Contents

14905851

This book is to be returned on or before the last date stamped below

DIGITAL
INTERFACING
WITH
ANALOG
WORLD

SECOND EDITION

JOSEPH J. CARR

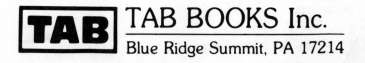

TAB BOOKS Inc.

Blue Ridge Summit, PA 17214

SECOND EDITION
FIRST PRINTING

Copyright © 1987 by TAB BOOKS Inc.
Printed in the United States of America

First edition copyright © 1978 by TAB BOOKS Inc.

Library of Congress Cataloging in Publication Data

Carr, Joseph J.
    Digital interfacing with an analog world.

    Includes index.
    1. Computer input-output equipment.   2. Transducers.
3. Analog-to-digital converters.   4. Digital-to-analog
converters.   I. Title.
TK7887.5.C37   1987   621.398   86-30018
ISBN 0-8306-0950-4
ISBN 0-8306-2850-9 (pbk.)

Questions regarding the content of this book
should be addressed to:
    Reader Inquiry Branch
    Editorial Department
    TAB BOOKS Inc.
    P.O. Box 40
    Blue Ridge Summit, PA 17214

Introduction to
the Second Edition

SINCE THE FIRST EDITION OF THIS BOOK WAS PUBLISHED, THE electronics field has experienced a lot of revolutionary changes. Since I first wrote this book in the mid-seventies, the microcomputer has all but taken over in circuit design. It is impossible to be a designer or user of electronic instrumentation, measurement apparatus, or control systems without knowing something of digital computers. In the mid-seventies, the microcomputer was just coming out; Intel did not invent their pathfinding 4004, 8008, and 8080 devices until after 1972. By the mid-seventies, there were a few groups of computer hardware hackers in the country, but the cost of the hardware was far too high for most hobbyists or small laboratories. Now you can buy a computer for less than $200 that is far more powerful than the IBM 1601/1620 that I used as a freshman engineering student.

Because of the explosion of the microcomputer, I have included chapters on microcomputer interfacing and on small computers. These computers, and the interfacing information presented, were selected for their applicability to a wide range of readers. These machines (Timex, Apple II, AIM-65, Commodore 64, etc.) are inexpensive and can be pressed into service for a wide variety of instrumentation/control system applications.

Other information in this book has been updated. When the first edition was written, the cat's meow was the ubiquitous operational amplifier. But today we can make use of integrated circuit instrumentation amplifiers (ICIA) that far exceed the performance of similar circuits built with discrete operational amplifiers. Companies like National Semiconductor, Burr-Brown, Analog Devices and others now have ICIA offerings.

The quality of the standard operational amplifiers has increased markedly. In the "old days," we had to put up with 0.5-volt dc offsets, 1-milliampere input bias currents, and frequency response limitations that, on some op amps, made them unusable even for communications speech amplifiers. But today we have operational amplifiers with picoampere level input bias currents, input impedances on the order of 1.5×10^{12} ohms, low drift, low noise, and many other heightened specifications. The analog operational amplifier that is offered today is much nearer the ideal described in textbooks than the device we had in the early to mid-seventies—at least in the low price range (one had to pay considerably for premium devices with specifications that are now commonplace).

If you enjoyed the first edition of this book, then I suggest that you will also enjoy this version. It's updated, expanded, and revised to reflect modern technology, but without losing anything of the old technologies that are still valid. It is my hope that you will keep this book at your workbench or, if you are an engineering or technology student, keep it tucked inside your laboratory notebook when you are working on the much dreaded "Senior Project."

Introduction to the First Edition

THE MICROPROCESSOR-BASED SMALL COMPUTER IS NO LONG-
er a new fad for the dilettante. Both professional and hobbyist
are past the "Oh Gosh! It really is . . ." stage, and many want to
learn how to put the machine to good use. Applications is the watch-
word today. This book is offered to fill an apparent gap in the literature. Many fine books have been written on programming, and
most fields, especially business programming, seem too well
covered. But to the user who wants to use the machine to measure
or control, little is available. Those I/O ports are essentially use-
less unless they are connected to a prefabricated peripheral. I explain what else you can do with them.

To the professional engineer, both hardware and software
types, learning to apply the microprocessor on the chip or controller
level has become an economic necessity; more and more employ-
ment advertisements list "microprocessor design experience"
among entry level requirements. A friend of mine is a nuclear engineer, and only deals with electronics as one of several tools, then
only on the black box level. But his first civilian employer, who
had been a Navy "nuke," advised him to become familiar with the
design of instrumentation and control systems based on the
microprocessor before reporting for work. A lot of night sessions
with a KIM-1 proved profitable, I am sure.

The scientist also finds work for the microcomputer. The machine will handle the analytical chores almost as well as any other

computer because it will number crunch with the best of them, only usually slower. But the microcomputer can also be used to make a low-cost data logger or control system for an experiment. Bare-bones microcomputers are available that will do the job for less than $100. Systems with lots of whistles and bells can be obtained for less than $1000—and even look like computers to the layman. Even retail hobby vendors such as Radio Shack are into this market. Radio Shack offers their TRS-80 line, which is rated a "best buy" in its class by some experts.

The computer hobbyist has become far more sophisticated. In the past, the hobbyist would assemble and debug a kit, then amuse friends and a dubious spouse with dozens of video games on the computer, including not less than a dozen versions of space war. But after shooting down the Klingon warships with your photon torpedoes for the six-hundredth time, many users want to explore some real, honest, practical uses for their toy.

This present book is designed to fill an apparent gap in the literature of the hobbyist (and professional for that matter). There are numerous books on computers, how they work, microcomputers/microprocessors, etc. There are also numerous books on BASIC programming, and even a dozen or so compendiums of BASIC programs useful to certain segments of the market. The business and scientific number-cruncher type of user may, in fact, be too well covered by existing texts; there may be too many to select from intelligently.

But for the user who views the microcomputer as a bit of hardware to be applied and software as either a simple set of instructions to make the machine go, or more importantly, a valid substitute for hardware, little has been written in book form. This book presents information, almost in handbook style, for those users of microcomputers who want to design a device or system with a microcomputer at its heart.

Applications is too broad a word on which to base a single book because the range of possible applications ideas is possibly much larger than the total number of readers. Your own imagination and design acumen are the limiting factors. If you supply the imagination, then I will do my part by helping you gain a little bit of the acumen part of the equation. You will learn certain aspects of computer interfacing, plus information about technology that will help you in solving a wide range of practical problems. The rest of this introduction is given to a discussion of the types of things that will be covered in the chapters to follow.

Chapter 1 discusses transducers or, more properly, transduction. Most of the discussion is on the strain gauge, but many of the concepts are applicable to almost any form of transducer.

A transducer is a device that will convert energy from a stimulus parameter (i.e., pressure, force, position, temperature, etc.) to an electrical voltage or current that is proportional to the value of the stimulus. In Chapter 2, various types of transducers are discussed. Among the types selected for coverage are position transducers (both voltage-output and digitally encoded styles), velocity transducers, acceleration transducers, fluid pressure transducers (i.e., liquids and gases), and several different types of temperature transducers. Note well that the nature of the discussion will lead you to other uses than those intended by the manufacturer. Both position (i.e., displacement) and pressure transducers, for example, can be configured to measure force.

The ubiquitous operational amplifier is covered in Chapters 3 and 4. Although the book is on interfacing with digital computers, it must be realized that many applications require a certain amount of analog signal processing before the digital computer can do its work. After all, it's an analog world for the most part. Most of the signals produced by the transducers used to sense this world produce analog voltages and currents. Furthermore, these analog signals will rarely be in a condition for application to the computer or A/D converter directly out of the transducer; at least some amplification will be needed, and in some cases some more elaborate signal processing will be required. While many of the signal-processing jobs can be performed in software (in fact some are better performed in software routines), it is often only those with the luxury of a fast, large-memory computer who can take advantage of signal-processing algorithms. For others, especially those with a slow-speed microcomputer or limited memory (some controllers have only 1K), the operational amplifier (analog signal processor) is the answer.

In some cases, you might want to use one of the commercial signal processors available instead of building an operational amplifier circuit. But that approach costs money. Anybody can solve problems by throwing money at them, but this book is intended primarily for those who cannot afford that approach. Note well that most of the operational amplifier circuits presented here are well enough behaved that even those people with a limited or nonexistent electronics background can make them work with a little effort. At this point, let me recommend two additional TAB books that are a must for people desiring to learn how to design and build electronic analog circuits:

1. *Linear IC/Op Amp Handbook—2nd Edition,* by Joseph J. Carr (TAB No. 1550).
2. *How To Design & Build Electronic Instrumentation,* by Joseph J. Carr (TAB No. 2660).

Two chapters of this book are devoted to analog function modules and certain digital circuits that have proven to be of especial usefulness. Note that neither of these chapters is an exhaustive study, but serve as guideposts for further study. Although I myself have found the circuits in these chapters to be useful, it is recommended that you consult the manufacturers' literature and catalogues for further ideas.

Chapters are also provided on subjects such as data transmission, including telephone interfacing and readout/display devices. Note that it is no longer either unusual or prohibitive to have a data link to a remote computer.

One chapter is devoted to controlling external devices (not ordinary computer peripherals) with your computer. I discuss turning ac loads on and off (safely), and controlling small dc motors in both open-ended and negative-feedback servo control applications.

By far, the largest signal portion of the book is devoted to digital data acquisition. After all, a digital computer can digest only that data presented in digital form. Digital computers simply do not know how to handle analog data, so conversion is a must.

The data conversion section leads off with a chapter on different types of digital codes used in computers. Although most microcomputers are formatted in straight binary, they can do code conversions in software. In fact, many are already equipped to do this type of job if they have an ASCII keyboard connected to one input port. This type of program will allow designers to use devices that produce other codes, or allow the ability to select a code that is most suitable for the application at hand. It may be, for example, that you want to use a Gray code position transducer, or take data from the BCD (i.e., display) lines of an instrument or device that uses seven-segment displays for readout.

In Chapter 10, I discuss the basic principles of data conversion circuits. The chapter covers the elementary digital-to-analog (D/A) converter and several methods for performing analog-to-digital

(A/D) conversions, including some techniques that use a D/A converter in a negative-feedback loop.

Chapters 11 and 12 complement Chapter 10 by providing information about real products available from leading suppliers. These chapters remove the discussion from the simply academic to the real world—that is, from the blackboard to the workbench.

Note that these chapters are not comprehensive; there are too many A/D and D/A converter products on the market for that to be the case. The criteria used for selecting the recommended products and manufacturers were availability and ease of application. There are many other products available, but these were not covered for any of five reasons: cost, difficulty of application, lack of space, ignorance of them, or lack of availability in low quantities.

If you do not see some particular favorite component in Chapters 11 and 12, please allow me to apologize in advance for my humble ignorance, and plead that I did the best I could for the greatest number. Various reasons kept popping up to keep some popular devices from being recommended. Some, for example, were rejected because they were not easily adapted for microcomputer service; they were originally developed for some other application. Some were too difficult to use, while others required too much external circuitry or parts that were either hard to obtain or costly. (Why is a $5 A/D converter desirable if it requires a $30 reference voltage chip?) Some very useful and practical devices were rejected because of difficulty encountered in actually obtaining these devices in low quantities. Some companies would happily sell 100, but turned their corporate noses up at an order for 1 or 2. Other companies had no local representatives or distributors (who usually will sell one or two on a "cash only" basis), or had too high a minimum factory order price. One attractive A/D product was rejected, for example, because it was not available through distributors, and the factory wanted a $250 minimum order (sigh). Precision Monolithics, Inc. (PMI) and Datel products are given seemingly excessive coverage, but I was able to buy the products in the onesy-twosy manner of hobbyists and low-volume professional users.

Data converter applications, as well as some highly specialized A/D converters, are covered in Chapter 13. This chapter is intended to make you aware of certain types of tricks that can be played using data converters, quite apart from their more easily recognized role of converting data. The multiplying D/A converter is especially useful in this manner and has many applications.

A short chapter on analog and digital multiplexing is included as part of the section on data converters and signal acquisition because these devices allow a single data converter to serve several channels. In many applications, the converter would idle much of the time, so it is available to perform other conversions. Although many of the circuits in Chapters 10 through 12 are low enough in cost to warrant the use of a separate converter for each channel, the use of a multiplexer will further reduce costs in many cases.

Prefabricated data acquisition systems are presented for: (1) those who do not care to design and build their own, but instead prefer a "plug 'n chug" system, and (2) those who wisely realize that the design and construction of a fast, multichannel, data acquisition system is a nontrivial matter. Systems by various manufacturers are discussed in both universal and peculiar-to-one computer (system) configurations.

The information given in this book is not intended to be followed step by step unless by some coincidence it ideally suits your needs. This book is intended to allow you to examine interfacing problems with a more practiced and knowledgeable eye. In fact, I would consider it the greatest form of flattery if a reader proves clever enough to modify the techniques presented here to more exactly fit specific situations. To paraphrase (i.e., steal from) a well-known ham radio publication: "Just like in Carr's book, except . . ."

Chapter 1

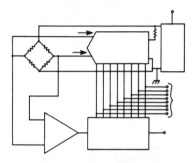

Transducers

A TRANSDUCER IS A DEVICE OR CIRCUIT THAT CONVERTS physical parameters such as position, force, pressure, temperature, velocity, acceleration, and so forth, to an electrical signal for purposes of measurement or control. The actual form taken by any given transducer depends a great deal on the intended function and the manufacturer's opinion as to what style is beautiful. Many different transducers can exist for any specific purpose or application, so the proper choice is sometimes obscured, making selection difficult. In other cases, there will be only one form that is applicable, so selection is easily done. In this chapter, we will consider some of the basic methods of transducer action—that is, transduction—leaving later chapters for the development of specific types.

Transducers must cause some electrical parameter to vary with the applied physical stimulus. To accomplish transduction we might make use of changes in electrical resistance, capacitance, inductance, or some combination of these to produce an electrical current, a voltage (ac or dc), a frequency, or a digital word that will be unique at each permissible value of the applied stimulus.

The most common forms of transducers use changes in ohmic resistance to indicate changes in the applied parameter. Various materials exist that will transduce through thermoresistance, photoresistance, piezoresistance (deformation), or simply the position of a potentiometer wiper.

1

Thermoresistance devices are also known as *thermistors* (i.e., thermal resistors). A thermistor will change its electrical resistance with changes in the applied temperature. All electrical conductors exhibit thermoresistance to some degree, but proper thermistors are designed to optimize and linearize this property.

The electrical resistance of almost any material that is an electrical conductor can be determined from:

$$R = \varrho \left(\frac{L}{A} \right) \tag{1.1}$$

where R is the electrical resistance in ohms, L is the length, A is the cross-sectional area of the conductor, and ϱ is the resistivity property of the particular material.

Resistivity, being a natural property of the material, is fixed in most (but not all) conductors, so we depend upon quantity L/A for transduction. This quotient can be changed only by deforming the material, and in most cases this is most easily accomplished by changing length L in either tension or compression.

When a material is in tension, the length will increase slightly, so the cross-sectional area must decrease because the overall volume is constant ($L \times A$). This action forces the numerator of Eq. 1.1 down and the denominator up, both factors tending to increase the value of quotient L/A, hence the resistance.

Similarly, when the material is in compression, the length decreases and the cross-sectional area increases making the electrical resistance lower.

Transducers of this type are called strain gauges. The key expression for evaluating strain gauge transducers is the gauge factor (K), which is defined as:

$$K = \frac{\Delta R/R}{\Delta L/L} \tag{1.2A}$$

$$K = \frac{\Delta R/R}{\epsilon} \tag{1.2B}$$

where R is the resistance in ohms, L is the length, K is the gauge factor, ϵ is the quotient $\Delta L/L$, and Δ is a symbol meaning a small change in . . .

Equation 1.2 in both versions will hold true only when ΔL is small compared with the length L and assumes that the change when in tension does not go past the limit of elasticity.

UNBONDED STRAIN GAUGES

The unbonded strain gauge uses a thin wire (the emphasis is on the word, "thin") stretched taut between two supports. The applied stimulus will either distend or compress the wire, changing its length. The unbonded strain gauge tends to be relatively fragile but is capable of superior precision under the correct set of circumstances.

BONDED STRAIN GAUGES

The bonded strain gauge consists of a thin wire, piece of thin metal foil, or a thin semiconductor slab cemented to a thin diaphragm. The applied stimulus will distend the diaphragm, thereby changing the length of the elements. The bonded strain gauge tends to be more stable and durable than unbonded types.

RESISTANCE CIRCUITS

In this discussion we will use the nomenclature of the piezoresistive strain gauges to explain circuit action, but most of the circuits are also valid with the other forms of resistance transducer.

Figure 1-1 shows the basic potentiometric circuit. Resistor $R1$ is a fixed resistor, while $R2$ represents the resistance of the strain gauge element. The current following in this circuit is given by:

$$I1 = \frac{E1}{R1 + R2} \qquad (1.3)$$

and ordinarily should not be greater than the current specified by the manufacturer as being the minimum that will cause self-heating of the strain gauge.

The output voltage, $E2$, is given by the ordinary voltage divider equation, namely:

$$E1 = E\ 1 \times \left(\frac{R2}{R1 + R2} \right) \qquad (1.4)$$

Differentiating Eq. 1.4 with respect to $R2$ gives us an appreciation of how the output voltage changes with stimulus-caused

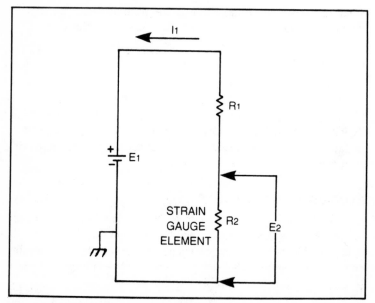

Fig. 1-1. Half-bridge transducer circuit.

changes in resistance $R2$. By the quotient rule for differentiation from basic calculus:

$$dE2 \;=\; E1 \;\times\; \left[\frac{(R1 + R2)\;\; dR2 - R2dR}{(R1 + R2)^2} \right] \qquad (1.5)$$

$$dE2 \;=\; E1 \;\times\; \left[\frac{dR2\,(R1 + R2 - R2)}{(R1 + R2)^2} \right] \qquad (1.6)$$

$$dE2 \;=\; \left[\frac{E1R1\;dR2}{(R1 + R2)^2} \right] \qquad (1.7)$$

By Eq. 1.2.

$$K \;=\; (\Delta R/R)/(\Delta L/L) \qquad (1.8)$$

which can be expressed as

$$K \;=\; dR2/dL \qquad (1.9)$$

$$KdL \;=\; dR. \qquad (1.10)$$

4

Substituting Eq. 1.10 into Eq. 1.7, we get

$$dE2 = \frac{E1\ R1\ K\ dL}{(R1 + R2)^2} \qquad (1.11)$$

Example 1-1

Assume the following:

$$R1 = 500 \text{ ohms}$$
$$R2 = 200 \text{ ohms (no strain conditions)}$$
$$dL = 0.001 \text{ in./in.}$$
$$E1 = 10 \text{ volts dc}$$
$$K = 4.0$$

Find the change in output voltage. From Eq. 1.11:
Solution:

$$dE2 = \frac{(10)(500)(4)(0.001)}{(500 + 200)^2} \qquad (1.12)$$

$$dE^2 = 4.08 \times 10^{-5} \text{ volts} \qquad (1.13)$$

$$dE2 = 40.8 \text{ microvolts } (\mu V) \qquad (1.14)$$

THE WHEATSTONE BRIDGE

A Wheatstone bridge circuit is shown in Fig. 1-2. Output voltage $E2$ is found from:

$$E2 = E1 \times (E_A - E_B) \qquad (1.15)$$

$$E2 = \frac{R2}{R1 + R2} - \frac{R4}{R3 + R4} \times E1 \qquad (1.16)$$

The bridge is said to be balanced when the output voltage is zero. For $E2$ to be zero, either $E1$ or the expression inside of the parentheses must be zero. $E1$, however, is fixed and always non-zero in practical circuits, so we can conclude that when the Wheatstone bridge is balanced:

$$\frac{R2}{R1 + R2} - \frac{R4}{R3 + R4} = 0 \qquad (1.17)$$

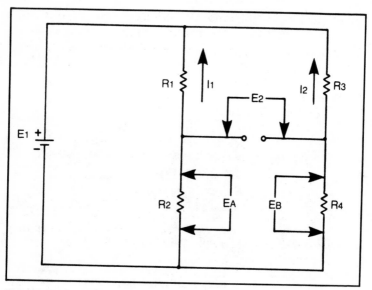

Fig. 1-2. Classic Wheatstone bridge.

So,

$$\frac{R2}{R1 + R2} = \frac{R4}{R3 + R4} \tag{1.18}$$

In the null condition;

$$I1\,R2 = I2\,R4 \tag{1.19}$$

and,

$$I1\,R1 = I2\,R3 \tag{1.20}$$

Dividing Eq. 1.19 into Eq. 1.20, gives us the sole necessary condition for balance in a Wheatstone bridge.

$$\frac{R2}{R1} = \frac{R4}{R3} \tag{1.21}$$

Equations 1.16 and 1.21 are the expressions usually given to describe the behavior of the Wheatstone bridge.

Example 1-2

Let us assume a resistive strain gauge in which

$R1 = R3 = 500$ ohms, and both $R1$ and $R3$ are fixed resistors. $R2$ and $R4$ are 200-ohm strain gauge elements similar to those used in Example 1-1. Assume the following parameters:

$$dL = 0.002 \text{ in./in.}$$

$$E1 = 10 \text{ volts dc}$$

$$K = 4.0$$

Find the change in output voltage $E2$.

Solution:

We will assume that the manufacturer arranged strain gauges $R2$ and $R4$ such that one is in compression under stimulus conditions and the other is in tension. This gives their respective dE terms opposite signs.

$$dE_A = \frac{E1 \; R1 \; K \; dL}{(700)^2} \tag{1.22}$$

and,

$$dE_B = -\frac{E1 \; R3 \; K \; dL}{(700)^2} \tag{1.23}$$

$$dE2 = dE_A - dE_B \tag{1.24}$$

$$dE2 = \left[\frac{E1 \; K \; dL}{(700)^2} \right] [R1 - (-R3)] \tag{1.25}$$

$$dE2 = \left[\frac{E1 \; K \; dL}{(700)^2} \right] (R1 + R3) \tag{1.26}$$

$$dE2 = \left[\frac{(10)(4)(0.002)}{(700)^2} \right] 500 + 500) \tag{1.27}$$

$$dE2 = 163 \text{ microvolts} \tag{1.28}$$

For a given stimulus, then, we get a lot higher output voltage. This is one principal advantage of the Wheatstone bridge over the so-called half bridge of Fig. 1-1.

Another advantage of the Wheatstone bridge is that the output voltage is zero when the stimulus is also zero, provided, of course, that the bridge is balanced under that condition. The half bridge, on the other hand, always produces an output voltage, so only changes in the applied stimulus can be noted with ease. Wheatstone bridge circuit transducers come in varieties with one, two, or four active strain gauges as elements.

CAPACITANCE TRANSDUCERS

A parallel plate capacitor is made by opposing two conductive metal surfaces parallel to each other. If these plates can be made to move relative to each other under the influence of an applied stimulus, then either the capacitance or the capacitive reactance can be used as a transduction property. A bridge such as Fig. 1-2 can be constructed with elements $R2$ and $R4$ replaced by variable capacitances. If $E1$ is an ac source, then we can use X_{C1} and X_{C2} to perform out transduction.

Alternatively, we can use a capacitive transducer to control the frequency of an oscillator. Transduction is by discrimination of the frequency in either a phase detector or a counter circuit.

Various forms of capacitive transducers exist, and their form will depend upon the nature of the job that is to be performed. Some will have the distance between the plates vary, as in a capacitor microphone, while in others the plates rotate relative to each other, so by their geometry either more or less area on each plate is shaded by the other plate.

INDUCTIVE TRANSDUCERS

We may also use inductance to produce transduction in the same manner as capacitance. Inductors can be used as reactance elements in a Wheatstone bridge, or to vary the frequency of an LC oscillator. Some transducers use the bridge method, but few use the oscillator circuit technique. There are also other inductive transducer techniques that are not similar to resistive or capacitive methods.

Most inductors used in transducers use variable cores to change the inductance, so are more properly called variable permeability devices. The applied stimulus will change the position of the coil's magnetic core material, thereby changing the inductance of the coil.

One of the most successful inductive transducers is the linear differential voltage transformer (LDVT), shown in Fig. 1-3. This transformer consists of a primary connected to an ac excitation

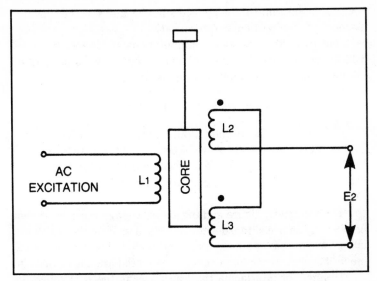

Fig. 1-3. Linear differential voltage transformer (LDVT).

source and two secondary windings. The secondary windings are cross connected so that the secondary currents generated by induction from the primary cancel each other.

When the core is inside both coils equally, then the two secondary currents cancel each other exactly; the net output voltage across the load is zero. If the core is displaced a little bit, one coil will have greater inductance than the other. This change will unbalance the relationship between secondary currents in the respective windings, making the cancellation less than total. This makes the output voltage greater than zero. Furthermore, the phase of the ac output voltage is determined by the direction of the core displacement. This situation means that we have both amplitude and directional displacement information.

SENSITIVITY

A transducer of the type which we have been discussing thus far will be excited by either an ac or dc voltage, and will produce an output voltage proportional to the value of the applied stimulus. The transfer function of this system is of the form

$$\frac{E_{\text{OUT}}}{E_{\text{IN}}} = \Psi X \qquad (1.29)$$

where X is the value of the applied stimulus and Ψ is the sensitivity factor of the particular transducer.

Sensitivity factor Ψ is a constant property and is specific to the transducer. It is usually given in units of output volts per volt of excitation per unit of applied stimulus. In other words:

$$\Psi = \text{volts/volt/unit stimulus} \qquad (1.30)$$

In most transducers the output voltage will be given in millivolts, or even microvolts. A well-known fluid pressure transducer used to measure arterial blood pressure in medical electronics, for example, is rated by its manufacturer to have a sensitivity of 10 μV/V/torr.

The sensitivity figure and excitation voltage must be known before you can use a transducer to encode data from the real world into a form usable by a computer or other digital electronic instrument. The sensitivity and excitation potential are used to design an amplifier for interfacing the transducer to the A/D converter at the computer.

Consider a system such as Fig. 1-4 where a physical stimulus is applied to a transducer. The transducer produces an output voltage ($E2$). This potential is then amplified to become $E3$, which is compatible with the input requirements of the A/D converter.

Example 1-3

Let us assume that it is necessary to interface a fluid pressure transducer to an eight-bit microprocessor. The A/D converter has an analog input range of 0 to 2.56 volts ($2^8 = 256$), and the instrument is to measure up to 100 torr of pressure. Transducer excitation is 10 volts dc and the sensitivity is known to be 10 μV/V/torr. How much amplifier gain is required?

Solution:

$$E2 = \Psi X \qquad (1.31)$$

$$E2 = \frac{(10\ \mu V)(10V)(100\ \text{torr})}{(V)\ (\text{torr})} \qquad (1.32)$$

$$E2 = 10,000\ \mu V \qquad (1.33)$$

$$E2 = 0.01\ \text{volts} \qquad (1.34)$$

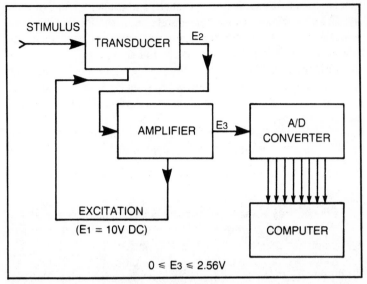

Fig. 1-4. Digital instrumentation system.

The gain required of the amplifier following the transducer is found by taking the quotient of the output voltage desired over the input voltage produced at either full range or some specified intermediate level. In this case:

$$A_V = \frac{2.56V}{0.01V} \qquad (1.35)$$

$$A_V = 256 \qquad (1.36)$$

The sensitivity property is also known by the term *transduction ratio*.

TRANSDUCER GLOSSARY

Linearity. Linearity is a measure of how well a transducer meets the ideal calibration curve. Ideally, when you plot the transducer output voltage against the applied stimulus, the graph will be a straight line. This line is usually constructed by connecting, with a straight edge, calibration points at both zero and full scale stimulus valves, although in some cases, a best-fit curve through these end points and several intermediate points is used.

11

Real transducers do not have such ideal properties in that their actual calibration curve might lie outside of the ideal curve by a considerable margin. Linearity is the deviation from ideal, expressed (usually) as a percentage of full scale.

Hysteresis. Some transducers will produce a different reading if a value is approached from below than if the same value were approached from above. The difference is called the hysteresis of the transducer.

Precision. Precision of the transducer measures the repeatability of a measurement. If the same identical stimulus is applied time after time, the output voltage should be the same with each trial. The precision is the measure of how well a given transducer meets this ideal condition.

Frequency Response. Frequency response of a transducer refers to a dynamic property that tells us how fast the transducer will track changes in the applied stimulus.

Any waveform can be defined mathematically as a Fourier series, which is a sum of assorted sine and cosine functions. The amplitude of the specific sine and cosine terms will determine the exact shape of the composite waveform.

The frequency response of a transducer used to measure an odd (nonsinusoidal) waveform stimulus must be high enough to pass the highest significant harmonic in the Fourier series of that particular waveform. Otherwise, the transducer will distort the shape, which could lead to erroneous measurement results. Similarly, if the transducer frequency response is too great, it may pass noise artifacts, which can lead to equally horrendous errors.

Chapter 2

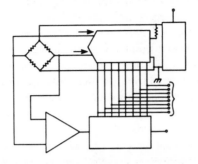

Types of Transducers

YOU COULD PROBABLY GUESS FROM THE MATERIAL IN
Chapter 2 that the word, "transducer," covers a wide range
of devices, all of which have different properties and characteristics. The job of a transducer is to look at some aspect of the parameter being measured, then deliver an output current or voltage proportional to its value. This output signal can then be processed to become a data signal in an instrumentation or control system.

TRANSDUCER EXCITATION

Most transducers are passive devices and, therefore, cannot create any electrical energy on their own. These transducers will require some type of excitation source in the form of an ac or dc voltage, depending upon the particular transducer. Regardless of the other excitation requirements, however, I find it necessary to have as much stability as possible to prevent errors and artifacts from creeping into the data record. I might, for example, have a nominal excitation potential on the order of $+7.5$ volts dc. It is often the case that there is sufficient latitude in the instrument to tolerate errors in the actual excitation potential, so I really do not care what the exact potential is, provided that it is somewhere between 7.0 volts and 8.0 volts. I do care, however, if the excitation voltage changes significantly during the time when the transducer is being used. Even very small changes can be significant if there is a lot of gain in the amplifiers following the transducer, so it is imperative that the stability of the excitation source be assured.

Part of the protocol for using transducers includes adequate warmup time. All transducers, amplifiers, and excitation sources can be expected to drift somewhat for a few minutes following a cold start. The wise user will turn on the electronics and transducer excitation (usually the same act accomplishes both) not less than 15 minutes prior to use. Some systems that are hypercritical in this respect must be allowed to warm up for several hours prior to use.

If the excitation source for any given transducer must produce a constant dc voltage, then using some sort of ordinary voltage regulator, even a simple zener diode should be sufficient. The exact form of circuit required will depend a great deal on whether a bipolar or single power supply is needed, and the exact level of excitation voltage that is required. In the discussion of excitation voltage to follow, I will assume that the transducer is in the form of a resistive Wheatstone bridge.

Figure 2-1 shows the most elementary form of excitation voltage source. A zener diode ($D1$) is used to keep the transducer voltage nearly constant, while $R1$ is used to limit the current drawn by the zener diode to a safe level.

This technique suffers from not less than two major problems. The first shortcoming is that zener diodes do not have a stable zener

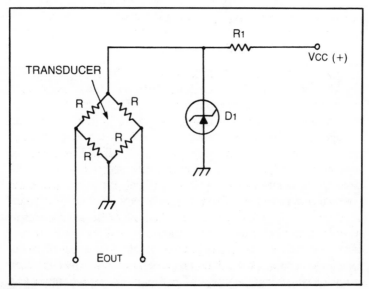

Fig. 2-1. Transducer excitation using a zener diode regulator.

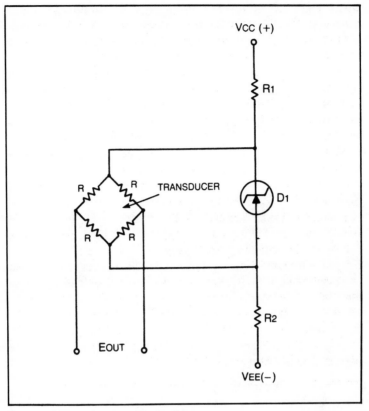

Fig. 2-2. Alternate zener diode regulator.

potential in all cases. The voltage can vary with temperature, especially if the looking-back resistance of the transducer (essentially R in a Wheatstone bridge where all elements are equal) is low. A partial solution is to use a reference-grade zener diode.

The second problem is that the supply is a single positive voltage, so I am limited to situations where the transducer output is positive with respect to its rest position where the bridge is balanced. This may not be a problem in all cases, such as rectilinear position transducers that operate in only one quadrant or Wheatstone bridges where a true differential amplifier follows the transducer, but in other cases it could seriously hamper my efforts.

An example of the second problem might be a pressure transducer that will read positive for gauge pressures, is balanced at atmospheric pressures, and should produce a negative output for vacuums. If I use a positive-going or negative-going excitation

source and a single-ended input amplifier in that situation, I will
have to make sure that the electronics to follow understand how
to interpret the results, or that the results are unimportant (hardly
likely).

A superior technique is shown in Fig. 2-2. In this method of
transducer excitation I place the zener diode between the V_{CC} (+)
and V_{EE} (–) power supplies.

A slightly different approach is shown in Fig. 2-3. Here the use
of a three-terminal integrated circuit voltage regulator, in this case
a 7805, LM340-5, or LM309K, produces a 5-volt dc output. This
series of IC regulators are offered by most of the major semiconductor manufacturers in three different packages and current
ranges. One package, designated by the letter H in suffix to the
type number, is the familiar TO-5 transistor case. Most of these
regulators are rated at 100 milliamperes of output current. The
LM309H, for example, produces a nominal 5-volt dc output potential and will source a current up to 100 milliamperes. This current
level can be used to drive transducers with a looking-back resistance of 50 ohms or greater.

The T package is the same as the TO-220 transistor package
and is made of plastic. Most T suffix voltage regulators will deliver up to 750 milliamperes of current in free air and up to 1 ampere if properly heat sinked.

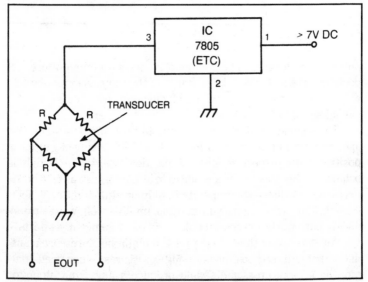

Fig. 2-3. Transducer excitation using a three-terminal IC voltage regulator.

The *K* package is the same as the diamond shaped TO-3 power transistor package. Most *K* suffix regulators will deliver up to 1 ampere in free air and 1.5 amperes if heat sinked. There are some regulators in this package, though, that will handle a lot more current. An LM323, for example, will source up to 3 amperes, while Lambda Electronics markets its LAS-1905 which is capable of delivering up to 5 amperes.

ELEMENTARY POSITION TRANSDUCERS

The simple potentiometer is the most common form of position transducer. Examples are shown in Fig. 2-4. The potentiometer shown in Fig. 2-4A (at the top) is useful for indicating position along the *X*-axis (in other words, along a line) in the first quadrant of a Cartesian plane. When the potentiometer's wiper is at the origin, it is electrically grounded so the output potential is 0 volts. As the wiper moves along the positive *X*-axis, the output voltage rises until it is at VCC (+), the highest permissible value in the *X*-domain.

If the position could be either minus or positive along the *X*-axis, then the circuit of Fig. 2-4B (at the bottom) must be used. In this circuit, one end of the potentiometer is connected to the *VCC* (+) supply as before, but the other end is connected to the *VEE* (−) instead of ground. When the potentiometer wiper is at the origin, it is in the center of its resistance range. The respective *VCC* and *VEE* contributions to the net output voltage will exactly cancel each other so the output voltage is zero. If the wiper displaces to the right along the positive *X*-axis, the *VCC* will predominate, so the output voltage is positive. Similarly, if the wiper were displaced to the left along the minus *X*-axis, the *VEE* would predominate producing a negative output voltage.

The form taken by such a potentiometer would depend on the type of motion that was being measured. If the motion was strictly rectilinear, that is to say in a straight line, then a slide type of potentiometer might be needed. If, on the other hand, the motion were angular or circular, then a rotary potentiometer might be the indicated choice. You could, for example, designate 0° of rotation as 0 volts output and 359° of rotation as *VCC*. Some potentiometers are available that will produce output voltages proportional to the sine, cosine, or sine and cosine of the angle of displacement.

The LDVT can also be used as a displacement transducer because the output voltage is zero when the core is equally inside of

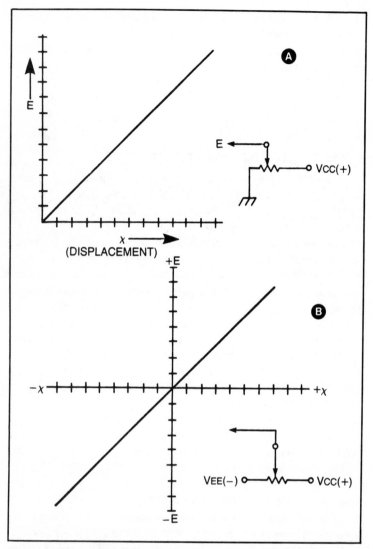

Fig. 2-4. Displacement transducers. (A) Single-quadrant type. (B) Two-quadrant type.

both secondary windings, yet will assume one polarity when the core is displaced in one direction and the opposite polarity when the displacement is in the opposite direction.

Regardless of the type of position transducer used, there might be a requirement for some linkage or other mechanics to reduce the throw of the mechanism to the throw of the potentiometer. Alternatively, you might want to use a gear reduction system for the rotary potentiometer.

DIGITAL CODE POSITION SENSORS

Optoelectronic circuits can be used to create a position sensor, and are particularly common in shaft encoders and other angular motion indicators. Figures 2-5 and 2-6 show an example of such a system.

Three different code wheels are shown in Fig. 2-5. These code wheels are mounted concentric to the axis of the shaft or through a gear train if that is appropriate for the particular application. Position information is given by the outer rim of each wheel, while a synchronization signal is given by the inner rim.

The information is coded onto the disc at Fig. 2-5A using holes drilled in the metal disc, while the disc in Fig. 2-5B uses light and dark shading, a technique amenable to low-cost production because the disc can be made photographically. The wheel in Fig. 2-5C is my salute to the past, in that it is a cam wheel driving a pair of microswitches. This system, though, is anachronistic but may be encountered even in equipment that is not really all that old.

Both of the wheels shown in Figs. 2-5A and 2-5B operate by creating an electrical pulse when light passes through the opening. A suitable read head is shown in Fig. 2-6A. The coded portions of the wheel passes through a light path between a light-emitting diode (LED) sender and a phototransistor (PT) receiver, or alternatively a photoresistor receiver. When the light path is blinded, no light reaches the phototransistor, so the output of the photo-transistor (see Fig. 2-6C) is low. When the path is not blinded, however, the base of the transistor is illuminated, so the output voltage is high. The alternating light and dark gives a pulse train at the emitter of the two phototransistors (see Fig. 2-6B) that can be used to infer position. Let me consider a hypothetical case using a code wheel such as Fig. 2-5B and a decoder such as Fig. 2-6C.

The heart of the decoder is an N-bit binary counter, where N is the number of unshaded spots on the outer rim of the code wheel. If I want to be able to resolve angular increments of $1°$, then I will need 360 unshaded spots on the outer rim of the wheel. Keep in mind that an eight-bit binary counter can only resolve 256 separate states, so I will need at least a nine-bit binary counter for this job. A nine-bit binary counter will resolve 2^9 or 512 different states, so it is more than adequate for this particular problem.

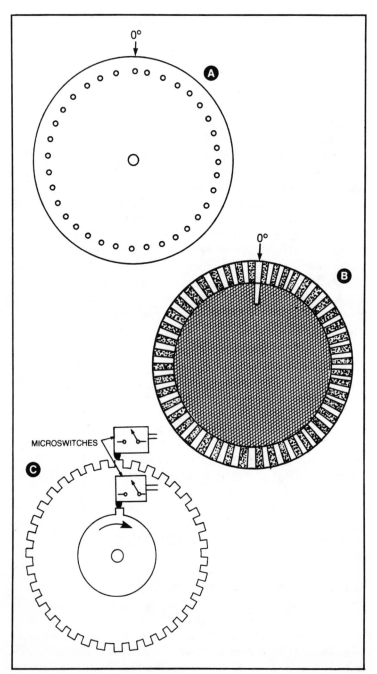

Fig. 2-5. Sequential code wheels.

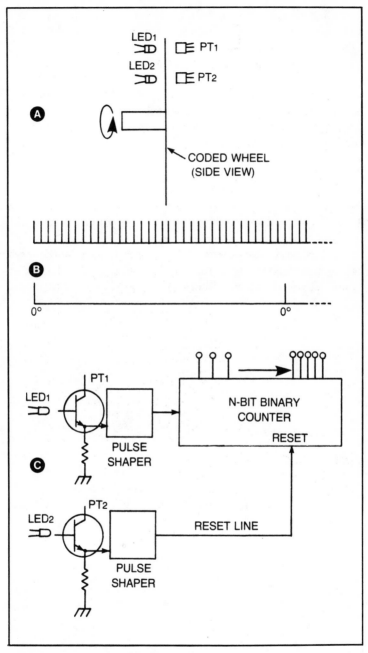

Fig. 2-6. Code wheel circuits. (A) Transducer configuration. (B) Waveforms. (C) Circuit block diagram.

21

The unblinded spot on the inner rim of the code wheel is used to indicate a specified reference position, usually defined to be 0°. One pulse will be created in *PT2* for every 360 pulses from *PT1*, so there is a means of recognizing the reference or 0°/360° position.

Pulses will begin arriving from *PT1* to the input of the binary counter just as soon as the wheel begins to turn, so a count will accumulate on the counter's output lines. When the wheel reaches the 0° position on its initial revolution, there will be a pulse generated in *PT2* that resets the binary counter to zero. From this point onwards, each pulse indicates a change in angular position of 1°. If, for example, the wheel goes through a rotation of one-quarter turn (i.e., 90°), then 90 pulses would have been generated, so the binary word at the output of the counter would be 001011010 in base 2 or 90 in decimal (the binary counter outputs in base 2).

The code wheel can also be used as a tachometer to indicate angular velocity, particularly if the reference pulse on the inner wheel is ignored. If the output of *PT1* were fed to a frequency counter I would find an output frequency of

$$F_{\text{hertz}} = \frac{360 \text{ (revolutions)}}{\text{second}} \tag{2.1}$$

Which, by a little algebra becomes

$$F_{\text{hertz}} = \frac{360 \text{ rev}}{\text{sec}} \times \frac{60 \text{ sec}}{\text{min.}} \tag{2.2}$$

should I want the data in that form.

A more sensible approach for a code wheel that is purely for tachometer use would be to use either 100 or 1000 unshaded spaces on the outer rim rather than 360. This would result in a readout in hertz proportional by a power of ten to the speed in revolutions per second. In that case:

$$F_{\text{hertz}} = \frac{100 \text{ rev}}{\text{sec}} \tag{2.3}$$

or,

$$F_{\text{hertz}} = \frac{1000 \text{ rev}}{\text{sec}} \tag{2.4}$$

Let me assume a case in which a code wheel has 1000 spaces, allowing me to apply Eq. 2.4. The wheel rotates at 40 revolutions per second. This yields an angular frequency of

$$F_{hertz} = \frac{(40 \text{ rev}) (1000)}{\text{seconds}} \tag{2.5}$$

$$F_{hertz} = 40,000 \text{ hertz} \tag{2.6}$$

If I wanted to display this on a frequency counter as 40.0 (kHz), I could read it as 40.0 RPS (revolutions per second) because it is numerically the same as my desired data.

But most motor shaft speeds are given in terms of revolutions per minute (RPM), so how do I deduce RPM from frequency? There are actually several methods for doing this job. I could use the computer as a frequency counter, or present the data from a hardware frequency counter to a computer input port, then multiply it in software by a factor of 60. This would take only a very short subroutine.

Another alternative would be to print more lines on the code wheel, such as to make the total 600 or 6000. The frequency counter in the above example would read:

$$F_{hertz} = \frac{(40 \text{ rev}) (6000)}{\text{second}} \tag{2.7}$$

$$F_{hertz} = 240,000 \tag{2.8}$$

But I know that 40 RPS is the same as 2400 RPM, so if I display the 240,000 Hz as 2400.00, I would have the tachometer in the form desired.

Still another alternative is to use the original 1000-space code wheel, then modify the frequency counter gate time. A frequency counter consists of a decimal counting assembly (DCA), a main gate, and a gate timing circuit. Pulses can only reach the DCA by passing through the gate; the gate's on time is set by the timing circuit. Most frequency, or events-per-unit-of-time (EPUT), counters have gate-open times of 1 second, 0.01 second, or some other power of ten multiple or submultiple of the basic 1-second period. This will give me a readout in hertz or one of the commonly recognized larger divisions of frequency based on the hertz, that is, the kilohertz or megahertz.

But what would happen if the gate were only open for a period of 0.60 second (600 millisecond) instead of the usual 1 second? With 1000 openings on the code wheel there would be, at 40 RPS or 2400 RPM:

$$\frac{(40 \text{ rev}) (1000)}{\text{second}} = 40,000 \text{ hertz} \qquad (2.9)$$

generated at the output of the phototransistor, as before. But if the gate were only open for 0.60 seconds at a time, the readout would not display 40,000 as before, but instead it would read:

$$(40,000) (0.60) = 24,000 \text{ events} \qquad (2.10)$$

This output would be interpreted as 2400.0 RPM by the judicious placement of the decimal point on the counter, a matter that is very easy to affect in hardware.

Code wheels can also be programmed to produce various recognized computer or other digital codes. These can be used to indicate position. An example is shown in Fig. 2-7. Ordinary base-2 binary code is weighted by position into powers of 2 according to the scheme:

	Bit No.		Binary Weight	Decimal Weight
(LSB)	1		2^0	1
	2		2^1	2
	3		2^2	4
	4		2^3	8
	5		2^4	16
	6		2^5	32
	7		2^6	64
	8		2^7	128
	9	(MSB)	2^8	256

In the code wheel segment shown in Fig. 2-7, the arrangement is to have bit 1, the least significant bit (LSB), on the outer rim and bit 9, the most significant bit (MSB), on the inner ring. Notice that the 6° position is unblinded at bits 2 and 3 making the digital

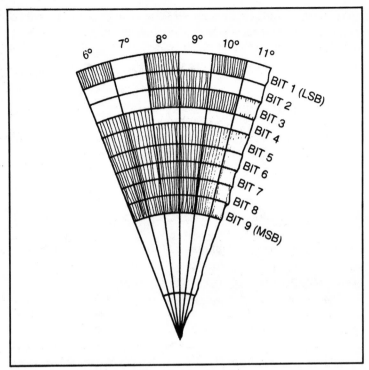

Fig. 2-7. Binary, gray, or BCD code wheel.

word at the output 000000110. This is deciphered in decimal form as:

$$0 + 0 + 0 + 0 + 0 + 0 + 2^2 + 2^1 + 0 = 4 + 2 = 6$$

Similarly, the 7° position of the wheel is unblinded at bits 1, 2, and 3 making the equivalent digital word 000000111, which is interpreted as:

$$0 + 0 + 0 + 0 + 0 + 0 + 2^2 + 2^1 + 2^0 = 4 + 2 + 1 = 7$$

Code wheels are also found in drum or cylindrical form. Some of these use the counter type of configuration in which the necessary code is around the circumference of the cylinder. Another form uses the binary code technique of Fig. 2-7 by arranging the blinded and unblinded squares along the length of the cylinder body. At least one high-performance phase-locked loop (PLL) communications receiver uses such a drum ganged to the tuning knob. The binary code is used to drive the PLL local oscillator and the electronic digital frequency indicator used as the frequency readout.

VELOCITY & ACCELERATION TRANSDUCERS

The analog position transducer can be used to generate both velocity and acceleration information. Recall from elementary physics that *velocity (v)* is the first time derivative of position, and that *acceleration (a)* is both the first time derivative of velocity and the second time derivative of position. In mathematic notation:

$$v = dx/dt \qquad\qquad (2.11)$$
$$a = dv/dt \qquad\qquad (2.12A)$$
$$a = d^2 \times /dt^2 \qquad\qquad (2.12B)$$

The most pressing implication of these facts is that I may often find it convenient to generate a position signal, and it is almost always cheaper to generate the position signal. I can then pass the position signal through an electronic differentiator to obtain a velocity signal. The velocity signal can then be passed through another differentiator to obtain an acceleration signal.

A popular angular velocity transducer is the ac alternator ganged to the rotating shaft. Most of these devices consist of an armature laden with a number of small pole pairs (see Fig. 2-8) and a set of coils. As the shaft is rotated, these magnets generate an alternating current in the coils, and this signal is used as the angular velocity signal. Most alternator transducers have a frequency output scalar specification giving the number of hertz per revolution that are produced. Typical specifications are on the order of 10, 60, or 100 hertz per revolution.

There are two ways this can be used to indicate angular velocity. In one method, I simply count the output frequency. If the specification is 60 hertz/revolution, what frequency would indicate a rotational speed of 1800 RPM?

$$f_{OUT} = \frac{60 \text{ cycles}}{\text{rev}} \times \frac{1800 \text{ rev}}{\text{min.}} \times \frac{1 \text{ min.}}{60 \text{ sec.}} \qquad (2.13)$$

$$f_{OUT} = 1800 \text{ cycles/sec} = 1800 \text{ hertz} \qquad (2.14)$$

It doesn't take a genius to figure out that a scale factor of 6, 60, or 600 cycles per revolution will result in a highly readable dis-

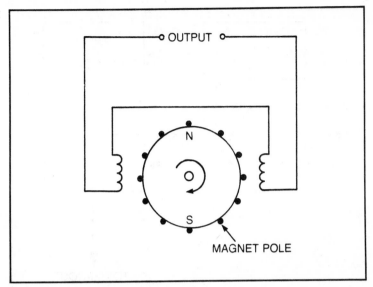

Fig. 2-8. Alternator tachometer.

play on the frequency counter because the frequency of the tachometer output is numerically the same as the shaft speed.

The tachometer signal will be ac, so unless the counter is equipped with a front end that is capable of handling sine waves, it will be necessary to provide some signal conditioning. A Schmitt trigger or zero-crossing detector between the tachometer output and the counter input will usually suffice.

Sometimes an analog voltage or current level is preferred or required. The tachometer is still usable in that case by either of two means: frequency-voltage conversion by a counter with a D/A converter or integration. If a frequency counter is employed, then a D/A converter can be used at its output lines to produce the analog signal. But counter circuits tend to be expensive and are often very slow. The cure, our second method, is shown in Fig. 2-9.

The output of the alternator is conditioned in a zero-crossing detector that is designed to produce an output pulse every time the ac input signal crosses the zero-volt baseline. An operational amplifier comparator with one input grounded and the ac signal applied to the other input will do this job. The signal is then differentiated and passed to a switching diode to obtain a negative-going spike.

The pulses from the zero-crossing detector are used to trigger a monostable (one-shot) multivibrator stage. The one-shot mul-

27

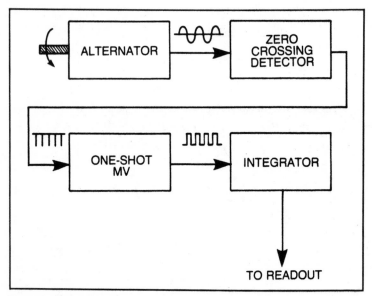

Fig. 2-9. Tachometer block diagram.

tivibrator will produce one output pulse every time it is triggered. These pulses will all have the same time duration and amplitude. When they are time averaged in the following integrator stage, a dc level is produced that is proportional to the frequency. This signal, then, is the angular velocity signal.

The analog velocity signal can be integrated to form a position signal:

$$x = C1 \int E(v)\, dt \qquad (2.15)$$

or differentiated to form an acceleration signal;

$$a = \frac{dE(v)}{dt} \qquad (2.16)$$

Neither of these derived signals, incidentally, is as accurate as the velocity signal because a certain distortion of the data occurs due to phase shifts inherent in operational integrator and differentiator circuits.

Accelerometers are sometimes built similar to the device shown in Fig. 2-10. Transduction occurs when a permanent magnet moves relative to a fixed inductor. In actual practice, the permanent magnet is cylindrical, and is mounted inside of and coaxially to the cylindrical inductor.

Two springs connected to rigid anchors hold the magnet and a calibrated mass in place. The force equation for a spring (in this case the two springs are taken to be a single unit) is:

$$F = -Kx \qquad (2.17)$$

I also know that Newton's second law defines force as

$$F = ma \qquad (2.18)$$

I can conclude that

$$ma = -Kx \qquad (2.19)$$
$$a = -Kx/m \qquad (2.20)$$

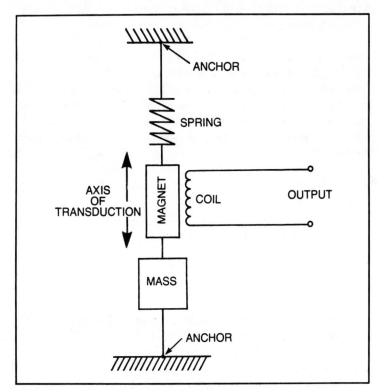

Fig. 2-10. Force-displacement accelerometer.

But the quantity $-K/m$ is a constant in any given transducer, so Eq. 2.20 can be written as

$$a = cx \qquad\qquad (2.21)$$

where a is the acceleration, x is the displacement along the axis of transduction, and c is a unique constant.

Acceleration can only exist when the position and velocity are changing, a fact that is inferred by the fact that acceleration is the first derivative of the latter and the second derivative of the former. Since the output potential of the inductor is proportional to the rate of change of position (x) and x is proportional to acceleration, we can conclude that the output potential is also proportional to the acceleration.

A vibration transducer is made almost exactly like Fig. 2-10, except that the mass is deleted and the $-K$ term is set for the size vibrations that the transducer is expected to handle. This class of transducers is used both in engineering applications and in seismological studies.

The basic transducer of Fig. 2-10 is also occasionally used in force measurements. Again by Newton's second law, I can infer the force applied from the core displacement. But this type of transducer works only for dynamic force situations, and will not respond at all to static forces and responds only poorly to slowly changing forces. A better solution, perhaps, is the transducer of Fig. 2-11. Here, a potentiometer is ganged to a rod that drives a spring. Again the force is proportional through a constant to the displacement of the spring along the axis of transduction. This type of transducer will yield an output signal regardless of whether the applied force is static or dynamic.

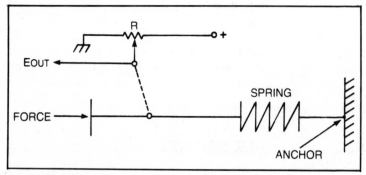

Fig. 2-11. Alternate accelerometer.

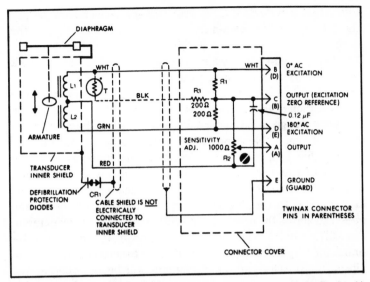

Fig. 2-12. Inductive fluid pressure transducer. (Courtesy of Hewlett-Packard.)

FLUID PRESSURE TRANSDUCERS

There are three forms of common fluid-pressure transducer which we will consider: resistive strain gauges, inductive strain gauges, and solid state. These are represented by the products of Statham, Hewlett-Packard, and National Semiconductor, respectively.

The resistive strain gauge type of pressure transducer uses a four-element Wheatstone bridge bonded to a thin metal diaphragm. The fluid under pressure is applied to the other side of the diaphragm, and this distends the diaphragm, thereby changing the resistance of the strain gauge bridge elements.

Figure 2-12 shows the circuit diagram for a popular inductive fluid-pressure transducer, the Hewlett-Packard Model 1280B/C. This is also a form of Wheatstone bridge, but requires an ac excitation source. The variable bridge elements are the reactances of coils $L1$ and $L2$, while resistors $R1$ and $R2$ form the fixed bridge arms. Resistor $R3$ and the thermistor are used in some models for temperature compensation.

Under zero pressure (open to atmosphere) conditions the core will be inside of both $L1$ and $L2$ equally. But under pressure, the diaphragm distends and drives the rod partially out of one coil and into the other, and this unbalances the system.

Few transducers accurately meet their sensitivity specifications

(Fig. 2-13), so in the H-P 1280 series, a large barrel connector (see Fig. 2-14) is used to contain the fixed elements of the bridge, the thermal compensation, and a sensitivity adjustment potentiometer. This feature allows you to apply a standard pressure, then adjust the sensitivity to a standard output level.

Figures 2-14 and 2-15 show the basic form of the pressure transducer. Although this is specifically the H-P 1280, it is applicable to a wider range of products, especially those in the medical arterial and venous blood pressure monitoring business.

The fluid is applied to the diaphragm, and is contained within a plastic pressure dome. There are actually two types of pressure domes. The conventional type of Fig. 2-14 and the disposable type of Fig. 2-15. In the disposable type there is a thin membrane stretched across the opening that is in contact with the diaphragm. A small drop of liquid is placed on the diaphragm to couple the membrane and diaphragm together. According to certain physical principles, the pressure on the diaphragm will be the same as the pressure on the fluid side of the membrane, provided that no air is in the system. The disposable dome finds a warm reception in

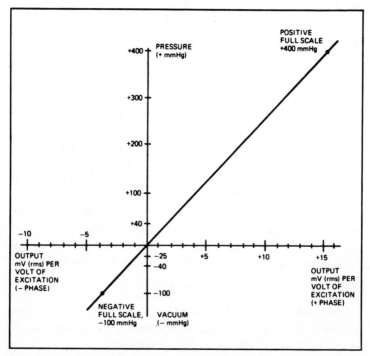

Fig. 2-13. Transfer function for Fig. 2-12. (Courtesy of Hewlett-Packard.)

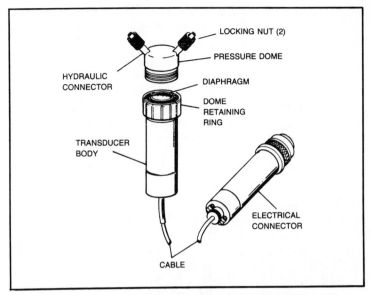

Fig. 2-14. H-P Model 1280 transducer. (Courtesy of Hewlett-Packard.)

medical applications because it is shipped sterile, thereby reducing or even eliminating the sterilization of the transducer between patients.

The medical configuration shown in the figures is also useful in other physical science and engineering applications, but there are numerous other configurations available for special purposes. For these let me recommend that you consult the manufacturer's literature.

National Semiconductor manufactures a clever line of temperature compensated integrated circuit pressure transducers (see Fig. 2-16). These contain the bridge, regulated excitation source, temperature compensation, and operational amplifiers needed to produce a high-level output voltage from a pressure applied to a fitting on the IC body. The transducers, designated by the maker as their LX series, are available in different configurations and ranges for different applications. For specific details consult National Semiconductor's assorted catalogues.

TEMPERATURE TRANSDUCERS

Three basic forms of temperature transducer are of nominal interest to us in this discussion: thermoresistive (i.e., thermistors), thermocouples, and semiconductor PN junctions.

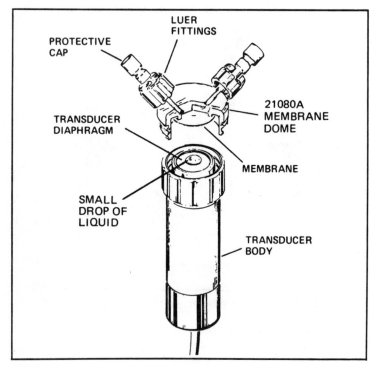

Fig. 2-15. Use of disposable pressure dome. (Courtesy of Hewlett-Packard.)

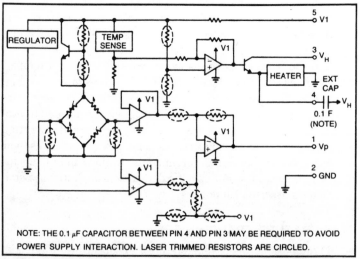

NOTE: THE 0.1 μF CAPACITOR BETWEEN PIN 4 AND PIN 3 MAY BE REQUIRED TO AVOID POWER SUPPLY INTERACTION. LASER TRIMMED RESISTORS ARE CIRCLED.

Fig. 2-16. National Semiconductor solid-state pressure transducer. (Courtesy National Semiconductor.)

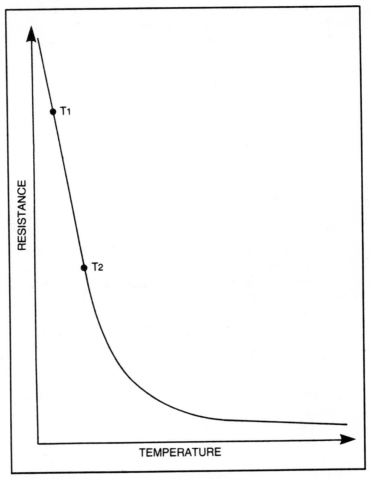

Fig. 2-17. Typical thermistor transfer function.

All electrical conductors will change resistance somewhat with changes in applied temperature. A thermistor is a device in which this property is optimized and predictable.

Figure 2-17 shows the graph of a typical thermistor with a negative temperature coefficient. This is the resistance-versus-temperature plot. The shape is decidedly nonlinear except for the region between temperatures $T1$ and $T2$. This region can be used directly for temperature measurement in ordinary bridge or half-bridge circuits. But beyond temperature $T2$, I must use one of several possible linerization techniques in order to avoid huge errors.

A *thermocouple* is formed by joining two dissimilar metals, usually wires, into a V (see Fig. 2-18). Every material has associated with it a natural property known as its work function. Differences in the respective work functions of material A and material B gives rise to a millivolt-range potential that is a function of temperature. When the junction is heated, a potential is found across the ends. One problem associated with the thermocouple, incidentally, is that other thermocouples are formed between the wires forming each leg and the copper wires connecting the device to its electronic circuit. This is of little consequence if both wires are of identical material, but it can be a problem otherwise.

Various types of thermocouple and combinations exist; and they have differing properties and applications (see Fig. 2-18B):

J-Type
This thermocouple consists of a positive iron wire and a negative constantan alloy wire. It can be used in chemically reducing atmospheres and at temperatures up to 1600 °F if appropriate wire diameters are obtained.

T-Type
This type, copper and constantan, is used in mildly oxidizing and reducing atmospheres at temperatures up to 750 °F.

K-Type
This type uses a positive chromel alloy wire and a negative alumel alloy wire. It can be used in some oxidizing atmospheres at temperatures up to 2300 °F.

E-Type
This thermocouple, chromel and constantan, is used in mildly reducing or oxidizing atmospheres and in vacuums up to temperatures of 1600 °F.

A PN junction of semiconductor material, such as a common diode, consists of N-type and P-type semiconductor material joined together in a bond. This type of junction has some excellent temperature transduction properties, which are of immense use even though limited to lower temperature ranges than the thermocouple.

The temperature transduction properties of an ordinary reverse-biased power supply rectifier can be demonstrated with an ordinary ohmmeter. Connect the terminals of the ohmmeter to the diode such that the ohmmeter battery reverse biases the diode junction. Note the resistance, then apply heat. A match, heat gun (i.e., a hair dryer), or cigarette lighter will give spectacular results, but

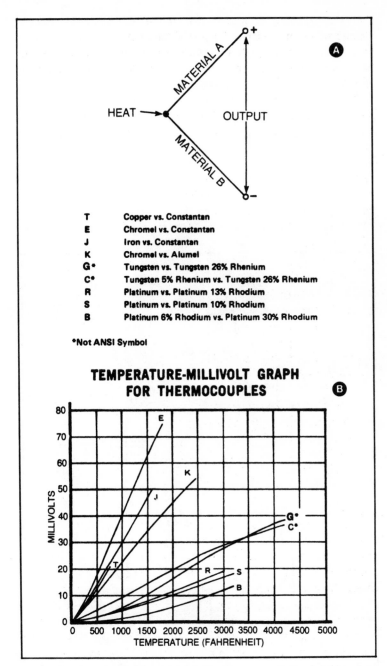

T	Copper vs. Constantan
E	Chromel vs. Constantan
J	Iron vs. Constantan
K	Chromel vs. Alumel
G•	Tungsten vs. Tungsten 26% Rhenium
C•	Tungsten 5% Rhenium vs. Tungsten 26% Rhenium
R	Platinum vs. Platinum 13% Rhodium
S	Platinum vs. Platinum 10% Rhodium
B	Platinum 6% Rhodium vs. Platinum 30% Rhodium

•Not ANSI Symbol

TEMPERATURE-MILLIVOLT GRAPH
FOR THERMOCOUPLES

Fig. 2-18. Thermocouples. (A) Physical configuration. (B) Transfer function. (Courtesy of Omega.)

even the heat in your fingers can cause a noticeable resistance change when you grasp the body of the diode.

A matched pair of diode-connected transistors offers the best PN junction temperature transducer (see Fig. 2-19). The base-emitter voltage (V_{BE}) in a transistor is given by:

$$V_{BE} = \frac{kT}{q} \ln \left(\frac{I_C}{I_S} \right) \quad I_C / I_S >> 1 \tag{2.21}$$

where k is Boltzmann's constant (1.38×10^{-23} J/°K), T is the absolute temperature in degrees Kelvin (°K), q is the elementary electric charge (1.6×10^{-19} coulombs), I_S is the reverse saturation current nominally taken to be 1.87×10^{-14} amperes in one tran-

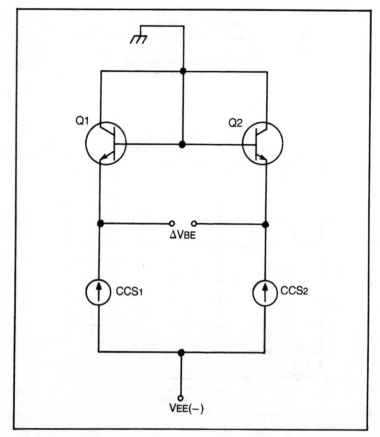

Fig. 2-19. Transistor pn junction temperature transducer.

38

sistor used for this purpose, and I_C is the collector current expressed in amperes.

If a circuit such as Fig. 2-19 is constructed using a matched-dual transistor (two identical transistors inside one case), I can calculate the *differential* base-emitter voltage (ΔV_{BE}) from the combined expression:

$$\Delta V_{BE} = \frac{kT}{q} \ln\left(\frac{I_{C1}}{I_{S1}}\right) - \frac{kT}{q} \ln\left(\frac{I_{C2}}{I_{S2}}\right) \quad (2.22)$$

$$\Delta V_{BE} = \frac{kT}{q}\left[\ln\left(\frac{I_{C1}}{I_{S1}}\right) - \ln\left(\frac{I_{C2}}{I_{S2}}\right)\right] \quad (2.23)$$

$$\Delta V_{BE} = \frac{kT}{q}\left[\ln\frac{(I_{C1}/I_{S1})}{(I_{C2}/I_{S2})}\right] \quad (2.24)$$

$$\Delta V_{BE} = \frac{kT}{q}\left[\ln\left(\frac{I_{C1}}{I_{C2}} \cdot \frac{I_{S2}}{I_{S1}}\right)\right] \quad (2.25)$$

$$\Delta V_{BE} = \frac{kT}{q}\left[\ln\left(\frac{I_{C1}}{I_{C1}}\right) + \ln\left(\frac{I_{S1}}{I_{S1}}\right)\right] \quad (2.26)$$

But I_{S1} and I_{S2} are very nearly equal in a monolithic matched pair, so Eq. 2.26 can be rewritten in the form

$$\Delta V_{BE} = \frac{kT}{q}\left[\ln\left(\frac{I_{C1}}{I_{C2}}\right) + \ln(1)\right] \quad (2.27)$$

and since the natural logarithm of 1 is 0,

$$\Delta V_{BE} = \frac{kT}{q}\left[\ln\left(\frac{I_{C1}}{I_{C2}}\right)\right] \quad (2.28)$$

In Eq. 2.28 the terms k and q are physical constants, and if the respective collector currents are made constant and nonequal (to prevent their ratio from being 1), I may lump these terms together to form a new constant, k' (the log of a constant is also a constant).

39

I can then rewrite Eq. 2.28 in the form:

$$\Delta V_{BE} = k' \, T \qquad (2.29)$$

and rearranging to obtain an equation that is a function of temperature:

$$T = \frac{\Delta V_{BE}}{k'} \qquad (2.30)$$

I can conclude, therefore, that voltage ΔV_{BE} is proportional to the absolute temperature of the junction. In fact, assuming that the ratio I_{C1}/I_{C2} is set equal to 2:1, and

$$k' = (k/q) \ln (IC1/ZI_{C2}) \qquad (2.31)$$
$$k' = (1.38 \times 10^{-23}(1.6 \times 10^{-19})/\ln 2 \qquad (2.32)$$
$$k' = 5.978 \times 10^{-5} \qquad (2.33)$$

By rearranging Eq. 2.30 we gain a conversion factor relating the output voltage to the temperature:

$$\frac{\Delta V_{BE}}{T} = k' = 5.978 \times 10^{-5} \qquad (2.34)$$

$$\frac{\Delta V_{BE}}{T} = 59.78 \ \mu V/°K \qquad (2.35)$$

I can amplify this potential to make it both larger and numerically equal to the temperature. The required amplification to give me a scale factor of 10 mV/°K is:

a. Convert Eq. 2.35 to mV/°K:

$$\frac{\Delta V_{BE}}{T} = 0.05978 \ mV/°K \qquad (2.36)$$

b. The amplification,

$$A_V = E_{OUT}/E_{IN} \qquad (2.37)$$
$$A_V = (10 \ mV)/(0.05978 \ mV)/°K \qquad (2.38)$$
$$A_V = 167.3 \qquad (2.39)$$

TEMPERATURE MEASUREMENT DEVICES

Since the first edition of this book was published, a number of semiconductor device manufacturers have offered temperature measurement/ control integrated circuits (TMIC). These devices are almost all based on the PN junction properties discussed earlier in this chapter, although at least one by Analog Devices, Inc. uses an external thermocouple. In this chapter, I will look at the semiconductor TMIC devices offered by National Semiconductor and Analog Devices, Incorporated. In addition, I will look at a method for converting temperature measurements to frequencies that can be transmitted along a communications link or recorded on a tape recorder.

National Semiconductor LM-335. The National Semiconductor LM-335 device shown in Fig. 2-20 is a three-terminal temperature sensor. The two main terminals are for power (and output), while the third terminal, shown coming out the body of the diode symbol is for adjustment and calibration. The LM-335 device is basically a special zener diode in which the breakdown voltage is directly proportional to the temperature. It has a transfer function of close to 10 millivolts per degree Kelvin (10 mV/°K).

The LM-335 device and its wider range cousins, the LM-135 and LM-235, operate with a bias current set by the designer. This current is not supercritical, but must be within the range of 0.4 to 5 milliamperes. For most applications, designers seem to prefer currents in the 1-mA range.

Accuracy of the device is relatively good, and is more than sufficient for most control applications. The LM-135 version offers uncalibrated errors of 0.5 to 1 °C, while the less costly LM-335 device offers errors of < 3 °C. Of course, clever design can reduce these errors even further if they are out of tolerance for some particular application.

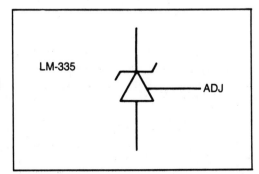

Fig. 2-20. LM-335 transducer diode.

41

One difference between the three devices is the operating temperature range, which are as follows:

Device Type No.	Temperature Range (Centigrade)
LM-135	− 55 to + 150
LM-235	− 40 to + 125
LM-335	− 10 to + 100

There are two packages used for the LM-135 through LM-335 family of devices. The TO-92 is a small plastic transistor case ("Z" suffix to part number, e.g., LM-335Z), while the TO-46 is a small metal can transistor package (smaller than the familiar TO-5 case). This case is identified with the suffix "H" or "AH" (for example, LM-335H or LM-335AH).

The simplest, although least accurate, method of using the LM-335 device is shown in Fig. 2-21A. The LM-335 is essentially a zener diode, and here it is connected as a zener diode. The series current-limiting resistor limits the current to around 1 milliampere. This value of $R1$ (i.e., 4700 ohms) is appropriate for + 5 volt power supplies as might be found in digital electronic instruments. The resistor value can be scaled upwards for higher values of dc potential according to Ohm's law (keeping I = 0.001 amperes):

$$R(ohms) = (V+) \times 1000 \qquad (2.40)$$

For example, when the power supply voltage is + 12 volts dc, the value resistor in series with the LM-335 is:

$$R(\text{ohms}) = (V+) \times 1000$$
$$R(\text{ohms}) = (12 \text{ volts}) \times 1000$$
$$R(\text{ohms}) = 12,000 \text{ ohms}$$

The output of the circuit in Fig. 2-21A is taken across the LM-335 device. This voltage has an approximate rate of 10 mV/°K. Recall from earlier that "degrees Kelvin" is the same as "degrees Centigrade," except that the zero point is at absolute zero (close to − 273 °C) rather than the freezing point of water. Using ordinary units conversion arithmetic will show us how much voltage to expect at any given temperature. For example, suppose I want to know the output voltage at 78 °C. The first thing I must do is convert the temperature to degrees Kelvin. This neat little trick

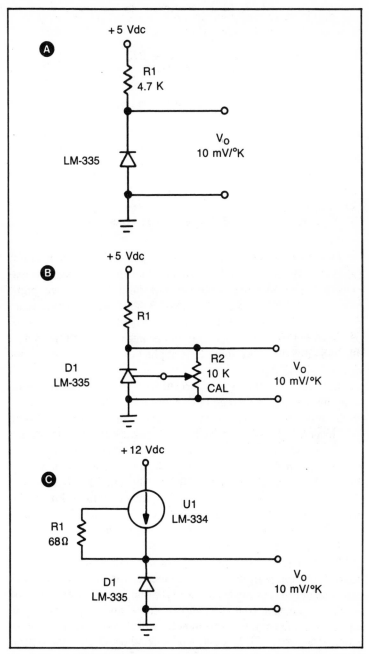

Fig. 2-21. (A) LM-335 simplest circuit. (B) LM-335 adjustable circuit. (C) Circuit using LM-334 current source diode.

is done by adding 273 to the centigrade temperature:

$$°K = °C + 273 \qquad\qquad (2.41)$$
$$°K = 78 \ °C + 273$$
$$°K = 351$$

Next, we convert the temperature to the equivalent voltage:

$$V = \frac{10 \ mV}{K} \times 351 \ K$$

$$V = (10 \ mV)(351)$$

$$V = 3510 \ mV = 3.51 \text{ volts}$$

One problem with the circuit of Fig. 2-21A is that it is not calibrated. While that circuit works well for many applications (especially where precision is not needed), for other cases we might want to consider the circuit of Fig. 2-21B. This circuit allows single-point calibration of the temperature. The calibration control is obtained from the 10-kohm potentiometer in parallel with the zener. The wiper of the potentiometer is applied to the adjustment input of the LM-335 device.

Calibration of the device is relatively simple. One only needs to know the output voltage (a dc voltmeter will suffice), and the environmental temperature in which the LM-335 exists. In some less than critical cases, one might take a regular glass mercury thermometer and measure the air temperature. Wait long enough after turning on the equipment for both the mercury thermometer and the LM-335 device to come to equilibrium. After that, adjust the potentiometer (R2) for the correct output voltage. For example, if the room temperature is 25 °C (i.e., 298 °K), then the output voltage will be 2.98 volts. Adjust the potentiometer for 2.98 volts under these conditions.

Another tactic is to use an ice-water bath as the calibrating source. The temperature 0 °C is defined as the point where water freezes, and is recognized by the fact that ice and water coexist in the same spot (the ice neither melts nor freezes, it is in equilibrium). A mercury thermometer will show the actual temperature of the bath. The potentiometer is adjusted until the output voltage is 2.73 volts (note: 0 °C = 273 °K).

Still another tactic is to use a warm oil bath for the calibration.

The oil is heated to somewhat higher than room temperature (maybe 40 °C), and stirred. Again, the mercury thermometer is used to read the actual temperature, and the potentiometer is adjusted to read the correct value. The advantage of this method is that the oil bath can be a constant temperature situation. There are numerous laboratory pots on the market that will keep water or oil at a constant preset temperature, a factor that avoids some problems inherent in the other methods.

Another connection scheme for the LM-335 is shown in Fig. 2-21C. In this variation, a National Semiconductor LM-334 three-terminal adjustable current source is used for the bias of the LM-335 device. Again, the output voltage will be 10 mV/°K.

Any application where the sensor is operated directly into its load suffers a potential problem or two, especially if the load impedance changes, or if it is lower than some limit. As a result, this sometimes justifies using the buffered circuit of Fig. 2-22.

A "buffer" amplifier is one that is used for one or both of two purposes: 1) impedance transformation or 2) isolation. The impedance transformation factor is used when the source impedance is high (not true of the LM-335). The isolation factor is somewhat more concern to us here. The operational amplifier in Fig. 2-22 places an amplifier between the sensor and its load. The gain of the amplifier in this case is unity (i.e., "1"), but a higher gain could be used if desired. In that case, simply substitute one of the gain amplifier circuits shown later in this book.

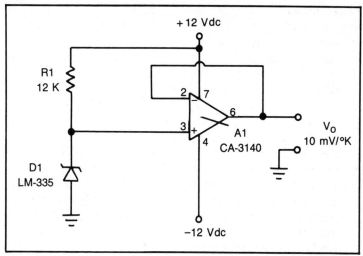

Fig. 2-22. LM-335 transducer diode used in operational amplifier circuit.

The operational amplifier shown here is an RCA CA-3140 device. This is simply for the freedom from bias currents exhibited by the BiMOS RCA operational amplifiers. The bias currents found on cheaper operational amplifiers could conceivably introduce error. The CA-4140 is not the only operational amplifier that will work, however; any low input bias current model will work nicely.

The noninverting input of the operational amplifier is connected across the zener diode-like LM-335. In this respect, this circuit looks somewhat like the typical voltage reference circuits seen elsewhere. The bias for the LM-335 is from a 12-kohm resistor, which is in keeping with our rule given earlier (Ohm's law, remember?).

Since there is no voltage gain in this circuit, the output voltage factor is the same as in previous designs, 10 mV/°K.

A circuit like Fig. 2-22 might prove useful in monitoring remote temperatures. If the operational amplifier is powered, a four-wire line is needed (V−, V+, ground and temperature). The advantage is that the line losses are overcome by the higher output power of the operational amplifier. The LM-335 is a rugged little low-impedance device, however, and, in many cases, such measures would not be needed.

TEMPERATURE SCALE CONVERSIONS

The Kelvin scale is used extensively in scientific calculations, but is not always the most popular in practical measurement situations. In fact, I suspect that most readers of this book will want to make their temperature measurements in either degrees Centigrade or degrees Fahrenheit. In this section, I will discuss the circuit methods used for both.

If the sensor is being input into a microcomputer, then it might be prudent to use the simplest circuit available. This is to measure in degrees Kelvin, and then let the computer do the neat trick of converting the units. The formulae below are useful for this purpose:

$$C = K - 273 \qquad\qquad (2.42)$$

and,

$$F = (1.8C) + 32 \qquad\qquad (2.43)$$

Of course, the first job will be to make the computer think it is seeing the correct kind of data. The analog-to-digital (A/D) converter will input (more than likely) a binary number between 00000000 and 11111111. This number must be scaled to the proper value that represents a temperature value. I will assume that I have an eight-bit A/D converter and a temperature range of 0 to 100 degrees Centigrade. The input voltage to the A/D will be 2.73 to 3.73 volts. If the A/D converter can provide offset measurements, I can set the maximum range for 1 volt, and then offset it to 2.73 volts. In that unlikely case, 00000000 would represent 0 °C and 11111111 would represent 100 °C. More likely, I will use a 5-volt unipolar input A/D converter to measure the narrow range of 2.73 to 3.73 volts and suffer a resolution loss. Of course, this loss is not what it may seem because, in many cases, it will still be less than the nonlinearity of the transducer/sensor. In such a scheme, the voltage represented by a 1-LSB change in the A/D output data word is approximately 20 mV, so would thus represent 2 °K. If all I need to measure is within two degrees, then I can use this system. Otherwise, some form of offset measurement is needed.

Figure 2-23 shows a scheme for converting the "degrees Kelvin" output of the LM-335 sensor (D1) into "degrees Centigrade." Since Centigrade degrees are the same size as Kelvin degrees, no change of slope in the output factor is needed: the output is 10 mV/°C, and the circuit gain is unity.

The basic circuit of Fig. 2-23 is a dc differential amplifier based on a common operational amplifier (741-family devices work fine). The gain is set by $R4/R2$, assuming $R2 = R3$ and $R4 = R5$. The noninverting input of the dc differential amplifier receives the temperature signal, while the inverting input receives a dc offset bias. This circuit is adjusted by using potentiometer $R6$ to set the voltage at point "A" to +2.73 volts (use a 3-1/2 digit or more digital voltmeter). The result is that the output will be 2.73 volts less than it would have were the offset not placed in the circuit—thus, the output potential is scaled in degrees Centigrade.

Figure 2-24 shows a circuit for converting degrees Centigrade to degrees Fahrenheit. The problem here is that the two types of degrees: 1) are offset from each other (like Kelvin and Centigrade, they have different zero references), and 2) have different sizes. Thus, the conversion amplifier must offer both an offset and a change of slope. Figure 2-24 shows both. The offset is provided by potentiometer $R5$, which is used to set the ice-point (zero degrees Centigrade) output level. The feedback potentiometer is used to set a calibration point at some higher temperature (for example, 25 °C, or room temperature, i.e., 77 °F).

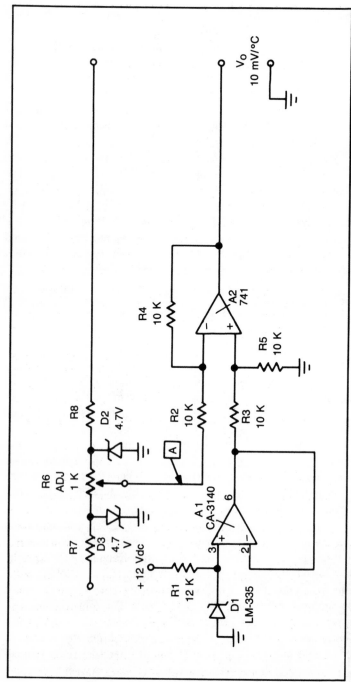

Fig. 2-23. Adjustable LM-338 circuit.

48

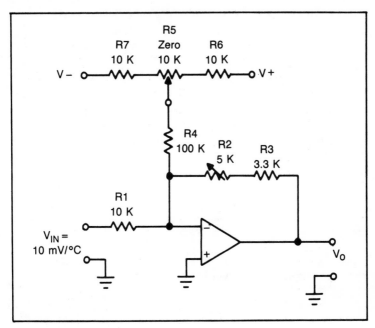

Fig. 2-24. F/C converter.

Calibration of the two points is performed in a manner similar to the above. The zero point is set using an ice bath (adjust *R*5); the higher point is probably best set at room temperature. In both cases, the actual temperature could be measured with an ordinary mercury thermometer. Of course, the best accuracy is obtained with a laboratory-grade mercury thermometer.

ANALOG DEVICES AD-590

The Analog Devices AD-590 is another form of solid-state temperature sensor. This particular device is a two-electrode sensor that operates as a current source with a one microampere-per-degree Kelvin (1 uA/°K) characteristic. The AD-590 will operate over the temperature range −55 to +150 degrees centigrade. It is capable of a wide range of power supply voltages, working with anything in the range +4 to +30 volts dc (this range is more than sufficient for most solid-state applications). Selected versions are available with linearity of −/+0.3 degrees Centigrade and a calibration accuracy of −/+0.5 degrees Centigrade.

The AD-590 comes in two different packages. There is a metal can (TO-52) that is recognized as the small-size transistor package (smaller than TO-5). There is also a plastic flat-pack available.

Being a two-terminal current source, the AD-590 is simplicity itself in operation. Figure 2-25 shows the most elementary circuit that can be calibrated for the AD-590. Since it is a current source producing a current proportional to temperature, I can convert the output to a voltage by passing it through a resistor. In Fig. 2-25, the resistance is approximately 1,000 ohms, and consisting of $R2$ (950 ohms) and $R1$ (a 100-ohm potentiometer). From Ohm's law, I know that 1 uA/K converts to 1 mV/K when passed through a 1000-ohm resistor. I can calculate the voltage output at any given temperature from the simple relationship below:

$$V_O = \frac{1 \ mV}{K} \times \text{TEMP} \qquad (2.44)$$

Thus, if I have a temperature of 37 °C, which is (37 + 273) or 310K, then the output voltage will be:

$$V_O = \frac{1 \ mV}{K} \times \text{TEMP}$$

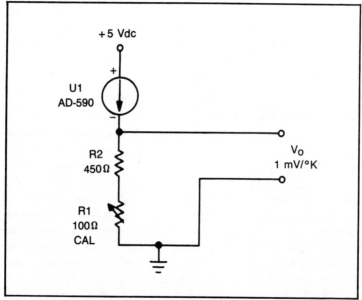

Fig. 2-25. Analog Devices AD-590 temperature transducer.

$$V_O = \frac{1 \; mV}{K} \times 310K$$

$$V_O = 310 \; mV$$

Potentiometer $R1$ is used to calibrate this system. You can make a quick and dirty calibration with an accurate mercury thermometer (laboratory grade recommended) at room temperature. Connect a digital voltmeter across the output, and allow the system to come to equilibrium (should take about ten minutes). Once the system is stable, adjust the potentiometer for the correct output voltage. For example, assume that the room temperature is 25 °C, which is 77 °F. This temperature converts to (272 + 25), or 298 °K. The output voltage will be (1 mV × 298), or 298 millivolts (0.298 volts). Using a 3-1/2 digit voltmeter is sufficient to make this measurement.

In some cases, it might be wise to delete the potentiometer and use a single 100-ohm resistor in place of the network shown. There might be several reasons for this action. First, the calibration accuracy is not critical for the application at hand. Second, potentiometers are points of weakness in any circuit. Being mechanical devices, they are subject to stress under vibration conditions and may fail prematurely. If the temperature accuracy is not crucial, and reliability is, then consider the use of a single fixed 1-percent tolerance resistor in place of the network shown in Fig. 2-25.

The circuit of Fig. 2-25 is sometimes used to make a temperature alarm. By using a voltage comparator to follow the network, and biasing the comparator to the voltage that corresponds to the alarm temperature, I can create a TTL level that indicates when the temperature is over the limit. A "window comparator" will allow me to have an alarm of either under- or over-temperature conditions. Some electronic equipment designers use this tactic to provide an overtemperature alarm. In one application, a commercial minicomputer generated a large amount of heat (it used a 35-ampere, + 5-volt dc power supply!). The specification called for an air-conditioned room for housing the computer. An AD-590 device was placed inside at a critical point. If the temperature reached a certain level (45 °C), then the comparator output snapped LOW and created an interrupt request to the computer. The computer would then sound an alarm and display an "overtemperature warning" message on the CRT screen.

The circuit of Fig. 2-25 suffers from one little problem: it allows calibration at only one temperature. Unfortunately, this situ-

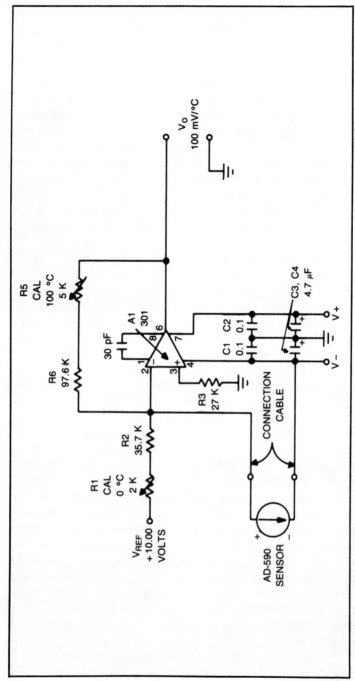

Fig. 2-26. Two-point calibration circuit using AD-590.

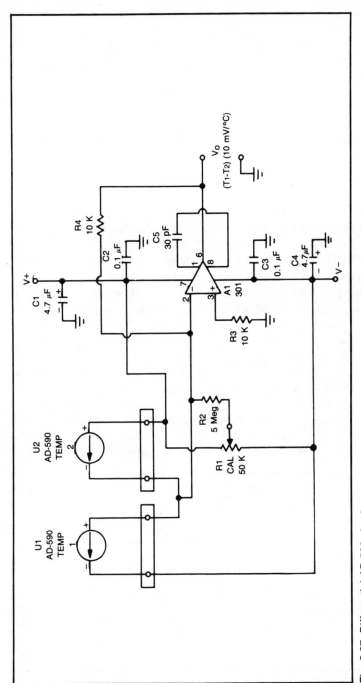

Fig. 2-27. Differential AD-590 circuit.

53

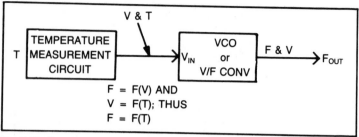

Fig. 2-28. Temperature-to-frequency converter.

ation does not allow for optimization of the circuit. I can, however, improve the situation using the two-point calibration circuit of Fig. 2-26. In this case, there is an operational amplifier in the inverting-follower configuration. The summing junction (inverting input) receives two different currents. One current is the output of the AD-590 (i.e., 1 uA/K), while the other current is derived from the reference voltage (V_{ref}) of 10.00 volts. Adjustment of this current provides the zero-reference adjustment, while the overall gain of the amplifier provides the full-scale adjustment.

The operational amplifier selected is the LM-301 device, although almost any premium operational amplifier will suffice. The RCA CA-3140 BiMOS device, or some of those by either Analog Devices or National Semiconductor will also work nicely. If the LM-301 or similar device is used, then be sure to use the 30-pF frequency compensation capacitor. See Figs. 2-28 and 2-29.

The V − and V + power supply lines are bypassed with 0.1-μF and 4.7-μF capacitors. The 0.1-μF capacitors are used for high-frequency decoupling, and must be mounted as close as possible

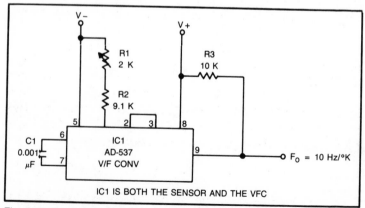

Fig. 2-29. Temperature-to-frequency converter using AD-537.

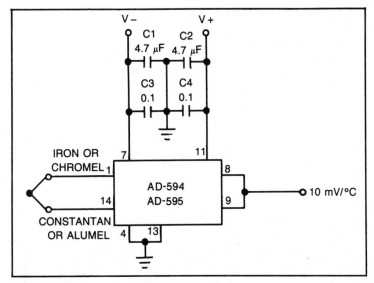

Fig. 2-30. Thermocouple signal processor IC.

to the body of the operational amplifier. The values of these capacitors are approximate, and they may be anything from 0.1-μF to 1-μF.

Calibration of the device is simple, although two different temperature environments are required. The zero-degrees Centigrade adjustment ($R1$) can be made with the sensor in an ice-water bath (as described above). The upper temperature value may be anything from 0.1-μF to 1-μF. See Figs. 2-30 and 2-31.

Calibration of the device is simple, although two different temperature environments are required. The zero-degrees Centigrade adjustment ($R1$) can be made with the sensor in an ice-water bath. The lines are bypassed with 0.1-μF and 4.7-μF capacitors. The 0.1-μF capacitors are used for high frequency decoupling, and must be mounted as close as possible to the body of the operational amplifier. The values of these capacitors are approximate. Bias current is derived from potentiometer $R1$. In the operational amplifier chapters to follow, you will learn that the output voltage of this circuit is proportional to the summation (or difference) of the input currents. Thus, the output voltage is given by:

$$V_O = (T1 - T2) \times (10 \ mV/°C) \qquad (2.45)$$

As you can see, the output voltage is proportional to the scale factor (10 mV/C) and the difference in temperatures.

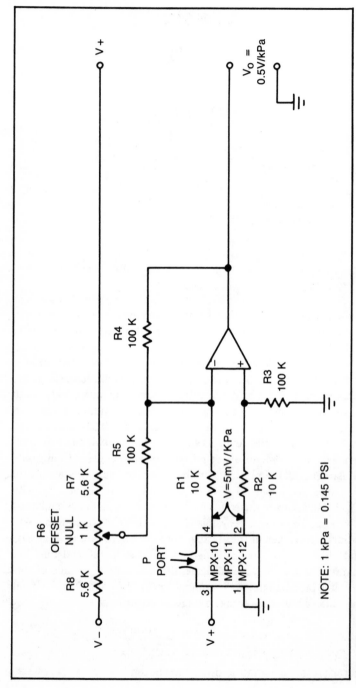

Fig. 2-31. Solid-state Motorola pressure transducer circuit.

Chapter 3

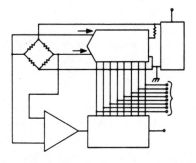

Operational Amplifiers

ELECTRONIC CIRCUIT DESIGN IS SOMETIMES RATHER ARCANE, so is usually left to the electrical engineer trained in electronics. If you wanted to design a transistor amplifier to use with the transducer in the previous chapter, which required a gain of 167.3, you would very rapidly find that it is a tough chore. But the introduction of the integrated circuit operational amplifier has changed all of that. In past decades, discrete operational amplifiers in self-contained chassis were often used by engineers and physical scientists to perform instrumental chores, because the operational amplifier responds to some simple design equations. The IC operational amplifier brings the application down to the circuit design level, and as one writer termed it, "the contriving of contrivances is a game for all."

The operational amplifier has certain properties which allow us to use simplified design and analysis techniques. Considering first, the ideal operational amplifier, there are the following basic properties:

1. Infinite open-loop voltage gain
2. Infinite input impedance
3. Zero output impedance ($Z_o = 0$)
4. Infinite bandwidth
5. Infinite common-mode rejection ratio

And since most IC operational amplifiers possess both inverting (−) and noninverting (+) inputs, property 1 can be amended to include an infinite open-loop differential voltage gain, and add one further property to the list:

6. Both inputs stay together, meaning that setting conditions at one input will effectively set the same conditions at the other input.

These properties define the operational amplifier and account for the elegant simplicity of operational amplifier circuit design. The implication of properties 1 through 3 is that the characteristics of an operational amplifier circuit follow directly from the characteristics of the negative-feedback loop. I may, therefore, design a circuit using operational amplifiers as the active element through consideration of the transfer function of the feedback network (sigh, ain't it exciting?).

INVERTING & NONINVERTING AMPLIFIERS

Figure 3-1 shows the symbol for an IC operational amplifier. The two inputs shown are the inverting (−) and noninverting (+) inputs. The inverting input produces an output signal that is 180°

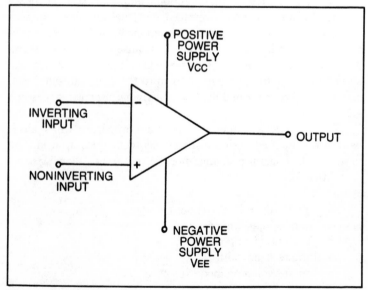

Fig. 3-1. Operational amplifier symbol.

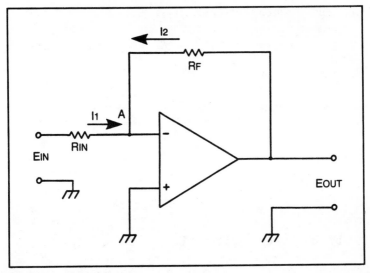

Fig. 3-2. Inverting follower.

out of phase with the input, while the noninverting input produces an inphase output signal.

Initially, I will consider the inverting amplifier configuration shown in Fig. 3-2. For the sake of simplicity, the power connections are deleted in our drawings.

In this circuit, there are two resistors, input resistor R_{IN} and negative-feedback resistor R_F. The noninverting input is grounded, so by property 6 we can legally treat point A (the junction of R_{IN}, R_F, and the inverting input) as if it were also grounded. This concept is usually called a "virtual ground" for lack of a word that says it better.

There are two methods for analyzing the operational amplifier circuit in order to obtain the transfer function equation: feedback theory and Kirchhoff's current law. Unless it is necessary to use a complex feedback network however, the latter is usually superior for making the circuit analysis and determining the overall circuit transfer function.

KIRCHHOFF'S LAW METHOD

Consider property 2: infinite input impedance. This implies that the input current is zero. The inverting input will neither sink nor source any current. The total current flowing into the summing junction at point A is $(I1 + I2)$, which by Kirchhoff's current law must sum to zero. Therefore:

$$-I1 = I2 \qquad (3.1)$$

But, by Ohm's law,

$$I1 = \frac{E_{IN}}{R_{IN}} \qquad (3.2)$$

and

$$I2 = \frac{E_{OUT}}{R_F} \qquad (3.3)$$

So, by substituting Eq. 3.2 and Eq. 3.3 into Eq. 3.1, I get

$$-\frac{E_{IN}}{R_{IN}} = \frac{E_{OUT}}{R_F} \qquad (3.4)$$

Solving Eq. 3.4 for E_{OUT} gives me the transfer function for an operational amplifier inverting follower:

$$E_{OUT} = \frac{-R_F E_{IN}}{R_{IN}} \qquad (3.5A)$$

By letting the voltage gain constant R_F/R_{IN} be represented by A_V I rewrite Eq. 3.5A in the form:

$$E_{OUT} = -A_V E_{IN} \qquad (3.5B)$$

The circuit for the noninverting follower will yield to a similar analysis. Because of property 6, I can claim that point A in Fig. 3-3A is at a potential equal to E_{IN}. In that case current $I1$ is

$$I1 = \frac{E_{IN}}{R_{IN}} \qquad (3.6)$$

and $I2$

$$I2 = \frac{E_{OUT} - E_{IN}}{R_F} \qquad (3.7)$$

60

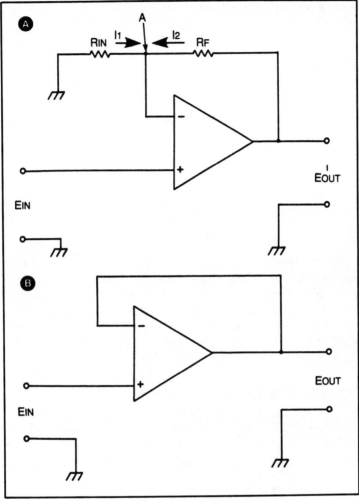

Fig. 3-3. Noninverting followers. (A) With gain. (B) Without gain.

In the noninverting amplifier, I must rewrite Eq. 3.1 as

$$I1 = I2 \qquad (3.8)$$

so, by plugging Eq. 3.6 and 3.7 into 3.8 I get

$$\frac{E_{IN}}{R_{IN}} = \frac{E_{OUT} - E_{IN}}{R_F} \qquad (3.9)$$

Solving for E_{OUT} yields the transfer function for a noninverting follower:

$$\frac{R_F E_{IN}}{R_{IN}} = E_{OUT} - E_{IN} \qquad (3.10)$$

$$\frac{R_F E_{IN}}{R_{IN}} + 1 = E_{OUT} \qquad (3.11)$$

Rearranging to satisfy a sense of order,

$$E_{OUT} = \left(1 + \frac{R_F}{R_{IN}}\right) E_{IN} \qquad (3.12)$$

Using the feedback-theory approach yields precisely the same results in both inverting and noninverting cases.

A special case of the noninverting follower is shown in Fig. 3-3B. This is the *unity-gain follower* configuration. Since the operational amplifier is connected directly to the $(-)$ input, the ratio $R_F/R_{IN} = 0$, so $A_V = 1$. This circuit is used mostly for buffering and impedance transformation where no gain is desired.

THE DC DIFFERENTIAL AMPLIFIER

The two inputs on the typical IC operational amplifier have equal but opposite effects on the amplitude of the output signal. Figure 3-4 shows the differential voltage situation as it appears to the two-input operational amplifier. Since the output voltage is de-

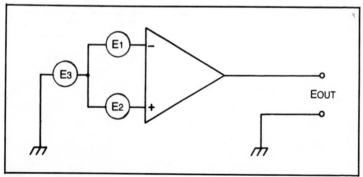

Fig. 3-4. Voltages affecting operational amplifier inputs.

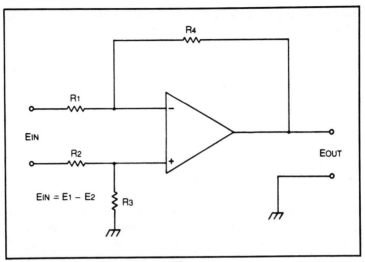

Fig. 3-5. Simple dc differential amplifier.

fined in terms of the input voltage and a gain factor, by property 6 I must rewrite Eqs. 3-5 and 3-12 in the form

$$E_{\text{OUT}} = \frac{-R_F}{R_{\text{IN}}}\,(E1 - E2) \qquad (3.13)$$

and,

$$E_{\text{OUT}} = 1 + \left(\frac{R_F}{R_{\text{IN}}}\right)(E1 - E2) \qquad (3.14)$$

respectively, whenever $E1 = E2$. In the case where $E1 = E2$, of course, $E_{\text{OUT}} = 0$.

Figure 3-5 shows a basic differential amplifier using a single, low-cost IC operational amplifier. If I require that $R1 = R2$ and $R3 = R4$, I can calculate the gain of the stage using Eq. 3.15:

$$A_V = \frac{R3}{R2} = \frac{R4}{R1} \qquad (3.15)$$

The transfer function is

$$E_{\text{OUT}} = \frac{R4\ E_{\text{IN}}}{R1} \qquad (3.16)$$

63

or

$$E_{\text{OUT}} = \frac{R3\ E_{\text{IN}}}{R2} \qquad (3.17)$$

where E_{IN} is the differential input voltage $E1 - E2$.

The balance of this type of circuit is critical if property 5 is to be preserved. The common-mode voltage ($E3$ in Fig. 3-4) should produce an output voltage of zero. If the resistor equalities $R1 = R2$ and $R3 = R4$ are not maintained, then common-mode rejection deteriorates.

The simple differential amplifier of Fig. 3-5 suffers from several problems, one of which is low input impedance. Because of this problem, there is often a modified input circuit. In some cases, particularly on equipment several years old, designers used high input impedance JFETs to buffer each input, but this becomes less attractive now that high-performance operational amplifier devices are available at low cost. The use of the JFET at each input was an open invitation to thermal drift. Another technique is to use a pair of unity-gain, noninverting followers—one at each input. Modern MOSFET input operational amplifiers work well at this job.

A superior alternative seems to be the instrumentation amplifier of Fig. 3-6, a technique that preserves the high impedance of the noninverting follower yet offers gain.

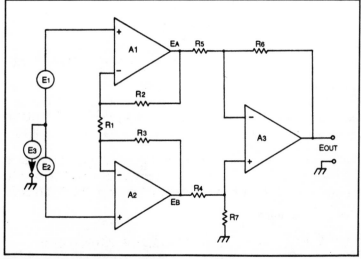

Fig. 3-6. Instrumentation amplifier.

I can treat input amplifiers $A1$ and $A2$ noninverting gain followers which produce output voltages of:

$$E_A = 1 + \left(\frac{R2}{R1}\right)E1 - \left(\frac{R2\ E2}{R1}\right) \qquad (3.18)$$

$$E_B = 1 + \left(\frac{R3}{R1}\right)E2 - \left(\frac{R3\ E1}{R1}\right) \qquad (3.19)$$

If I assume initially that the voltage gain of amplifier $A3$ is set to unity and that $R2 = R3$, then

$$E_{OUT} = E_B - E_A \qquad (3.20)$$

and if this operation is actually carried out using Eqs. 3.18 and 3.19 in place of E_A and E_B, the result will be

$$E_{OUT} = (E2\ -\ E1)\left(1 + \frac{2R2}{R1}\right) \qquad (3.21)$$

In the case where the gain of amplifier $A3$ is greater than unity, I must multiply Eq. 3.21 by the voltage gain of $A3$. If $R4 = R5$ and $R6 = R7$:

$$A_{V(A3)} = \frac{R6}{R5} \qquad (3.22)$$

So the transfer function of the instrumentation amplifier as a whole is given by:

$$E_{OUT} = (E2 - E1)\ 1 + \left(\frac{2R2}{R1}\right)\left(\frac{R6}{R5}\right) \qquad (3.23)$$

Oddly enough, mismatching resistors $R2$ and $R3$ will not create a significant common-mode rejection problem, as would be true if $(R6/R5) = (R7/R4)$, but it does cause an error in the differential voltage gain.

LIMITATIONS OF REAL OPERATIONAL AMPLIFIERS

Real IC operational amplifiers vary considerably from the ideal case. These limitations conspire to constrict the designer's freedom. For example, I have made the claim that the operational amplifier inputs will neither sink nor source current, but in real devices there is an input offset current. The input stage of the operational amplifier is a two-input transistor differential amplifier of rather ordinary configuration. Both transistors in the differential pair will require bias current, denoted I_{B1} and I_{B2}, and the offset current is their difference, namely:

$$I_{OFF} = I_{B1} - I_{B2} \qquad (3.24)$$

The offset current produces an output voltage artifact equal to $R_F \times I_{OFF}$. In the inverting configuration of Fig. 3-2, I find that bias current I_B flowing from the inverting input produces a voltage drop across the input and feedback resistors. This will create an output voltage aircraft equal to

$$\pm (R_F \, R_{IN}) I_B A_V \qquad (3.25)$$

In most cases, this artifact will be larger than the I_{OFF} artifact.

A good design practice to reduce this source of error is the application of an equal magnitude potential at the noninverting input. Such a potential is generated in Fig. 3-7 by resistor R_C, which has a value equal to the value of the parallel combination of R_F and R_{IN}.

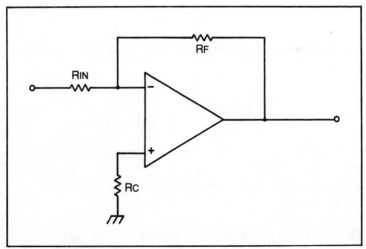

Fig. 3-7. Use of a compensation resistor.

66

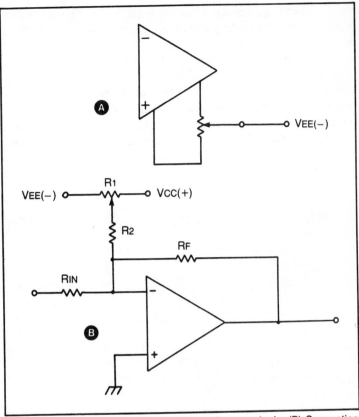

Fig. 3-8. Offset null techniques. (A) Using offset terminals. (B) Summation junction.

One other offset is a voltage term that seems to have no origin when considered solely from the viewpoint of external components—it is internally generated. This potential will be quoted in operational amplifier specification sheets as an "input offset voltage." It is defined as the voltage existing at the output when the term $E2 - E1$ is forced to be zero and when there is no resistance to create a voltage drop from bias currents.

Two general techniques are used to suppress all forms of output voltage offset artifact; these are shown in Fig. 3-8. That shown in Fig. 3-8A uses a special pair of null terminals found on some IC operational amplifiers. Take care, though, because not all operational amplifiers have these terminals.

The second technique, shown in Fig. 3-8B, offers applicability to almost all operational amplifier circuits. In this case, use a coun-

67

tercurrent through resistor $R2$ to cancel the offset. The potentiometer is connected to V_{CC} and V_{EE}, so offsets of either polarity can be accommodated.

The design of operation circuits for most simple applications is relatively easy, so simple in fact, that it removes the design of circuitry down from the exclusive realm of the engineer to the world populated by more ordinary people. Although some additional applications will be offered in this book, let me immodestly suggest that if you would like to pursue the topic further, refer to my other TAB books: *Linear IC/Op Amp Handbook—2nd Edition* (TAB No. 1550), and *How To Design and Build Electronic Instrumentation—2nd Edition* (TAB No. 2660).

Chapter 4

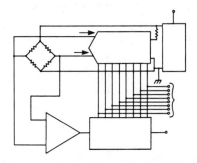

Operational Amplifier
Circuit Design

THIS CHAPTER COVERS DESIGN OF ACTUAL OPERATIONAL AM-
plifier circuits from the procedures point of view. Circuits are
not usually designed by simply plugging in values to the formulas
presented in books. There is some judgement required and a lot
of trade-offs.

INVERTING FOLLOWER CIRCUITS

The inverting follower circuit (an example of which is shown
in Fig. 4-1) uses just two resistors, input resistor $R1$ and feedback
resistor $R2$, to form an amplifier that obeys the transfer equation:

$$\frac{E_{\text{OUT}}}{E_{\text{IN}}} = \frac{R_F}{R_{\text{IN}}} \qquad (4.1)$$

The value of resistor R_{IN} effectively sets the input impedance,
and one does not ordinarily want the value of the feedback resistor to exceed 1 megohm unless a premium operational amplifier
or one with a JFET or MOSFET input stage is used.

Example 4-1
I need an inverting amplifier with a voltage gain of 10. The
source impedance is 1000 ohms.

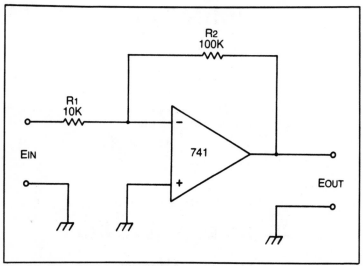

Fig. 4-1. Gain-of-10 inverting follower.

Solution: Since Z_S = 1000 ohms, a common rule of thumb is to make R_{IN} (in this case designated $R1$) 10 × 1000 ohms, or 10,000 ohms minimum. Therefore, let $R1$ = 10K. Hence,

$$A_V = R_F / R_{IN} \qquad (4.2A)$$
$$A_V = R2/R1 = 10 \qquad (4.2B)$$

Substitute in the value of $R1$ and solve for the value of $R2$.

$$10 = R2/10K \qquad (4.3)$$
$$10 \times 10K = R2 \qquad (4.4)$$
$$100,000 = R2 \qquad (4.5)$$

Example 4-2

Design an inverting amplifier with a voltage gain of 56 with an input impedance of 100K.
Solution:
1. To meet the input impedance specification set $R1$ (in Fig. 4-2) to 100K.
2. Compute the value of feedback resistor $R2$.

$$A_V = 56 = -R_F / R_{IN} \qquad (4.6A)$$
$$56 = -R2/R1 \qquad (4.6B)$$
$$56 = -R2/100K \qquad (4.7)$$

$$R2 = (56)(100K) \tag{4.8}$$
$$R2 = 5{,}600{,}000 \text{ ohms} \tag{4.9}$$
$$R2 = 5.6M$$

This value is uncomfortably high for use with the 741 and other low-cost operational amplifiers with bipolar input stages because the transistor bias current when dropped across such a high resistance will create a voltage offset of considerable magnitude at the output. It is recommended, then, that an RCA CA3140/CA3160 or some similar device is used for the amplifier element.

Example 4-3

I want an amplifier stage to produce a gain of 0.20 with an input impedance of 100K. Determine the value of the required resistors.

Solution:

1. Let $R_{IN} = 100K$ (Fig. 4-3).
2. Compute R_F.

$$R_F / R_{IN} = 0.20 \tag{4.10}$$
$$R_F / 100K = 0.20 \tag{4.11}$$
$$R_F = 0.2 \times 100K \tag{4.12}$$
$$R_F = 20{,}000 \text{ ohms} \tag{4.13}$$

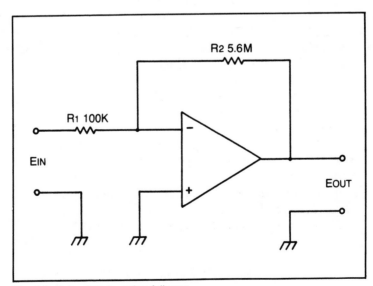

Fig. 4-2. Gain-of-56 inverting follower.

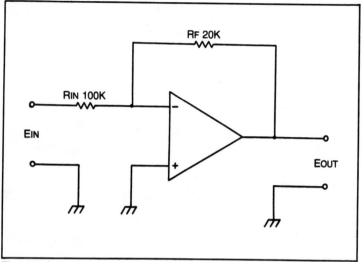

Fig. 4-3. Gain-of-2/10 inverting follower.

Example 4-4

Design an amplifier with a gain of 1200. The phase of the output signal is required to be positive with respect to the input, and the input impedance should be at least 10K.

Solution:

It is usually considered good practice when using low-cost operational amplifiers to limit the gain required of any one stage to something less than 500. So, I will want a two-stage or more amplifier chain to solve this problem. Since the output phase must be noninverted, it will be best to use two inverting amplifier stages.

1. Drawn an appropriate circuit diagram (see Fig. 4-4). The gain of this amplifier is given by:

$$A_V = 1200 = (R2/R1)\ (R4/R3) \qquad (4.14)$$
$$A_V = 1200 = (R2R4)/R1R3) \qquad (4.15)$$

2. Let $R1 = 10K$ in order to meet the input impedance specification.

3. The 741 sheet lists $Z_0 = 75$ ohms, so $R3$ should be greater than 10×75 ohms, or 750 ohms. Therefore, let $R3 = 1000$ ohms.

4. Rewrite Eq. 4.15 using the values determined in steps 2 and 3.

$$1200 = \frac{R2R4}{(10^4)(10^3)} \qquad (4.16)$$

$$1200 = \frac{R2R4}{10^6} \qquad (4.17)$$

To get a voltage gain of 1200 I merely factor 1200 to find more reasonable gain figures for $A1$ and $A2$. Possible factors:

$$
\left.
\begin{array}{c}
2 \times 600 \\
3 \times 400 \\
4 \times 300
\end{array}
\right\} \quad \text{one factor too high}
$$

$$
\left.
\begin{array}{c}
30 \times 40 \\
24 \times 50 \\
10 \times 120
\end{array}
\right\} \quad \text{more reasonable to obtain}
$$

Try all of these combinations of acceptable factors to find values of $R2$ and $R4$ that are near standard resistor values.

Trial	$A_{V(A1)}$	$A_{V(A2)}$	R1	R4	Comments
1	30	40	300K	40K	nonstandard values
2	40	30	400K	30K	nonstandard values
3	24	50	240K	50K	okay
4	50	24	500K	24K	okay
5	10	120	100K	120K	okay
6	120	10	1.2M	10K	okay

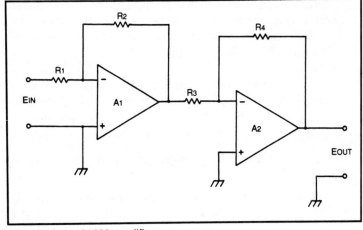

Fig. 4-4. Gain-of-1200 amplifier.

Trials 1 and 2 resulted in values for $R4$ and $R2$ that are not standard resistances, and trials 3 and 4 resulted in resistances that, while technically standard, are usually hard to obtain. It is best, then, to use trial 5 or 6.

Rule: Use the lowest values for all resistors in an operational amplifier circuit that will result in achieving the goal. In general, keep the feedback resistor less than 500K, or 1M at the outside, when using low-cost operational amplifiers.

If I follow the above rule, I would accept trial 5, which gives me 100K for $R2$ and 120K for $R4$. The required values then are:

$$R1: 10K \qquad R3: 1K$$
$$R2: 100K \qquad R4: 120K$$

Does the total gain check with the specifications of 1200? To find out plug the above values into Eq. 4.15 and see if the equation balances.

$$A_V \;=\; (R2R4)/(R1R3) \qquad (4.18)$$
$$\;=\; (100)\,(120)/(10)\,(1) \qquad (4.19)$$
$$A_V \;=\; 1200$$

The answer checks out.

Example 4-5
Design an inverting amplifier that will deliver an output signal of exactly 2.56 volts when the input voltage is 100 millivolts. Assume that the input impedance will be 10K or greater. This particular amplifier specification is common for those used with 8-bit microprocessors because they will recognize and resolve up to 256 different voltage levels.
Solution:
1. First convert the two voltages to the same units, so that the required voltage gain can be determined. Volts and millivolts are two different units, so will not mix in the same formula without conversion of one to the other.

$$100 \text{ mV} \times 1V/1000 \text{ mV} = 0.1V \qquad (4.20)$$

2. Compute the required voltage gain using $A_V = E_{OUT}/E_{IN}$

$$A_V = E_{OUT}/E_{IN} = 2.56/0.1 = 25.6 \qquad (4.21)$$

3. Set R_{IN} to 10K in order to meet the input impedance specification.
4. Compute the value of the feedback resistor.

$$R_F / R_{IN} = 25.6 \qquad (4.22)$$
$$R_F / 10K = 25.6 \qquad (4.23)$$
$$R_F = (25.6)\,(10K) \qquad (4.24)$$
$$R_F = 256K \qquad (4.25)$$

5. There are two tactics you may follow at this juncture. First, you *could* order 10K and 256K precision resistors, but you will find that suppliers of precision resistors are reluctant to sell them in quantities less than 50 each. The alternate solution, assuming that you don't want to live with the error inherent in using the nearest standard value, is to make the feedback resistor variable (see Fig. 4-5) across the 256K value that is required. In this case, set $R1 = 240K$ (a standard value) or 220K (also standard, but easier to obtain) at a tolerance of 5%. Then make $R2$ a potentiometer rated at 100K. This potentiometer should be a ten-turn trimmer pot for precision adjustment of the overall gain. You can then apply a potential of exactly 100 mV to the input and adjust $R2$ for precisely 2.56 volts at the output.

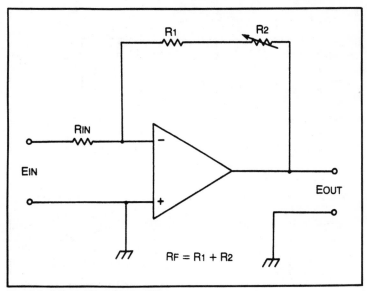

Fig. 4-5. Adjustable gain amplifier.

NONINVERTING FOLLOWERS

The noninverting follower (see Fig. 4-6) uses the noninverting input of the operation amplifier, so the output will be in phase with the input. The noninverting follower obeys the relationship:

$$A_V = 1 + \frac{R_f}{R_{IN}} \tag{4.26}$$

Note that negative feedback is still used, but the input resistor has one end grounded.

The input impedance of the noninverting follower is extremely high, and that is probably the single most important reason for using this circuit. The input impedance is *theoretically* infinite, but when I go from the ideal to the real I find that *very high* is a little nearer the truth. In low-grade operational amplifiers, the input impedance of the typical noninverting follower is on the order of 1 megohm or more, while certain operational amplifiers that have MOSFET input stages boast input impedances on the order of 1.5 teraohms (10^{12} ohms).

Example 4-6
Design a noninverting amplifier with a voltage gain of 11.

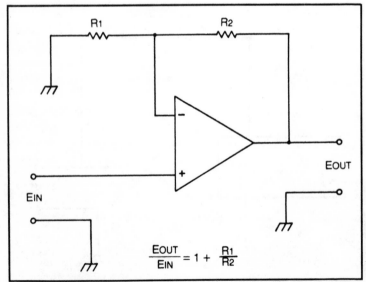

Fig. 4-6. Noninverting follower.

Solution:

1. Pick a value for the input resistor ($R1$). A good value is 1000 ohms since it lets us obtain respectable gain figures without using high values of feedback resistor. Let $R1$ = 1000 ohms.
2. Solve the general transfer equation for noninverting followers for the value of the feedback resistance assuming A_V = 11 and $R1$ = 1000 ohms:

$$A_V = 11 = 1 + R2/R1 \tag{4.27}$$

$$11 = 1 + R2/1000 \tag{4.28}$$

$$10 = R2/1000 \tag{4.29}$$

$$(1000)(10) = R2 \tag{4.30}$$

$$10,000 = R2 \tag{4.31}$$

So $R2$ will equal 10K.

Example 4-7

I want a preamplifier with a gain of 10 to be used as the front end of a voltmeter. The amplifier must, therefore, have a high input impedance so as to not load the circuit.

Solution:

1. A noninverting gain-of-ten follower is indicated because of the input impedance requirement. Furthermore, it should be a gate-protected MOSFET-input operational amplifier that is used. An example is the RCA CA3130, CA3140, and CA3160 series of devices.
2. Draw an appropriate circuit diagram (see Fig. 4-6).
3. Calculate trial values by setting $R1$ to some standard values, then calculating appropriate values of $R2$. Let A_V = 10, so

$$R2 = (A_V - 1)R1 = 9R1$$

Trial	R1	R2	Use	A_V	Error
1	680	6120	6.8K	11	+10%
2	820	7380	7.5K	10.1	+1%
3	1000	9000	9.1K	10.1	+1%
4	1200	10.8K	10K	9.3	-7%
5	1500	13.5K	15K	11	+10%
6	2200	19.8K	20K	10.1	+1%

Trial	R1	R2	Use	A_V	Error
7	3300	29.7K	27K	13	+30%
8	3900	35.1K	33K	9.5	−5%
9	4700	42.3K	39K	9.3	−7%
10	5600	50.4K	50K	9.9	−1%

If precision is desired, none of these trials is very reliable, making $R2$ a series combination of a fixed resistor and a ten-turn pot will give me the desired accuracy. Therefore, use the values given in trials 2, 3, 6, or 10. I will arbitrarily pick trial 6, and let $R1$ = 2.2K and $R2$ = 20K. A good trick for $R2$ is to use an 18K fixed resistor and a 5K ten-turn potentiometer.

Example 4-8
Design a noninverting follower with a gain of 76.
Solution:
1. Let $R1$ = 1000 ohms.
2. Solve the transfer equation for $R2$.

$$A_V = 76 = (R2/R1) + 1 \qquad (4.32)$$
$$76 = (R2/1000) + 1 \qquad (4.33)$$
$$75 = R2/1000 \qquad (4.34)$$
$$75,000 = R2 \qquad (4.35)$$

So $R2$ will be equal to 75K.

Example 4-9
An inverting amplifier is required that has a gain of 550 and a very high input impedance. Draw an appropriate circuit diagram (see Fig. 4-7), and calculate the resistor values.

A noninverting amplifier is used as the input stage so that the input impedance specification is met. An inverting amplifier follows this stage to provide inversion.
Solution:
1. Factor the gain specification (550) into possible gains for $A1$ and $A2$. Let $R1$ = $R3$ = 1000 ohms for convenience of calculation. Note that $R4$ = 1K × $A_{V(A2)\,and\,R2}$ = 1K × ($A_{V(A1)}$ − 1). Possible gain factors are as follows:

Trial	$A_{V(A1)}$	$A_{V(A2)}$	R2	R4
1	110	5	109K	5 K
2	5	110	4K	110 K

Trial	$A_{V(A1)}$	$A_{V(A2)}$	R2	R4
3	11	50	10K	50 K
4	50	11	49K	11 K
5	20	27.5	19K	27.5K

There may be, of course, innumerable factors, so select only a few.

Of the factors shown, the third has the best chance unless a potentiometer is used in place of either R2 or R4, or unless a gain error can be tolerated. The gain of this circuit is given by:

$$A_V = A_{VA1} \, A_{VA2} \qquad (4.36)$$
$$A_V = [(R2/R1) + 1] \, (R4/R3) \qquad (4.37)$$

I will look at each trial value to determine just what kind of gain errors might exist if the nearest standard value resistor is used in each case.

Trial	R2	R4	A_V	Error
1	110K	5K	555	0.9%
2	3.9K	110K	539	2.0%
3	10 K	50K	550	0 %
4	50 K	10K	510	7.3%
5	20 K	27K	567	3.1%

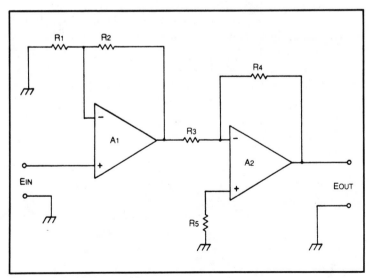

Fig. 4-7. Gain-of-550 amplifier.

If gain errors are not too important, then I would be quite free to accept any of these trials, although one might think that trial 4 with its 7.3% error is too far off for any practical purpose.

The value of $R5$ is found from

$$R5 = (R3R4)/(R3 + R4) \qquad (4.38)$$

Example 4-10

Design an amplifier to boost the output of a single-ended 20K transducer from 1 millivolt to 1 volt for display on an oscilloscope. Provide offset null, a sensitivity control (zero to full scale), and a position control capable of shirting the baseline of the output stage over the range -5 volts to $+5$ volts.

Solution:

1. Compute the overall gain that is required.

$$A_V = E_{OUT}/E_{IN} \qquad (4.39)$$
$$A_V = 1000 \text{ mV}/1 \text{ mV} \qquad (4.40)$$
$$A_V = 1000 \qquad (4.41)$$

2. Draw a suitable, if only tentative, circuit diagram (see Fig. 4-8). Of course, I must make some basic assumptions. In this case, I will want resistors $R3$ and $R5$ between 200 ohms and 15K, and I select gain factors for $A1$ and $A2$ that seem reasonable. In this case, I select 20 and 50 as suitable factors of 1000 for $A1$ and $A2$, respectively. These choices, incidentally, are made from trials as before, and are almost arbitrary. Also, to meet the specification of an input impedance high enough for a 20K source impedance it is best to use a noninverting follower as the input amplifier stage.

3. Compute the resistor values for $A1$.

$$A_{VA1} = 50 = (R2/R1) + 1 \qquad (4.42)$$
$$49 = R2/R1 \qquad (4.43)$$

I next try setting $R1$ to various standard values, then make trial computations for $R2$ for each. I will then select the value of $R2$ closest to a standard resistor value.

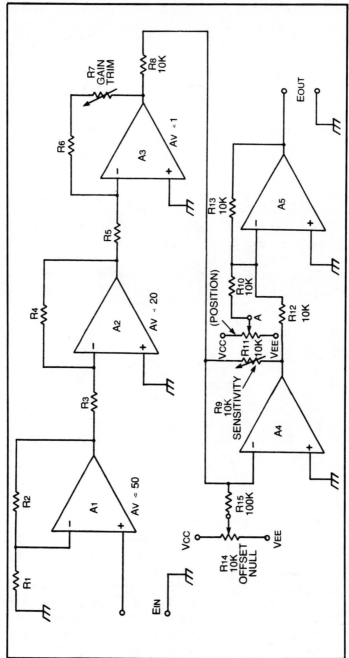

Fig. 4-8. Gain-of-1000 amplifier.

81

Trial	R1	R2	Use	A_v	Error
1	220	10.8K	10K	46.5	7%
2	270	13.2K	15K	56.6	13%*
3	330	16.2K	15K	46.0	8%*
4	390	19K	20K	51.3	2.6%
5	470	23K	24K	51.0	2%
6	560	27.4K	27K	48.0	4%
7	680	33.3K	33K	49.5	1%
8	1000	49K	49K	49.0	2%
9	1.5K	73.5K	75K	51.0	2%
10	1.8K	88K	82K	47.0	6%*
11	2.2K	108K	110K	51.0	2%

*Usually considered too much error.

Of these, the values in trial 7 are both easy to obtain and result in a gain of nearly 50. The resultant error is only − 1%. So let $R1$ = 680 ohms and $R2$ = 33K.

4. To compute the gain for $A2$, I use the same procedure as in step 3.

Trial	R3	R4	Use	A_v	Error
1	220	4.4K	4.7K	21.4	7%
2	270	5.4K	5.6K	20.7	3.5%
3	330	6.6K	6.8K	20.6	3%
4	390	7.8K	8.2K	21.0	5%
5	470	9.4K	9.1K	19.4	2%
6	560	11.2K	12 K	21.0	7%
7	680	13.6K	15 K	22.0	10.5%
8	820	16.4K	15 K	18.0	8.5%
9	1000	20 K	20 K	20.0	0%

Clearly, trial 9 in this case is the best from the gain point of view, so let $R3$ = 1000 ohms and $R4$ = 20K.

Amplifier $A3$ should have a nominal gain of unity, but I want to be able to trim out gain errors in amplifiers $A1$ and $A2$ caused by the use of standard resistor values. A good selection, considering the low errors of my choices, would be to let the gain of amplifier $A3$ vary over the range 0.9 to 1.1. If I let $R5$ = 10K and $R6$ = 9.1K (both are also standard values, but 9.1K can be made by paralleling 10K with 100K), then let $R7$ be a 2K ten-turn potentiometer, the gain of $A3$ will vary over the following range:

when $R7 = 2K$,

$$A_V = (R6 + R7)/R5 \qquad (4.44)$$
$$A_V = (9.1 + 2)/10 \qquad (4.45)$$
$$A_V = 11.1/10 = 1.1 \qquad (4.46)$$

when $R7 = 0$ ohms,

$$A_V = R6/R5 \qquad (4.47)$$
$$A_V = 9.1/10 \qquad (4.48)$$
$$A_V = 0.91 \qquad (4.49)$$

Both of these limits are approximately correct, so I will use them. If the range of $R7$ proves insufficient change it to a 5K potentiometer.

The last two stages are both unity gain, with amplifier $A5$ providing position control and $A4$ the control over sensitivity. I want to be able to shift the output baseline ± 4 volts; this is accomplished by a potentiometer $R11$ and resistor $R10$.

The output voltage due to $R11$ must shift between ± 4 volts, and is equal to

$$E_{\text{OUT}} = E_A (R13/R10) \qquad (4.50)$$

E_A will vary between VCC (+12 volts) and VEE (−12 volts), and $R13$ will be set to 10K. Find the value of $R10$.

$$E_{\text{OUT}} = E_A (10K/R10) \qquad (4.51)$$
$$R10 = E_A (10K)/E_{\text{OUT}} \qquad (4.52)$$
$$R10 = (12) (10K)/4 \qquad (4.53)$$
$$R10 = 30K \qquad (4.54)$$

This is a standard value, but is sometimes a little hard to find (not all standard values are stocked). If a 27K resistor is used instead:

$$E_{\text{OUT}} = (12V) (10K)/(27K) \qquad (4.55)$$
$$E_{\text{OUT}} = \pm 4.4V$$

which should present no problems in this application.

Provision for offset nulling and sensitivity control are made in amplifier $A4$. It is always good practice to null the cumulative effects of the offsets prior to the sensitivity control. Otherwise, the baseline will shift an amount approximately equal to the offset as the sensitivity control is varied through its range. The gain of $A4$ is:

when $R9 = 10K$,

$$A_V = R9/R8 = 10K/10K = 1 \qquad (4.56)$$

when $R9 = 0$ ohms,

$$A_V = R9/R8 = 0/10K = 0 \qquad (4.57)$$

Adjust $R14$ incrementally until $R9$ can be run from zero to full value with no dc shift in the output voltage. This is essentially a dc balance control.

DIFFERENTIAL AMPLIFIERS

A differential amplifier is a circuit that will deliver an amplified output voltage that is proportional to the difference between two ground-referenced input potentials, $(E1 - E2)$, known collectively as E_{IN}. The true differential amplifier will issue a zero output if the two input voltages are equal $(E1 = E2)$. Such voltages are known as common-mode potentials.

Example 4-11

Design a differential amplifier with a gain of 100. The amplifier should be suitable for amplifying the output of a source that has a looking-back resistance of 200 ohms.

Solution:

1. Resistors $R1$ and $R2$ (in Fig. 4-9) must be equal, and should have a value that is not less than five or ten times the looking-back resistance of the source. In other words:

$$(5 \times 200) \quad (R1 = R2) \quad (10 \times 200) \qquad (4.58)$$

$$1K \qquad R1,R2 \qquad 2K \qquad (4.59)$$

Personally, I prefer the ten times rule, so let $R1 = R2 = 2K$. Note that this is a minimum resistance.

2. Calculate $R3$

$$R3 = R4 \qquad (4.60)$$

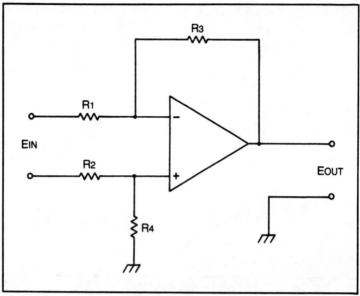

Fig. 4-9. Dc differential amplifier.

$$100 = R3/R1 = R4 \qquad (4.61)$$
$$100 = R3/2K = R4 \qquad (4.62)$$
$$200K = R3 = R4 \qquad (4.63)$$

The values, then, for a gain of 100 are:

$$R1 = 2K$$
$$R2 = 2K$$
$$R3 = 200K$$
$$R4 = 200K$$

Example 4-12

Design a differential amplifier with an input impedance of 10K and a gain of 256.

Solution:

1. Since the gain is greater than 100, it is good practice to use at least two operational amplifier stages, or a single stage with a premium operational amplifier as the active element. See Fig. 4-10 for an example of a suitable two-stage circuit.
2. Find several reasonable factors of 256.

   1. $2 \times 128 = 256$

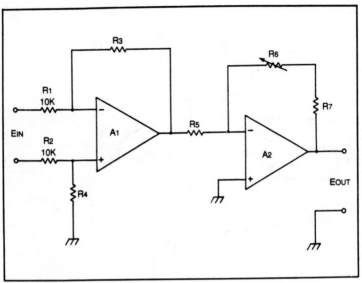

Fig. 4-10. Gain-of-256 differential amplifier.

2. $4 \times 64 = 256$
3. $8 \times 32 = 256$
4. $16 \times 16 = 256$

None of these factors are particularly exciting because they will not result in easy to obtain standard values of resistor, especially in amplifier $A1$, where all of the values are fixed. I will compromise, then, and modify factor 4 to 15×17. Furthermore, let me assign $A1$ a gain of 15 and $A2$ a gain of 17. Amplifier $A2$ is variable, so gain errors can be easily trimmed out.

3. Let $R1 = 10K$, and solve for $R3$ assuming that $A_V = 15$ for $A1$.

$$A_{V(A1)} = 15 = R3/R1 \qquad (4.64)$$
$$15 = R3/10K \qquad (4.65)$$
$$(15)(10K) = R3 \qquad (4.66)$$
$$150K = R3 \qquad (4.67)$$

4. Pick a standard value for $R5$ and compute $R6 + R7$ assuming that A_V is 17 for $A2$. Arbitrarily set $R5 = 2.2K$.

$$A_{V(A2)} = 17 = (R6 + R7)/R5 \qquad (4.68)$$

$$17 = (R6 + R7)/2.2K \qquad (4.69)$$

$$(17)(2.2K) = R6 + R7 \qquad (4.70)$$

$$37.4K = R6 + R7 \qquad (4.71)$$

A good move at this point would be to specify that $R7$ be set to 33K, a common standard value, and that $R6$ be made a 10K ten-turn potentiometer. This would allow the series combination $R6 + R7$ to vary over the range 33K to 43K. The values, then are:

$$R1 = 10K$$
$$R2 = 10K$$
$$R3 = 150K$$
$$R4 = 150K$$
$$R5 = 2.2K$$
$$R6 = 10K \text{ ten-turn pot}$$
$$R7 = 33K$$

Example 4-13

Find the correct resistor values for an instrumentation amplifier with a gain of 500.

Solution:

1. Make some assumptions using Fig. 4-11.
   a. Let $R1 = 1K$
   b. Resistances $R2 = R3$, $R4 = R5$, and $R6 = R7$.

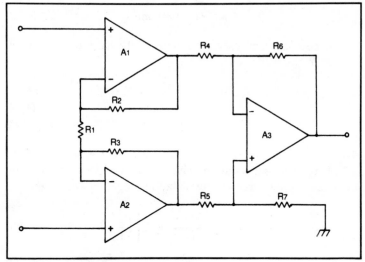

Fig. 4-11. Instrumentation amplifier.

c. Resistances $R4$ and $R5$ will be between 1K and 10K.

d. The voltage gain of $A3$ will be approximately 10, while that of $A1$ and $A2$ will be 50.

2. Compute $R2$ and $R3$

$$A_{V(A1 + A2)} = 50 = (2R2/R1) + 1 \qquad (4.72)$$

$$50 = (2R2/1000) + 1 \qquad (4.73)$$

$$49 = 2R2/1000 \qquad (4.74)$$

$$(49)(1000)/2 = R2 \qquad (4.75)$$

$$24.5\text{K (use 24K)} = R2 \qquad (4.76)$$

This results in a stage gain of:

$$A_{V(A1 + A2)} = (2R2/R1) + 1 \qquad (4.77)$$

$$= [(2)(24)/(1)] + 1 \qquad (4.78)$$

$$= 49 \qquad (4.79)$$

I will want to try various combinations of resistors for amplifier $A3$ to find if any will yield a precise gain of 500/49, or 10.2, for $A3$. Let $R7 = 10$K (an arbitrarily selected starting point). Find the value for $R4$.

$$10.2 = R7/R4 \qquad (4.80)$$

$$R4 = 10\text{K}/10.2 \qquad (4.81)$$

$$R4 = 980 \text{ ohms} \qquad (4.82)$$

Since a 980-ohm resistor might be a little hard to obtain, I will let $R7 = 12$K;

$$10.2 = R7/R4 \qquad (4.83)$$

$$R4 = 12\text{K}/10.2 \qquad (4.84)$$

$$R4 = 1.18\text{K} \qquad (4.85)$$

This result comes pretty close to what is available, actually, since a 1.2K resistor is a standard value. If this value were used, then the gain of the overall instrumentation amplifier would be:

$$A_V = \left[\frac{(2)(24)}{1} + 1 \right] \left(\frac{12}{1.2} \right) \qquad (4.86)$$

$$A_V = 490 \qquad (4.87)$$

For many applications this gain will suffice, being only a 2% error. If precision is required, then make resistor $R1$ a series network consisting of an 860-ohm resistor and a 200-ohm ten-turn potentiometer. This network will work as a gain control.

Example 4-14

Consider the circuit in Fig. 4-12. What will it do?

Solution:

1. With switch $S1$ closed $R2$ forms a load for E_{IN} and $R4$ forms a load for E_{OUT}. They do not, therefore, affect the gain of the stage. I can conclude, then:

$$-I1 = I2 \qquad (4.88)$$

$$\frac{-E_{IN}}{R1} = \frac{E_{OUT}}{R3} \qquad (4.89)$$

Since $A_V = E_{OUT}/E_{IN}$ for any stage,

$$A_V = -R3/R1 \qquad (4.90)$$

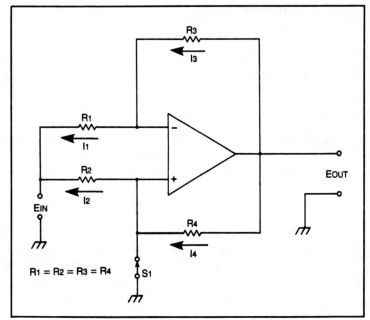

Fig. 4-12. ± Unity gain amplifier.

But $R1 = R3$, so

$$A_V = -1 \qquad (4.91)$$

2. When switch $S1$ is open, on the other hand, I may conclude:

$$I1 = I2 \qquad (4.92)$$
$$-I1 = I3 \qquad (4.93)$$
$$I2 = I4 \qquad (4.94)$$

Therefore,

$$I1 = I4 \qquad (4.95)$$
$$I1 = E_{IN}/R1 \qquad (4.96)$$
$$I4 = E_{OUT}/R4 \qquad (4.97)$$

By substitution of Eq. 4.97 and 4.96 into Eq. 4.95:

$$E_{IN}/R1 = E_{OUT}/R4 \qquad (4.98)$$
$$E_{OUT} = (E_{IN})(R4)/R1 \qquad (4.99)$$

Solve for the voltage gain,

$$A_V = R4/R1 \qquad (4.100)$$

But $R1 = R4$, so

$$A_V = +1 \qquad (4.101)$$

The circuit in Fig. 4-12 is then a $\pm$ unity follower. This type of circuit is very handy for inverting signals from their current polarity under switch control. If $S1$ is actually a solid-state switch that is driven by a TTL-type signal, this can be done automatically under control of a computer or other instrument.

Example 4-15

Design a high-impedance voltmeter using a milliammeter with a range of 0 to 1 mA as the readout. The full-sale potential to be displayed will be 10 volts when the input voltage is 100 millivolts. Specify appropriate V_{CC} and V_{EE} minimums.
Solution:
1. Use an RCA CA3160 operational amplifier, or some equiva-

lent, to meet the input impedance requirement of a voltmeter. This device has an input resistance of 1.5 teraohms (10^{12}).

2. I require a 10-volt output. Note in the CA3160 spec sheet that the maximum output voltage can swing to $+12$ volts in the VCC direction and -15 volts in the VEE direction, at power supply potentials of ± 15 volts dc. This means that a 3-volt difference exists between the maximum allowable output and VCC, and essentially no difference for VEE and the maximum negative output. I can conclude, then, that the minimum VCC potential is 13 volts and the minimum VEE is -10.5 volts. For convenience, though, both are set to 15 volts.

3. Draw an appropriate circuit (see Fig. 4-13).

4. Find the ratio $R2/R1$.

$$E_{\text{OUT}}/E_{\text{IN}} = A_V = (R2/R1) + 1 \tag{4.102}$$
$$10/0.1 = (R2/R1) + 1 \tag{4.103}$$
$$100 = (R2/R1) + 1 \tag{4.104}$$
$$99 = R2/R1 \tag{4.105}$$

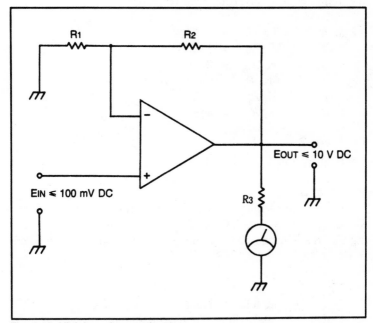

Fig. 4-13. High-impedance voltmeter.

91

5. Test standard values to find a combination in which this ratio holds nearly true.

    a. Let $R2 = 100K$.

$$R1 = 100K/99 = 1.01K \qquad (4.106)$$

    b. Let $R2 = 110K$.

$$R1 = 110K/99 = 1.11K \qquad (4.107)$$

    c. Let $R2 = 120K$.

$$R1 = 120K/99 = 1.21K \qquad (4.108)$$

Trial c is very nearly a standard value for $R1$, so it is used.

EXERCISES

Problem 1

Compute the gain in each of the following cases. Assume that the amplifier is an inverting follower stage.

a. $R_f = 470K$, $R_{IN} = 56K$
b. $R_F = 1M$, $R_{IN} = 120K$
c. $R_F = 10K$, $R_{IN} = 2.2K$
d. $R_F = 5.6K$, $R_{IN} = 560$ ohms
e. $R_F = 10K$, $R_{IN} = 10K$
f. $R_F = 10K$, $R_{IN} = 56K$

Problem 2

Find R_F and E_{IN} in Fig. 4-14.

Problem 3

In a noninverting follower, I have the following values: $R_F = 100K$, $R_{IN} = 2.2K$, and $E_{OUT} = + 3.5$ volts dc. Assuming that there are no offset problems, what is the voltage applied to the input?

Problem 4

A noninverting follower has $E_{IN} = 0.02$ volts, $E_{OUT} = 1.25$ volts, and $R_F = 68K$. Find the actual value of R_{IN}.

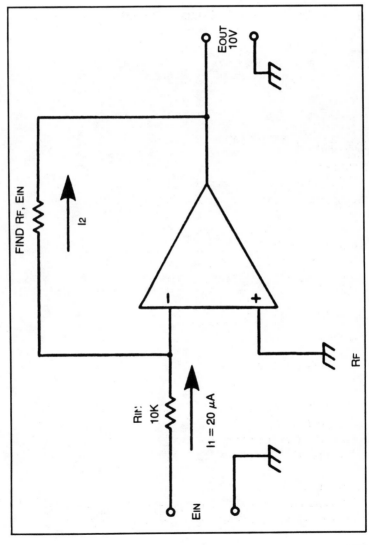

Fig. 4-14. Inverting follower. Find R_F and E_{IN} (see Problem 2).

93

ANSWERS TO PROBLEMS

1.

    a.   8.4
    b.   8.3
    c.   4.6
    d. 10.0
    e.   1.0
    f.   0.2

2. R_F = 500K, E_{IN} = 0.2 volts or 200 mV

3. E_{IN} = 75 mV, or 0.075 volts

4. R_{IN} = 1,106 ohms

Chapter 5

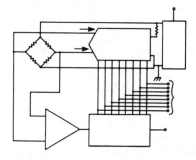

Isolation and Other Amplifiers

THE OPERATIONAL AMPLIFIER REVOLUTIONIZED ANALOG computer design about a million years ago, but early models were not well suited to general use in analog circuits. In the midsixties when the uA-709 I.C. operational amplifier was developed, it became possible to use this wonderful invention for a wide range of applications where such use was not feasible before.

Figure 5-1 shows the four basic forms of op amp circuit: inverting follower (Fig. 5-1A), noninverting follower with gain (Fig. 5-1B), unity gain noninverting follower (Fig. 5-1C), and dc differential amplifier (Fig. 5-1D).

The inverting follower circuit shown in Fig. 5-1A produces an output signal that is 180 degrees out of phase with the input signal—which is where it gets its name "inverting" follower. The gain of the inverting follower is simple $-R2/R1$, where the minus sign indicates the inversion. Thus, when a circuit like Fig. 5-1A has an input resistor ($R1$) of 10 kohms, and a feedback resistor ($R2$) of 100 kohms, the gain is $(-100 \text{ k})/(10 \text{ k})$, or -10. The input impedance of this circuit is equal to R1.

The noninverting follower with gain (Fig. 5-1B) provides no phase reversal between input and output. The input impedance of this circuit is very high, and is not related to the value of any resistors in the circuit. The gain of the circuit is $[(R2/R1) + 1]$, so, in the example above, when $R1 = 10$ kohms and $R2 = 100$ kohms, the gain will be 101.

The unity gain noninverting follower (Fig. 5-1C) is a special case of Fig. 5-1B, in which the feedback resistor is zero ohms: the output is connected directly to the inverting input. This circuit provides a voltage gain of one (or "unity" gain). There are two main purposes for the unity gain noninverting follower: buffering and impedance transformation. "Buffering" means that the circuit provides isolation between input and output. Thus, variations in load will not affect the input circuit, an ideal situation where oscillators and other such sensitive circuits must drive unstable loads. Impedance transformation is high to low: a high input impedance reduces to a low output impedance. Since the voltage remains the same, and the impedance drops at the output, it is obvious that the unity gain noninverting amplifier provides a power gain greater than unity, while providing a voltage gain of one.

Finally, there is the dc differential amplifier (Fig. 5-1D). This type of amplifier is used to provide a single-ended (i.e., unbalanced) output from a differential (i.e., balanced) input signal source. The output voltage V_O is proportional to the differential voltage gain (A_{vd}) and the difference between input voltages $V1$ and $V2$ (i.e., $V1 - V2$):

$$V_O = A_{vd} \times (V1 - V2) \qquad (5.1)$$

The gain is calculated in much the way of the inverting amplifier, and is equal to either $R3/R1$ or $R4/R2$, provided that $R1 = R2$ and $R3 = R4$ (these capabilities are important!).

The circuits above are based on single op amps. With two or more op amps however, even more complex circuits are possible. In the remainder of this Chapter, I will deal with I.C. versions of the instrumentation amplifier (I.A.) circuit shown in Fig. 5-2. The I.A. provides an extremely high input impedance (similar to the noninverting follower circuits), a high possible gain, and easy design. The gain equation for this circuit is shown in Fig. 5-2.

The I.A. circuit shown in Fig. 5-2 consists of two sections: $A1$-$A2$ and $A3$. Amplifier $A3$ forms a simple dc differential amplifier (such as Fig. 5-2), and obeys the same rules. The $A1$-$A2$ amplifier is a differential noninverting-input-with-differential-output stage. By cascading these two forms of amplifier, I obtain the instrumentation amplifier.

In many cases, where variable or adjustable gain is required, I leave all resistors constant except $R1$. I must be careful, however, because $R1$ appears in the denominator of the equation in Fig.

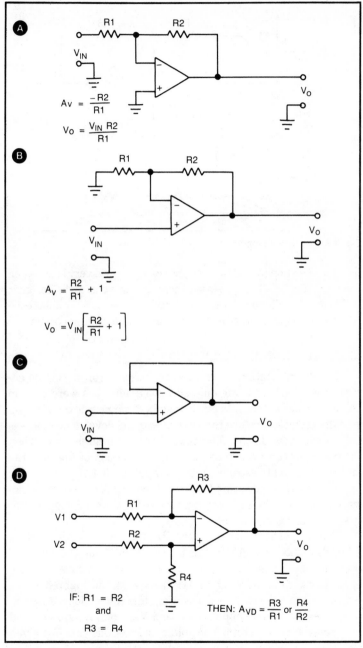

Fig. 5-1. (A) Inverting follower. (B) Noninverting follower with gain. (C) Unity gain noninverting follower. (D) Dc differential amplifier.

97

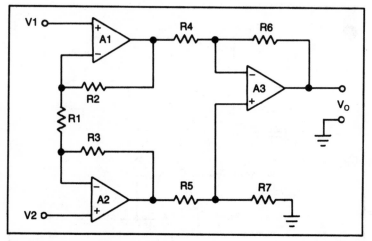

Fig. 5-2. Instrumentation amplifier.

5-2. This location means that the gain can get very, very large when the resistance of $R1$ drops close to zero. In some cases, the designer will place a small-value fixed resistor in series with a variable resistor (potentiometer) to adjust gain, but limit it to a maximum.

IC OPERATIONAL AMPLIFIERS

The operational amplifier truly revolutionized analog circuit design. For a long time, the only additional advances were that op amps became better and better (they became nearer the ideal op amp of textbooks!). While that was an exciting development, it was not a really new device. The next big breakthrough came when the analog device designers made an I.C. version of Fig. 5-2, the integrated circuit instrumentation amplifier (ICIA). Today, the manufacturers are offering better and better ICIA devices; it can truly be said with an early op amp textbook, "the contriving of contrivances is a game for all."

Figure 5-3 shows one popular ICIA, the Precision Monolithics AMP-01 device. The AMP-01 is housed in an 18-pin DIP package (Fig. 5-3A).

The basic circuit for the AMP-01 is shown in Fig. 5-3B. Notice how simple the circuit is! There are few connections: differential inputs, dc power supplies (V − and V +), output, ground, and two gain-setting resistors. The voltage gain of this circuit is given by:

$$A_{vd} = 20 \ R_s/R_g \qquad (5.2)$$

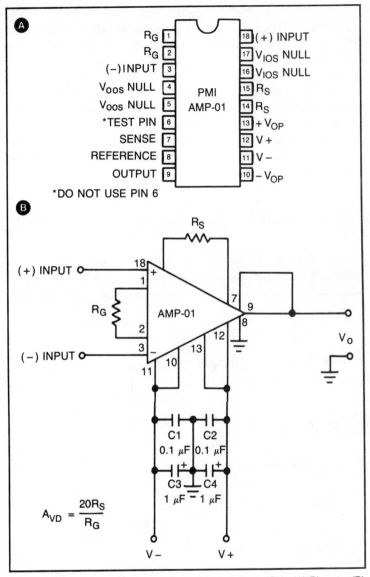

Fig. 5-3. Precision Monolithics, Incorporated AMP-01 ICIA. (A) Pinouts. (B) Circuit.

Suppose I want to make a differential voltage amplifier with a gain of 1000. I need to make a resistor ratio of 1000/20, or 50:1. Thus, if R_s is set to 100 kohms, and R_g is 2 kohms, I will have the required gain of 1000. The permissible gain range is 0.1 to 10,000.

The dc power supply voltages are up to $-/+$ 18-volts dc. Notice in Fig. 5-3B that the dc power supply lines are heavily bypassed. The 0.1-uF units are used to bypass high frequencies, while the 1-uF units are for low frequencies. The 0.1-uF units must be mounted as close as possible to the body of the amplifier.

The maximum operating frequency depends upon the gain. At a gain of 1, the maximum small-signal input frequency is 570 kHz, while at a gain of 1000 it reduces to 26 kHz.

The Burr-Brown INA-101 (Fig. 5-4) is another new ICIA device. This amplifier is also simple to connect. There are only dc power connections, differential input connections, offset adjust connections, ground, and an output. The gain of the circuit is set by:

$$A_{vd} = (40 \ k/R_g) + 1$$

The INA-101 is basically a low-noise, low-input, bias-current-integrated-circuit version of the I.A. of Fig. 5-2. The resistors la-

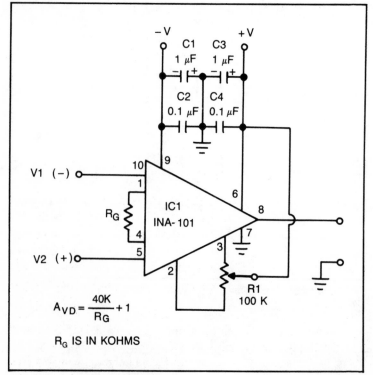

Fig. 5-4. Burr-Brown INA-101 ICIA.

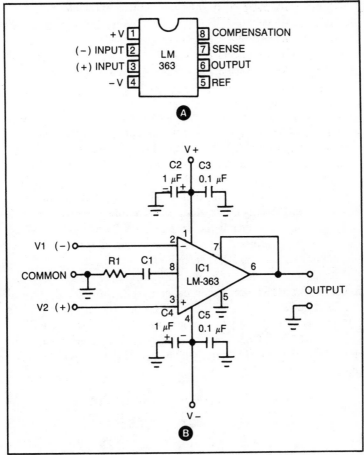

Fig. 5-5. National Semiconductor LM-363. (A) Pinouts. (B) Circuit.

beled $R2$ and $R3$ in Fig. 5-2 are both 20 kohms, hence the "40 k" term in Fig. 5-4.

Potentiometer $R1$ in Fig. 5-4 is used to null the offset voltages appearing at the output. An offset voltage is a voltage that exists on the output at a time when it should be zero (i.e., when $V1 = V2$, so that $V1 - V2 = 0$). The offset voltage might be internal to the amplifier, or a component of the input signal. It is true that dc offsets in signals are common, especially in biopotentials amplifiers such as ECG and EEG.

Still another ICIA is the LM-363 device shown in Fig. 5-5. The miniDIP version is shown in Fig. 5-5A (an 8-pin metal can is also available), while a typical circuit is shown in Fig. 5-5B. The LM-363 device is a fixed-gain ICIA. There are three versions:

Designation	Gain
LM-363-10	10
LM-363-100	100
LM-363-500	500

The LM-363-xx is useful in places where one of the standard gains is required and there is minimum space available. Two examples spring to mind. I could use the LM-363-x as a transducer preamplifier, especially in noisy signal areas. The LM-363-x can be built onto (or into) the transducer to build up its signal before sending it to the main instrument or signal acquisition computer. The other example is inbioamplifiers. The biopotentials are typically very small, especially in lab animals. The LM-363-x can be mounted on the subject and a higher level signal sent to the main instrument—a little exotic, but none the less, useful.

A selectable gain version of the LM-363 device is shown in Fig. 5-6. The 16-pin DIP package is shown in Fig. 5-6A, while a typical circuit is shown in Fig. 5-6B. The type number of this device is LM-363-AD, which distinguishes it from the LM-363-x devices. The gain can be 10, 100 or 1000 depending upon the programming of the gain setting pins (2, 3 and 4). The programming protocol is as follows:

Gain Desired	Jumper Pins
X10	(All Open)
X100	3 & 4
X1000	2 & 4

Switch S1 in Fig. 5-6B is the GAIN SELECT switch. This switch should be mounted close to the IC device, but is quite flexible in mechanical form. The switch could also be made from a combination of CMOS electronic switches (e.g., 4066).

The dc power supply terminals are treated in a manner similar to the other amplifiers. Again, the 0.1-uF capacitors need to be mounted as close as possible to the body of the LM-363-AD.

Pins 8 and 9 are guard shield outputs. These pins are a feature that makes the LM-363-AD more useful for many instrumentation problems than other models. By outputting a signal sample back to the shield of the input lines, I can increase the common mode rejection ratio (CMRR). This feature is frequently used in biopoten-

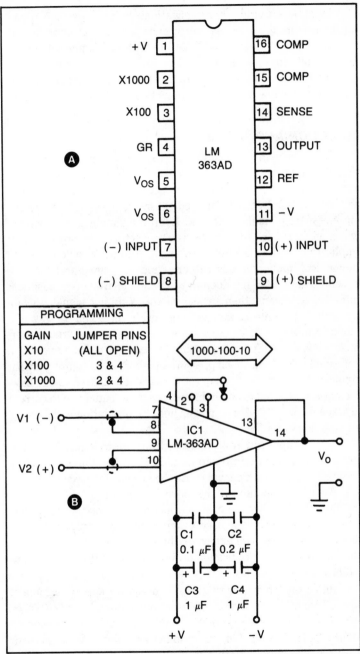

Fig. 5-6. National Semiconductor LM-363AD. (A) Pinouts and programming protocol. (B) Circuit.

tials amplifiers and in other applications where a low-level signal must pass through a strong interference (high noise) environment.

The LM-363 devices will operate with dc supply voltages of $-/+5$ volts to $-/+18$ volts dc, with a common mode rejection ratio of 130 dB. The 7 nV/(SQR(Hz)) noise figure makes the device useful for low noise applications (a 0.5 nV model is available at a premium cost).

ISOLATION AMPLIFIERS

There are many applications for the instrumentation amplifier that are dangerous for either the circuit or the user. In biomedical applications, the issue is patient safety. There are numerous signals acquisition needs in biomedical instrumentation where the victim—errrr patient—is at risk. Even the simple ECG machine, which measures and records the heart's electrical activity, was once implicated in patient safety problems. Another problem area in biomedical applications is catherization instruments. There are several tests where the doctors insert an electrode or transducer into the body, and then measures the resulting signal: the intracardiac ECG places an electrode inside the heart by way of a blood vein, the cardiac output computer uses a signal from a thermistor inside a catheter placed in the heart (also through a vein), and simple electronic blood pressure monitors use a transducer that connects to an artery. In all of these cases, it is undesirable to have the patient exposed to small differences of potential due to current leakage from the 60 Hz ac power lines. The solution is the use of an isolation amplifier.

Another application is signals acquisition in high-voltage circuits. Here it is undesirable to mix high voltage sources with low voltage electronics which will cause the low voltage circuits to blow out. Again, the solution is the isolation amplifier.

Figure 5-7 shows the basic symbol for the isolation amplifier. The break in the triangle used to represent any amplifier denotes the fact that there is an extremely high impedance (typically 10^{12} ohms) between the inputs and output terminal of the isolation amplifier.

Notice that there are two sets of dc power supply terminals. The V – and V + terminals are the same as found on all ICIA or op amp devices. These dc power supply terminals are connected to the regular dc supply of the equipment where the device is used. Such a power supply derives its dc potentials from the ac power mains by way of a 60-Hz transformer. The isolated dc power sup-

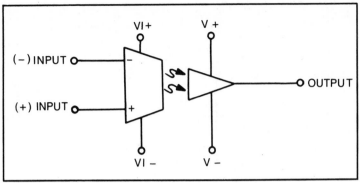

Fig. 5-7. Isolation amplifier symbol.

ply inputs (VI – and VI +) are used to power the input amplifier stages and must be isolated from the main dc power supply of the equipment. The VI – and VI + terminals are usually either battery powered, or powered from a dc-to-dc converter that produces a dc output from the main power supply by using a high frequency (50 to 500 kHz) oscillator. The high frequency "power supply" transformer does not pass 60-Hz signals well, so the isolation is maintained.

Figure 5-8 shows the circuit of an isolation amplifier based on the Burr-Brown 3652 device. This isolation amplifier is not generally available to hobbyists, but would be used even in small, "one-of-a-kind" professional labs.

The dc power for both the isolated and nonisolated sections of the 3652 is provided by the 722 dual dc-to-dc converter. This device produces two independent – / + 15 Vdc supplies that are each isolated from the 60-Hz ac power mains and from each other. The 722 device is powered from a + 12 Vdc source that is derived from the ac power mains. In some cases, the nonisolated section (which is connected to the output terminal) is powered from a bipolar dc power supply that is derived from the 60-Hz ac mains, such as a – / + 12 Vdc or – / + 15 Vdc supply. In no instance, however, should the isolated dc power supplies be derived from the ac power mains.

There are two separate ground systems in this circuit, symbolized by the small triangle and the regular three-bar "chassis" ground symbol. The isolated ground is not connected to either the dc power supply ground/common or the chassis ground. It is kept floating at all times and becomes the signal common for the input signal source.

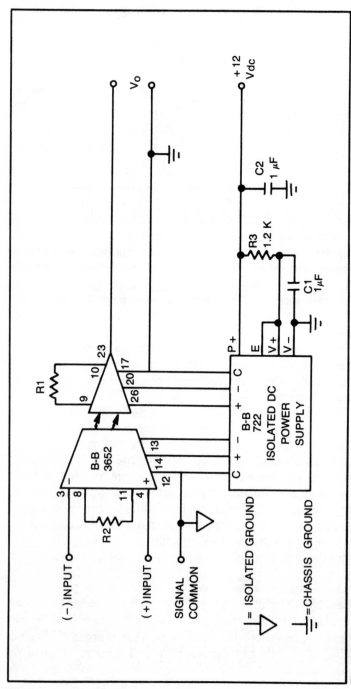

Fig. 5-8. Burr-Brown isolation amplifier circuit.

The gain of the circuit is approximately:

$$GAIN = \frac{R1 + R2 + 115}{1,000,000}$$

In most design cases, the issue is the unknown values of the gain-setting resistors. I can rearrange the equation above to solve for $(R1 + R2)$:

$$(R1 + R2) = \frac{1,000,000 - (115 \times GAIN)}{GAIN}$$

Where:

R1 and R2 are in ohms
GAIN is the voltage gain desired

Let's work an example. Suppose I need a differential voltage gain of 1,000. What combination of R1 and R2 will provide that gain figure?

If GAIN = 1000

$$(R1 + R2) = \frac{1,000,000 - (115 \times 1000)}{1000}$$

$$(R1 + R2) = \frac{1,000,000 - (115,000)}{1000}$$

$$(R1 + R2) = \frac{885,000}{1000}$$

$$(R1 + R2) = 885 \text{ ohms}$$

In this case, we need some combination of R1 and R2 that adds to 885 ohms. The value, 440 ohms, is "standard," and will result in only a tiny gain error if used.

CONCLUSION

The IC instrumentation amplifier and the isolation amplifier open new applications that the simple op amp cannot match. Digital electronics fans should be aware that analog is not dead—it lives in even more sophisticated manifestations.

Chapter 6

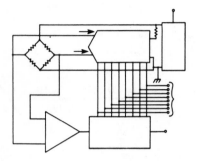

Some Useful Digital Circuits

ALTHOUGH IT HAS BEEN A BASIC ASSUMPTION IN WRITING
this book that you are familiar with basic electronic circuitry,
it is felt that it will be useful to include a section on certain types
of useful digital circuitry.

ONE-SHOT MULTIVIBRATORS

Monostable multivibrators, or one shots, are circuits that will
deliver a single output pulse of constant amplitude and duration
every time a trigger pulse is received at the input terminals.

One shots are used extensively in electronic instrumentation,
as well as in computer and other digital circuits. They are used,
for example, to clean up noisy waveforms, debounce switch contacts, form digital delay lines, and ensure the proper initial conditions in some instrument circuits.

In circuits where the signal information is contained in the repetition rate of irregularly shaped pulses, then a one-shot circuit can
be used to recover the data. Similarly, when decoding a tape-
recorded audio-FM data signal, a pulse-counting detector can be
used; this requires the same type of circuit as the case of the irregularly shaped pulse.

In both situations, a one shot is used to produce pulses of constant amplitude and constant duration, only the repetition rate varies
with the input signal. Since all individual one-shot output pulses
have constant area (i.e., duration x amplitude), then integrating the

pulse train will yield a dc voltage that is proportional to the repetition rate. This technique is used in many different types of scientific instruments.

The TTL chips in the 74121 through 74123 series are high-speed one-shot circuits, and that fact alone makes them a little nasty to tame. Many different problems result from improper layout, and those chips are difficult enough to use that one should try to employ a slower device such as the 555 timer if the pulse duration is between 1 microsecond and 10 seconds. The 555 is a bipolar-technology device, although *not* TTL, and will operate at supply voltages between + 4.5 volts dc and + 15 volts dc. This wide power supply range makes it easy to interface the 555 with TTL, CMOS, or discrete circuits.

The 555 output terminal (pin 3) will *sink* or *source* up to 200 milliamperes of current, making it easy to use with lamps, relay coils, and a variety of loads that are difficult to drive with either TTL or CMOS devices.

Figure 6-1 shows the basic monostable circuit using the 555 timer chip. The stages inside of the block are the 555 internal circuitry, while the rest is external.

The 555 uses two comparators. Comparator 1 is biased to a potential of 2/3 V_{CC}, while comparator 2 is biased to a potential of 1/3 V_C.

The circuit in Fig. 6-1 is triggered by bringing the voltage on the trigger terminal from a positive value down to less than 1/3 V_{CC}. This condition is met by applying a negative-going pulse to pin 2.

The trigger pulse causes the output of comparator 2 to set the control flip-flop, which in turn causes output terminal (pin 3) to snap high.

Prior to the trigger pulse, the discrete transistor inside of the 555 has been turned on, which kept capacitor $C1$ discharged. But now the transistor is turned off, so $C1$ can begin charging.

When the charge voltage across $C1$ reaches 2/3 V_{CC} the output of the comparator snaps high, resetting the control flip-flop. This turns the transistor on, discharging $C1$ rapidly, which causes the output terminal to drop low again. The duration of the output-high condition is given by:

$$T = 1.1\ R1C1 \qquad (6.1)$$

where T is the time in seconds, $R1$ is in ohms, and $C1$ is in farads.

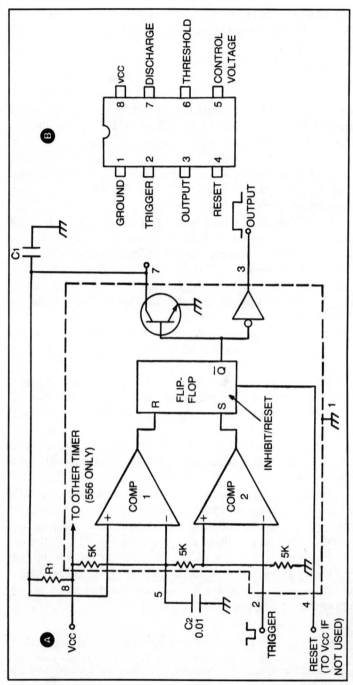

Fig. 6-1. One-shot using the 555 timer IC. (A) Circuit. (B) Pin layout.

110

Example 6-1

What is the period of a 555 one-shot circuit (Fig. 6-1) if $R1 = 470K$ and $C1 = 0.05\ \mu F$?

Solution

From Eq. 6,1,

$$T = (1.1)(4.7 \times 10^5)(5 \times 10^{-8}) \text{ seconds}$$
$$= (1.1)(23.5 \times 10^{-3}) \text{ seconds}$$
$$= 2.59 \times 10^{-2} \text{ seconds} = 26 \text{ ms}$$

The value of $R1$ can be almost anything between 10K and 12M, while $C1$ can be 100 pF to 10 μF.

The unique advantage of the 555 over some other one-shot circuits is that the output period is almost independent of supply voltage variations, because the period is set by the trip points of the comparators relative to the $C1$ potential. These trip points are $2/3\ Vcc$ and $1/3\ Vcc$, respectively, so if Vcc drifts, the one-shot will track the change.

Figure 6-2 shows two methods for triggering the 555. Remember that the trick is to drop pin 2 from a positive potential close to Vcc, down to a potential that is *less than* 1/3 Vcc. In Fig. 6-2A external transistor $Q1$ is normally turned off, so its collector (pin 2 of the 555) is at Vcc. When a trigger pulse is applied to the base of $Q1$, the collector will drop to ground potential, thereby triggering the 555.

Figure 6-2B shows a manual method for triggering the 555. Under rest conditions capacitor $C1$ will see the same voltage at both ends, Vcc. But when $S1$ is closed, one side of the capacitor begins to charge through $R2$. For a brief instant, until the capacitor is partially charged, the voltage on pin 2 drops very nearly to ground, triggering the 555.

The 555 is only usable to periods of approximately 10 seconds. For greater time durations, a device such as the Exar 2240, 2250, or 2260 (also Intersail 8240, 8250, and 8260, respectively) is a better selection.

The block diagram to the XR-2240 is shown in Fig. 6-3. The time base is a stage built similar to the 555, although different voltage divider resistor values are used, allowing the period to be found using the RC time constant, namely $R1 \times C1$. The significant difference between the XR-2240 and the 555 is the internal binary counter. This eight-bit stage uses open-collector outputs that can be connected in a wired-OR configuration through a single 10K pull-

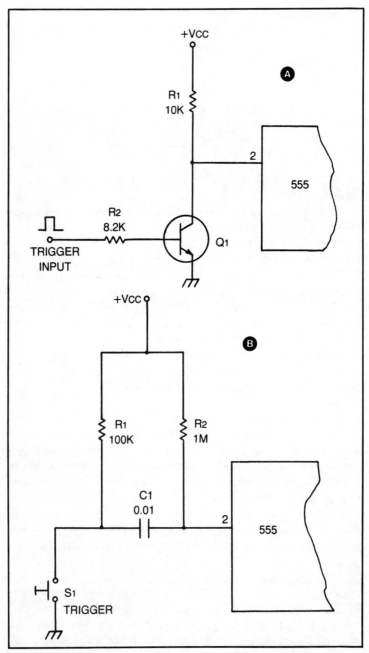

Fig. 6-2. Type 555 triggering circuits (A) Using an external transistor. (B) Manual method.

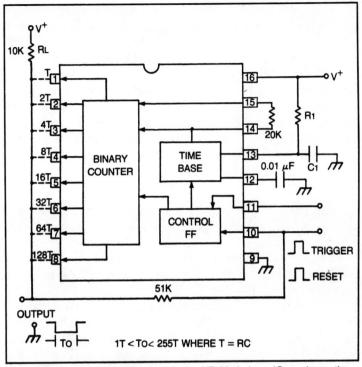

Fig. 6-3. One-shot multivibrator using the XR-2240 timer IC produces times up to $255R_1C_1$. (Courtesy of Exar Integrated Systems.)

up resistor to V_{CC}. As long as any OR-connected output is low, then the output from the entire stage is low.

The total time that the output is low following a trigger pulse is a function of the $R1C1$ time constant and the sum of the binary weights of the counter outputs that are ORed together. This period can be anything between $1R1C1$ and $255R1C1$ (the counter is an eight-bit device).

For example, if $R1C1$ is 1 second, and I want a 200-second time delay, I would OR-connect pin 8 ($128T$), pin 7 ($64T$), and pin 4 ($8T$), which totals $(128 + 64 + 8)T$, or $200T$, where T is the 1-second time constant $R1C1$.

Extremely long durations can be programmed by cascading two or more XR-2240 units. The output of one becomes the time base of the next. If two units are used, the total period is (256) (256) $R1C1$, or $65,536$ $R1C1$. If $T = 1$ second incidentally, then two XR-2240 timers in cascade yield a duration of 65,536 seconds, or over 18 hours.

113

In Fig. 6-3, the feedback switch determines whether or not the timer operates in astable or monostable modes. When the switch is closed, the device becomes self-retriggering at the end of each period; such behavior results in astable operation.

An external timer can be used in those cases where the computer inputs data only after a long process has taken place, or where the sample rate is measured in *events per minute* or *hour*. It would be wasteful to have the computer loop meaningless while awaiting the end of the period, so an external timer would allow it to perform other jobs in the meantime.

74100 LATCH

Data typically appears on a microprocessor/microcomputer data bus for less than a millisecond. A means must be provided to grab and hold that data; that is, to *latch* the data.

There are a number of data latches available as auxiliary or peripheral chips from microprocessor chip companies, but these tend to be relatively costly.

The 74100 is a TTL chip, and it costs $1.50, where some of the special purpose data latch chips cost over $12. The 74100 is defined as a *dual quad-latch* type, and essentially contains eight D-type flip-flops arranged in two banks of four each. The data and output terminals D, Q, and $\overline{Q}$ (*not Q*) on each flip-flop are independent, but all of the *clock* terminals in each bank are tied together.

In Fig. 6-4 the clock terminals from the two banks are connected together forming what can be called a *strobe* or *chip enable* terminal.

Data applied to the input pins will be transferred to the output only when the strobe terminal is brought high. When the strobe is low, the output terminals hold the data that existed at the last time the strobe was high.

The 74100 is a low-cost TTL device, so is a cheaper device to use in data latch applications. It will operate at speeds up to several megahertz, making it compatible with most microprocessors.

FLIP-FLOPS

There are several different types of flip-flops: RS, D-type, and JK. All of these flip-flops have different properties, so are useful under differing circumstances. All flip-flops can be made from various combinations of NAND, NOR, AND, OR, and exclusive-OR

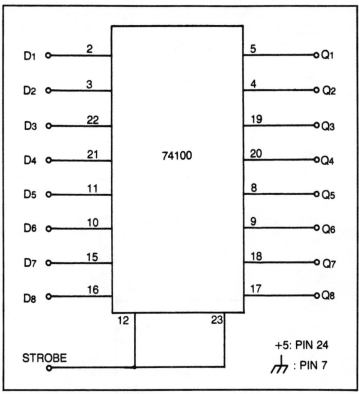

Fig. 6-4. Type 74100 quad-latch.

gates, but very few designers bother doing that anymore because all but the RS flip-flop are readily available in several forms of TTL and CMOS devices.

The RS flip-flop, also called the *set-reset* flip-flop, comes in two varieties that have inverse truth tables. The example shown in Fig. 6-5A uses NAND gates (i.e., 7400). The rules for operation of this flip-flop (see also the truth table) are summarized below.

1. If both R and S are high, then there is *no change* in output state.
2. If S is momentarily made low, then the output state goes to the condition where Q is high and $\overline{Q}$ is low.
3. If the R terminal is momentarily made low, then the output state goes to the condition where Q is low and $\overline{Q}$ is high.
4. If both R and S are low, then this is a disallowed state and should be avoided.

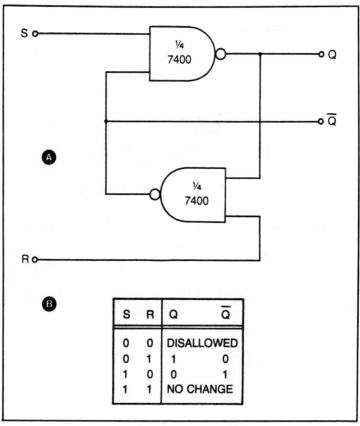

Fig. 6-5. NAND gate RS flip-flop. (A) Circuit. (B) Truth table.

S	R	Q	$\overline{Q}$
0	0	DISALLOWED	
0	1	1	0
1	0	0	1
1	1	NO CHANGE	

The related circuit of Fig. 6-6A uses NOR gates (i.e., 7402), and obeys the rules below:

1. Both R and S HIGH: disallowed state.
2. If S is brought high momentarily the output state goes to the condition where Q is high and $\overline{Q}$ is low.
3. If R is brought high momentarily, the output state goes to the condition where Q is low and $\overline{Q}$ is high.
4. Both inputs low: no change.

The D-type flip-flop is an example of a synchronous or clocked flip-flop: it will change state only when the clock terminal is high.

The circuit symbol for a D-type flip-flop is shown in Fig. 6-7. The data applied to the D input (i.e., a 1 or 0) will be transferred

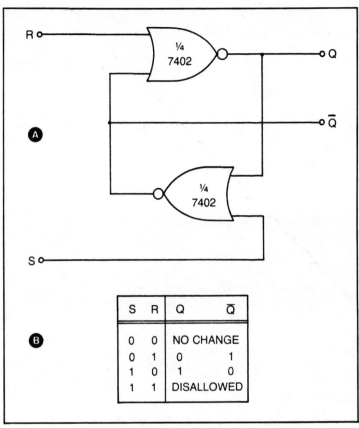

Fig. 6-6. NOR gate RS flip-flop. (A) Circuit. (B) Truth table.

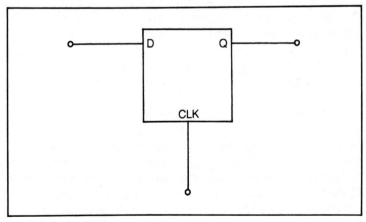

Fig. 6-7. D-type flip-flop.

117

to the output only when CLK is high. Between clock pulses the Q output holds the last data present at the input before the clock pulse drops low again.

The 7474 is a TTL and dual D-type flip-flop. The 7475 device is a quad D-type flip-flop, but is in the form of a latch which has two banks of two flip-flops each. The CLK terminals of the flip-flops in a bank are tied together. As previously noted, the 74100 is a dual quad-latch type containing eight D-type flip-flops in two banks of four each.

The JK flip-flop (e.g., 7473, 7476) is probably the most complex of the flip-flops. There are actually two sets of rules for the JK; one for clocked operation and another for direct (i.e., unclocked) operation.

In clocked operation (Fig. 6-8) the *preset* and *preclear* inputs are tied high, and the output responds to the states of J and K inputs, but only during negative-going transitions of the clock pulse.

If J is low, and K is high during the clock transition, then the output goes to the condition where Q is low and $\overline{Q}$ is high.

Exactly the opposite happens when J is high and K is low: the output condition following the clock pulse will be Q high and $\overline{Q}$ low.

In direct operation (Fig. 6-8C) the preset and preclear inputs control operation and are independent of the clock pulse. The rules of operation are summarized in the truth tables.

CLOCK CIRCUITS

Many of the most useful digital circuits operate in a synchronous manner, i.e., they require a digital clock to make things happen in a coherent manner. The clock signal is merely a chain of square waves (alternating high and low states) that continue indefinitely, or at least as long as the circuit is turned on. Even certain devices such as analog-to-digital converters are synchronized to a clock circuit; a computer could not operate were it not for the synchronicity provided by the clock circuit. In this section, we will discuss clock circuits of several different species.

Accuracy is only occasionally an issue in digital circuits. Very few oscillators made from digital logic integrated circuits (which work best with other digital IC logic circuits) make precision-frequency clocks. Of course, RC timers are inherently less accurate than crystal oscillators, and are more subject to thermal drift (change of frequency with temperature change). RC timers suffer from thermal defects because of the nature of the resistors and capacitors used in the timer network. Both elements, especially the resis-

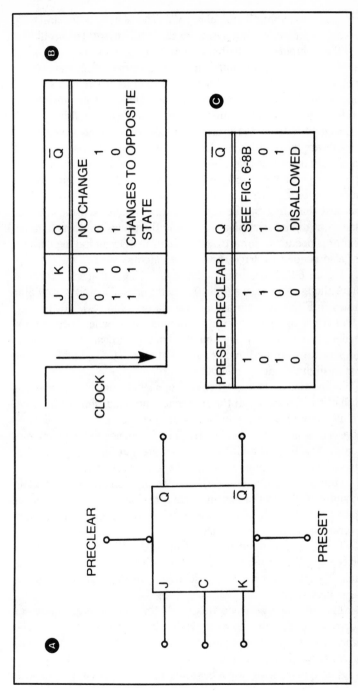

Fig. 6-8. JK flip-flop. (A) Circuit diagram. (B) Clocked operation truth table. (C) Direct operation truth table.

(B)

J	K	Q	Q̄
0	0	NO CHANGE	
0	1	0	1
1	0	1	0
1	1	CHANGES TO OPPOSITE STATE	

(C)

PRESET	PRECLEAR	Q	Q̄
1	1	SEE FIG. 6-8B	
0	1	1	0
1	0	0	1
0	0	DISALLOWED	

CLOCK

PRECLEAR

J

C

K

PRESET

Q

Q̄

(A)

119

tors, tend to change value slightly as the temperature shifts. Of course, a change in value will alter the frequency of the oscillator.

The temperature drift problem is of no importance in most cases because only an approximate frequency is important. Even drift will not bother most circuits as long as it is slow. If some specific application requires a precision clock circuit however, it might be wise to use a transistor oscillator circuit making sure it is designed with temperature compensation built in. Even when crystals are used with digital IC devices to make an oscillator, the results are often less than could be achieved with transistors as the active element. Not that transistor crystal oscillators are inherently more stable or more accurate than those made from digital ICs, it is just that they can be temperature compensated more easily than most digital versions. Yet, while the precision of such an oscillator is needed in some cases, TTL or CMOS digital logic levels are also needed to drive the logic elements.

Figure 6-9 shows two methods for using a transistor oscillator. Although there is a simple transistor crystal Colpitts oscillator shown in Fig. 6-9A, almost any oscillator could be used. In fact, if you are designing a circuit for an application where precision is needed, then it is likely that you will want to purchase a ready-built, temperature-compensated crystal oscillator (TCXO) and interface it with digital logic elements.

Transistor Q1 in Fig. 6-9A is an npn silicon unit that is selected so that it will oscillate at the frequency intended. For most cases, the unit should have a gain-bandwidth product (F_t) of 200 MHz or more for oscillators up to 20 MHz. The frequency determining element in Fig. 6-9A is a piezoelectric crystal cut for the desired frequency. The oscillating frequency of a crystal is partially dependent upon the circuit capacitance. By connecting a small series variable capacitor ($C3$) into the circuit, I can provide a certain amount of control over the frequency. It is possible to change the frequency several kilohertz in some cases.

The circuit is identified as a Colpitts oscillator by the capacitor voltage divider network $C1/C2$, which is used to provide feedback. This network is semi-critical, but, for most cases, I can use the values shown.

The interface between this oscillator and the digital devices it drives is accomplished by an LM-311 voltage comparator. The signal provided to the comparator is developed across emitter resistor $R2$ through capacitor $C5$.

A voltage comparator is basically an operational amplifier with no feedback resistor. This lack of feedback means that the gain seen by the inputs is the open-loop gain of the amplifier, which is at least 20,000 in even junk-grade operational amplifiers, and possibly over 1,000,000 in premium-grade devices. As a result, the output will be zero when the voltages applied to the inverting (–) and noninverting (+) inputs are equal. If the voltage applied to the inverting input is higher however, then the amplifier responds as though a positive potential were applied to the inverting input, so the output will be negative. Because of the gain of the device, the output will saturate against the negative supply voltage rail when the input difference is more than a few millivolts. Similarly, when the voltage applied to the noninverting input is greater, then the output will be trying to go positive. Again, because of the immense gain of the amplifier, the output voltage will be hard against the positive supply rail when the input potentials are more than a few millivolts different from each other.

The LM-311 contains an operational amplifier, but is specially configured for comparator service. When used in the monopolar mode (i.e., both pins 1 and 4 grounded) the output will be low when the two input terminals are the same, and high when they are at different potentials, so that the inverting input is higher than the noninverting. The output terminal of the LM-311 device is the open-collector version. To make the LM-311 TTL-compatible I must connect a 2.2-kohm pull-up resistor between the output terminal and the + 5-volt dc power supply. The rest of the LM-311 circuitry can operate from + 5 volts also, but there is no reason why I cannot always make the oscillator work nicely at + 5 volts dc. If I want, it is permissible to power the rest of the LM-311 from some positive potential higher than + 5 volts dc, upscale $R4$ (in Fig. 6-9A) proportionally (i.e., approximately 5.6 kohms for + 12 volts). If bipolar CMOS devices are used, then operate the LM-311 in the bipolar mode also.

The LM-311 device in Fig. 6-9A is connected so that the noninverting input is grounded at zero potential. The output will drop low every time the input voltage applied to the inverting input is either zero or negative; it will snap high whenever the input voltage is positive. So, for one half of each cycle produced by the oscillator, the output of the LM-311 is high, and for the other half it is low. As a result, the output of the LM-311 is a square wave with the same frequency as the input sine wave. I have the stability and precision of the transistor oscillator or commercial TCXO, and the logic levels required by digital circuits.

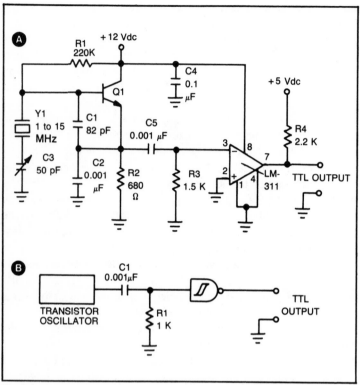

Fig. 6-9. (A) 1 to 15 MHz crystal oscillator circuit. (B) Converting output to TTL levels.

A variation on the theme is shown in Fig. 6-9B. The interface device here is a Schmitt trigger. A Schmitt trigger is a circuit which will snap high when a positive-going input signal crosses over one threshold point, and snaps low again when the signal then crosses over a lower threshold point. This operation is shown graphically in Fig. 6-9C for the standard 7414 TTL hex Schmitt trigger device. The upper threshold of the 7414 device is 1.7 volts, while the lower threshold is 0.9 volts; the hysteresis is therefore (1.7 – 0.9), or 0.8-volts. There is also a CMOS Schmitt trigger device, the 4093. That device has trip points that differ for different supply voltages, but are 2.9 volts and 2.3 volts for +5 volt dc supplies (i.e., TTL-compatible levels).

The signal from the oscillator in Fig. 6-9B is coupled to the Schmitt trigger via an RC network. Although the values of the components may be different for certain frequency ranges, those shown

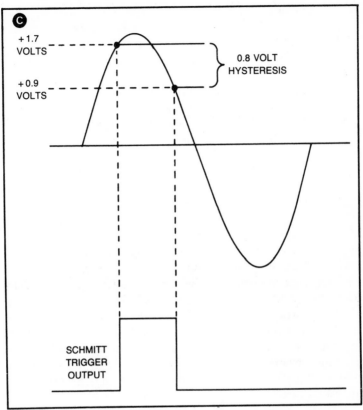

Fig. 6-9. (C) Waveforms for "B."

in Fig. 6-9B are usable for most frequencies between 1 and 20 MHz. The voltage is developed across resistor $R1$, causing the output of the Schmitt trigger to toggle if the amplitude exceeds the limits for the particular device selected. If necessary, use another transistor amplifier state in order to ensure that this level is achieved. If both thresholds are not crossed, then the Schmitt trigger will not toggle.

Figure 6-10 shows three basic clock circuits using TTL digital logic elements. Although two of these circuits appear to use 7400 NAND gate devices, they are connected as inverters. These circuits can, like the third of these circuits, also be built from TTL inverters. Keep in mind that both the NAND and the NOR gates will operate as inverters if both inputs are tied together. Therefore, we will consider all three circuits of Fig. 6-10 to be constructed from inverter elements.

123

The circuit of Fig. 6-10A is an RC-timed ring oscillator. The frequency of oscillation is set by resistors $R1$ and $R2$, and capacitor $C1$. For most popular frequencies, i.e., those most frequently used as clocks in digital circuits, the capacitor will be 0.001-μF. The circuit shows a variable resistor, $R1$. The potentiometer will allow variation of the oscillation frequency, but is not strictly necessary when the frequency is fixed. I can, therefore, replace resistors $R1$ and $R2$ with a single fixed-value resistor. The circuit of Fig. 6-10A will oscillate (with the values as shown) at frequencies between approximately 2 and 10 MHz, assuming that $C1$ is 0.001-μF. The frequency will not be stable because of the RC elements used to time the oscillations; the circuit is still useful for some limited applications where such high stability and precision is wasted.

A pair of crystal-controlled oscillations are shown in Fig. 6-10B and 6-10C. The version of Fig. 6-10B is well-known, but is not always the best circuit to actually use in practical circuits. In this circuit, the active elements of the oscillator are two inverters made from 7400 NAND gate sections; a third NAND gate inverter is used as a buffer to isolate the oscillator from the outside world. The crystal element is in the feedback loop from the output of inverter B to the input of inverter A. Since we have two inverters, each providing complementing (i.e., the binary equivalent of 180-degrees of phase shift), the feedback will be in-phase with the signal at the input of inverter A.

The frequency of oscillation is set by the interaction of crystal $Y1$ and variable capacitor $C1$. Since the frequency at which a crystal oscillates depends in part upon the circuit capacitance, varying $C1$ will also vary the frequency.

A second capacitor is used in the circuit to block the dc between the output of A and the input of B. This isolation allows us to bias each inverter with fixed resistors, $R1$ and $R2$. The values of these resistors are not supposed to be critical, but experience shows that some amount of experimentation is, indeed, worthwhile.

The circuit of Fig. 6-10B is not trustworthy enough for practical consideration. It is popular, but it tends to fail to start on occasion. I once had a kit-form microcomputer put together by a small outfit that went out of business a short while after its founding. The main system clock was one of these circuits, so on occasion the darn thing would fail to work. The outward appearance was a dead computer, but the root cause was that the main system clock was inoperative. A change in value for the feedback capacitor ($C1$

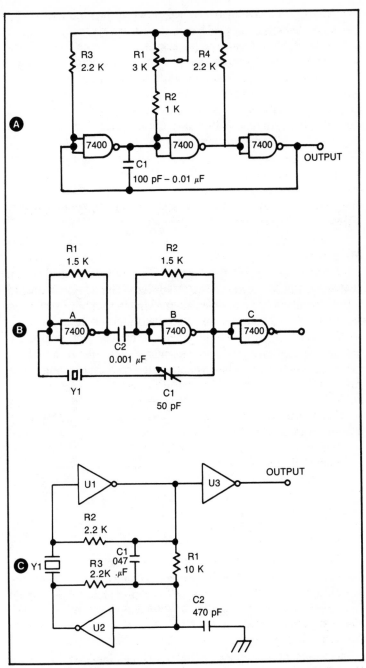

Fig. 6-10. Digital ''clock'' oscillators. (A) RC-timed. (B) Crystal. (C) Crystal.

125

was fixed), and a small capacitance between the output of B and ground, all but cured the problem . . . but still occasionally the clock fails to start or turn on.A better circuit is shown in Fig. 6-10C.

Later I bought a more reputable microcomputer, this one a Rockwell AIM-65, which always starts when I turn it on. My computer is in my office, which is located in an alarmed backyard shed that is heated with electric space heaters. During the winters, that shed is cold for the fist half-hour or so; in fact, it is freezing! Yet the little AIM-65 keeps on humming along nicely, and starts every time I turn the power on. The clock in that computer is shown in Fig. 6-10C. This circuit uses inverters, but could easily be built from inverter-connected NAND or NOR gates. Crystal $Y1$ is a 4-MHz unit in the computer, but the circuit should work well anywhere in the 1 to 10 MHz range.

MC-4024P

Perhaps the best solution for the TTL-compatible signal source problem is the Motorola MC-4024P device. The IC is now able to boast a standard "74xxx" number. It operates from + 5 volt dc and has two TTL voltage-controlled oscillators in one package, each with separate + 5 volt dc and ground terminals. Note that there are also package power (+ 5 volt) terminals on the device, and these must be connected whenever the device is operating. The MC-4024P pinouts are as follows:

Function	Osc A	Osc B
Capacitor or Crystal	10,11	3,4
Output	8	6
Control Voltage	12	2
+ 5 Volts	13	1
Ground	9	5

Package + 5-volts: 14
Package Ground: 7

It is possible to shift the oscillating frequency of the MC4024P over a range of approximately 3:1 by changing the control voltage from + 3 to + 5 volts. If this control voltage is an ac or varying dc signal, then the output signal will be frequency modulated by the signal. One application for this type of "FM" was once used in an analog modem that converted a $-/+$ 1-volt human electrocardiogram signal into a varying tone (FM). An operational amplifier with a gain of -1 was connected to the control input, and an offset potentiometer connected to the summing junction so that the "normal" output potential with no input signal was +4 volts—right in the middle of the MC-4024P control voltage range. The ECG signal would then vary −/+ 1 volt for a total range of +3 to +5 volts, and a frequency change of 3:1. It is important, by the way, that you not confuse the MC-4024P device with a CMOS 4024. The CMOS 4024 is a completely different device and bears no relationship to the MC-4024P.

Figure 6-11A shows one popular circuit. Here only one oscillator (OSC B) is used, but the pin numbers can be changed should you want to use OSC A. The package +5 volts terminal, the OSC B +5 volts terminal and the control voltage terminal are all connected together at +5 volts; no frequency control is possible with this circuit. The package ground and OSC B ground are both connected to ground, so the device will operate. It is possible to use the OSC A and OSC B grounds to control the on/off operation of each oscillator, provided that the package ground is kept at zero volts potential. I can, therefore, lift pin no. 5 off ground in order to turn off this oscillator.

Frequency of oscillation is set by either a capacitor (see inset) or a piezoelectric crystal, $Y1$. If the crystal is used, it is necessary that the frequency be at least 2.5 MHz and less than 25 MHz in the fundamental mode. Crystals of lower frequency do not always oscillate in this circuit. If the capacitor is used, then the frequency is set very roughly according to the formula:

$$F_{Hz} = 300/C_{\mu F} \tag{6.2}$$

It will be necessary to experiment with the capacitor value in the circuit. Using a frequency counter and a handful of capacitors on my breadboard, I found that the formula given in another text on this same subject was in error by a factor of approximately ten. Even so, this formula will not yield any better than ballpark results, and some tweeking will be in order if the exact frequency is important to you.

The recommended circuit is shown in Fig. 6-11B. Here everything is exactly the same as before, except that the control voltage applied to pin no. 2 is variable, rather than fixed. This circuit tends to work better with low frequency crystals, and is capable of providing a rather large degree of control. Note that pin no. 2 is bypassed to ground. Some people also like to use a 0.1-μF capacitor in parallel with the 10-μF unit because the electrolytic capacitor is not able to work at high frequencies.

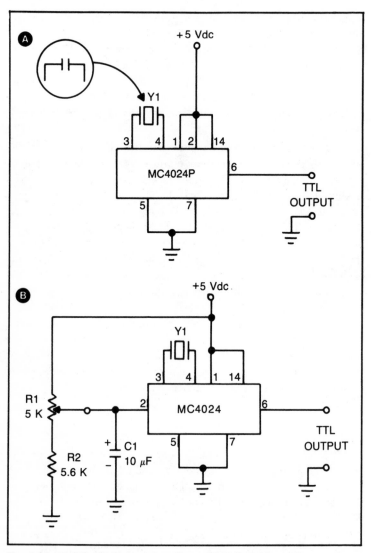

Fig. 6-11. (A) MC-4024P crystal oscillator. (B) Adjustable version.

CMOS CLOCK CIRCUITS

In the previous sections of this chapter, I have been discussing TTL compatible oscillator/clock circuits. Let me digress a bit and consider some of the CMOS versions. The circuit shown in Fig. 6-12A is a square wave oscillator based on the 4093 quad-CMOS NAND gate with Schmitt trigger inputs. This IC will operate as a normal NAND gate in that a low on either input will force the output high, and a high on both inputs will force the input low. The difference between the 4093 NAND gates and ordinary NAND gates is that the inputs operate as Schmitt triggers. For +5 volt supplies, the positive-going trip point is approximately +2.9 volts, while the negative-going trip point is about +2.3 volts.

The oscillator can be turned on and off by manipulating one of the inputs. One input is connected to V+ through a pull-up resistor (i.e., 10 kohms to 100 kohms). If I ground that input by closing

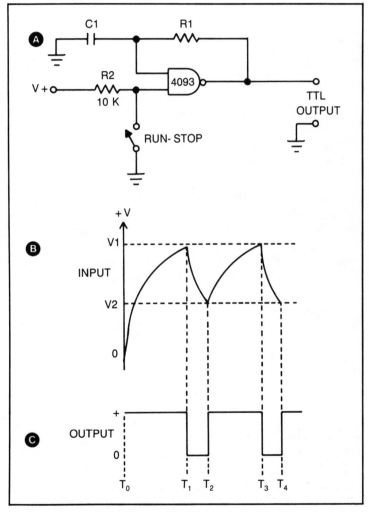

Fig. 6-12. (A) ON-OFF control of RC-timed oscillator. (B) Timing waveform.

switch S1, then the oscillator will stop. If however, I leave S1 open thereby keeping the input high, the oscillations continue. For applications where this control is not desired, simply strap the input high permanently through the pull-up resistor.

The operating frequency of this circuit is set by the combination of R1 and C1. The operating frequency is given very approximately by the following expression:

$$F = 2.7/R1C1 \qquad (6.3)$$

Where:

F is the frequency in hertz

R1 is the resistance in ohms

C1 is the capacitance in farads

Constraints: $V+ = 5$ volts dc, $V- = $ ground, C1 is greater than or equal to 0.001-μF

The timing waveform is shown in Fig. 6-12B. When the circuit is first turned on, the voltage across capacitor C1 (i.e., the voltage at the Schmitt trigger input) will be zero. This means the input is low, so the output will be high. A high on the Schmitt trigger output means that C1 will charge through resistor R1 at a rate determined by the Schmitt trigger output potential and the RC time constant. For practical reasons, the oscillation will not start instantly, but requires a delay of a few cycles. Eventually, though, the capacitor voltage climbs to the positive-going threshold V (at time T1). When this occurs, the output will snap low, allowing the capacitor to discharge through R1. When the capacitor voltage discharges down to the negative-going threshold V2, which occurs at time T2, then the output of the Schmitt trigger will snap high again. This alteration occurs periodically from then on, and we have a square wave oscillator with an unequal duty cycle.

Three other RC-timed CMOS oscillator/clock circuits are shown in Fig. 6-13A through C. The simplest version is shown in Fig. 6-13A, and consists of a pair of 4049A inverters and an RC timing network. Note that the 4049A will drive TTL-compatible loads when the package voltage is +5 volts dc, even though the fan-out is limited to two.

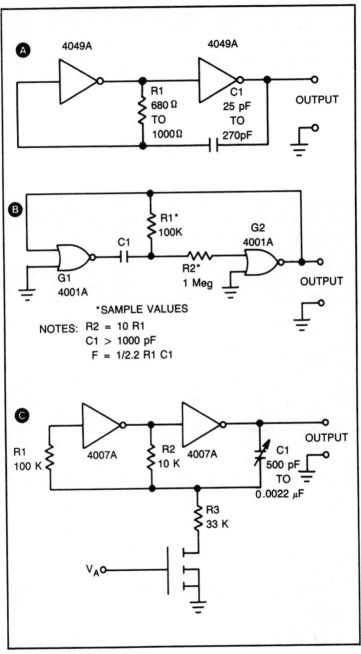

Fig. 6-13. (A) CMOS RC-timed oscillator. (B) Improved version. (C) Voltage-controlled version.

131

The preferred CMOS RC oscillator is shown in Fig. 6-13B. This circuit uses a second resistor, and this permits the duty cycle to be nearer 50 percent. In most cases, you will want to make resistor $R2$, which is in series with one input of the 4001A device, about ten times the resistance of $R1$. If capacitor $C1$ is greater than or equal to 0.001-μF, then the frequency of oscillation is approximately:

$$F = 1/2.2R1C1 \qquad (6.4)$$

The 4001A NOR gates are connected in an inverter configuration. The NOR gate will act as an inverter under two circumstances: a) when both inputs are tied together, and b) when one input is permanently grounded. I can, therefore, make this circuit a gated oscillator by lifting one or both grounded 4001A inputs ($G1$ and $G2$). If the input is grounded, then the circuit will oscillate, but if it is held high, then it will stop running.

A voltage-controlled, pulse-width version of this same circuit is shown in Fig. 6-13C. Here two additional components are added: a resistor ($R3$) and a MOSFET transistor. The transistor can be one section of the 4007 device. Voltage V_a applied to the gate of the transistor will vary its channel resistance; hence, it will affect the charge/discharge cycle of the RC network. I therefore gain pulse-width control as a function of voltage V_a.

The RC oscillators shown thus far will operate to frequencies of several megahertz, if you are lucky, but under 1 MHz is a bit more realistic, especially with the Schmitt trigger version. Selected devices will allow operation at frequencies greater than 1 MHz, but don't count on it for run of the mill CMOS devices.

Our last circuit is the crystal oscillator of Fig. 6-14. This circuit will operate to frequencies of 2 MHz, or thereabouts, with good precision and reliable operation. It uses a pair of 4001A NOR gates connected as inverters. As in the previous circuits, these devices can be replaced with either inverter-connected NAND gates or actual inverters. A variable capacitor allows some control over operating frequency. At some settings of $C2$ however, starting might tend to be a little flaky, so you will have to set the capacitance to a value that will ensure good starting 100 percent of the time.

OTHER CLOCKS

I have discussed a number of approaches to providing TTL and CMOS clock signals with an assortment of oscillators and astable multivibrators. There are other circuits and devices as well, but

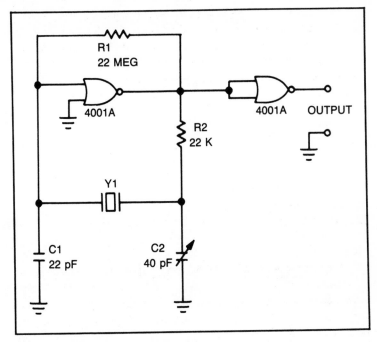

Fig. 6-14. CMOS crystal oscillator.

time and space simply do not permit covering all of them. I recommend, therefore, that you examine the most recent data books for devices most suited to your own needs. If you are designing a microcomputer or anything else around one of the popular microprocessor chips, then be aware that most semiconductor makers in that business offer special purpose clock chips that are especially suited for that particular application.

Chapter 7

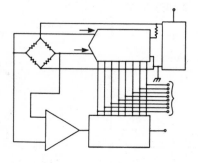

Analog Function Modules

A FUNCTION MODULE IS A PREPACKAGED ELECTRONIC CIR-cuit block that is designed to perform a special function, or a limited range of related functions, with a minimum of external circuit connections. This category of devices does not formally include special-purpose monolithic integrated circuits, but I think that a case for their inclusion can certainly be made.

But what is a function module? In most cases, the circuit required to do the job may be composed of operational amplifiers, other linear or digital integrated circuits, resistors, capacitors, diodes, and individual transistors much like a regular electronic circuit. The integrated circuits used are usually obtained in chip form from the semiconductor manufacturers rather than in the familiar packaged versions of ordinary electronic applications. These are mounted on a master ceramic substrate along with the other components to form a hybrid circuit. The substrate can be multilayered, and can contain printed circuit tracks in addition to angel hair wires from the "outside world" pins to the tracks or from the tracks to the integrated circuit chips. A cutaway view of a function module is shown in Fig. 7-1.

Several different common packages are used for housing analog function modules. The two types shown in Fig. 7-2 are probably the most evident. The package in Fig. 7-2B is a hermetically sealed metal package, while that in Fig. 7-2A is an epoxy-fiberglass "potted" version. The pin layouts and actual physical dimensions

Fig. 7-1. An analog function module.

tend to vary not only from one manufacturer to another, but between modules of the same manufacturer, so nothing meaningful can be said about them.

Most analog function modules are designed to be mounted and soldered directly to a printed circuit board. A few manufacturers, however, offer sockets at nominal cost, and these are especially useful in wire-wrapped applications.

The range of actual functions performed by the various modules is broad, and includes both exotic data acquisition systems and individual data converters (covered separately in Chapters 10 through 15 of this book) down to ordinary power supplies and isolated dc-to-dc converters.

Other common monolithic and hybrid function modules include active filters, voltage-to-frequency converters, frequency-to-voltage converters (and at least one module that will do both), logarithmic and antilog amplifiers, A/D converters, D/A converters, sample-and-hold circuits, high-performance operational amplifiers, and at least one multifunction converter module (the Burr-Brown 4301).

Figure 7-3 shows a printed circuit board manufactured by Datel using several Datel function modules along with a number of monolithic integrated circuits. This data acquisition product is shown here to emphasize to you what a function module is not. It is not a printed circuit (PC) board filled with components. The PC board is more properly called a special-purpose subassembly, built with function modules and other discrete or monolithic components as needed.

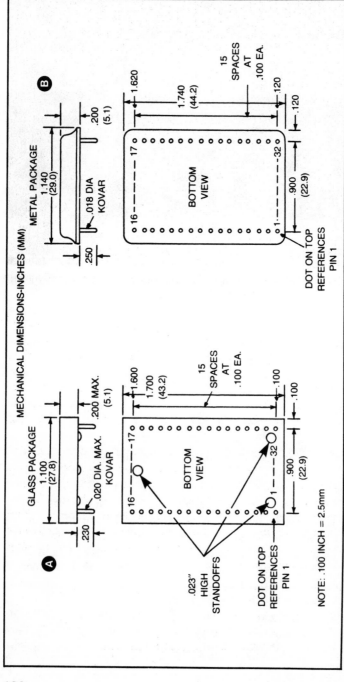

MECHANICAL DIMENSIONS-INCHES (MM)

A

GLASS PACKAGE

1.100 (27.8)

.200 MAX. (5.1)

.020 DIA. MAX. KOVAR

.230

16 -- 17

BOTTOM VIEW

1 -- 32

15 SPACES AT .100 EA.

1.600

1.700 (43.2)

.100

.100

.900 (22.9)

.023" HIGH STANDOFFS

DOT ON TOP REFERENCES PIN 1

NOTE: .100 INCH = 2.5mm

B

METAL PACKAGE

1.140 (29.0)

.200 (5.1)

.018 DIA KOVAR

.250

16 -- 17

BOTTOM VIEW

1 -- 32

15 SPACES AT .100 EA.

1.620

1.740 (44.2)

.120

.120

.900 (22.9)

DOT ON TOP REFERENCES PIN 1

Fig. 7-2. Function modules packages. (A) Metal. (B) Glass. (Courtesy of Datel.)

136

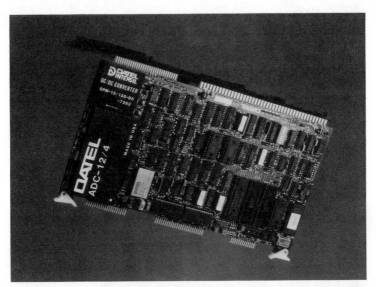

Fig. 7-3. Several Datel function modules used on a printed circuit. (Courtesy of Datel.)

I am not going to describe how they make an analog function module, even though the topic is potentially very interesting, but will describe several commercially available products and give their applications.

Figure 7-4 shows the block diagram for a Datel universal active filter function module, model FLT-U2. The internal circuit diagram is shown in Fig. 7-4A, and the external connections in Fig. 7-4B.

Operational amplifiers $A1$ through $A3$ form a "state variable" active filter. Although this class of filter is more complex than some of those given in Chapter 5, it offers the advantage of providing high-pass, low-pass, and bandpass properties *simultaneously* from different terminals. The FLT-U2 also includes an uncommitted operational amplifier ($A4$) that can be configured by the user to meet specific needs. The transfer functions for the state variable filter are:

Low-Pass Function

$$H(s) = \frac{k1}{s^2 + (\omega/Q)s + \omega^2{}_0} \qquad (7.1)$$

137

Bandpass Function

$$H(s) = \frac{k2s}{s^2 + (\omega/Q)s + \omega^2 o} \qquad (7.2)$$

High-Pass Function

$$H(s) = \frac{k3s^2}{s^2 + (\omega/Q)s + \omega^2 o} \qquad (7.3)$$

where ωo is the natural frequency of the circuit $2\pi f_o$ and $k1$, $k2$, and $k3$ are constants.

The external circuitry for the filter is shown in Fig. 7-4B. This module sees two different inputs, inverting and noninverting, but Table 7-1 shows which to use in order to give inverting or noninverting characteristics to the overall circuit.

Design Procedure

1. Determine desired f_o, Q, and configuration (Table 7-1).
2. Compute the quantity f_oQ. (At f_oQ less than 10^4 the Q of the finished circuit will closely approximate the calculated Q, at $f_oQ = 10^4$ there will be a Q error of approximately 1%, and at f_oQ greater than 10^4 there will be about 20% Q error.)
3. Find $R1$, $R2$ and $R3$ from the appropriate table. Use Table 7-2A for inverting configurations, and Table 7-3 for noninverting configurations.
4. Set $R4 = R5$ and compute:

$$R4 = R5 = 5.03 \times 10^7/f_o$$

Table 7-1. Datel FLT-U2 Active Filter Input/Output Phases.

Input Used	Output Phase		
	BP	HP	LP
Noninverting	Inverted	Noninverted	Noninverted
Inverting	Noninverted	Inverted	Inverted

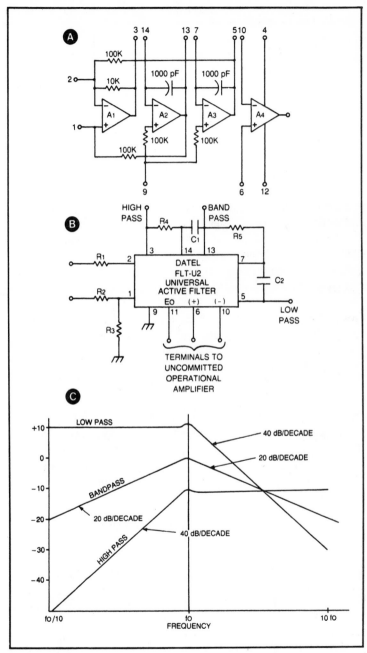

Fig. 7-4. Function module state-variable filter. (A) Block diagram. (B) Circuit. (C) Response plot. (Figs 7-4A and B, Courtesy of Datel.)

139

Table 7-2. Datel FLT-O2 Inverting Configuration Resistor Values.

Form	R_1	R_2	R_3
LP	100K	—	100K/(3.8Q − 1)
HP	10K	—	10K/(6.64Q − 1)
BP	Q × 31.6K	—	10K/(3.48Q)

(Note: $R4$ and $R5$ need not actually be equal so long as $(R4R5)^{1/2}$ remains constant. If $R4$ is made fixed, then $R5$ can be made variable to trim the resonant frequency f_o to an exact value.)

Capacitors $C1$ and $C2$ are used if the filter is operated at f_o below 50 hertz, in which case,

$$R4 = R5 = 5.03 \times 10^7/f_o C$$

assuming that the C term, given in picofarads, is the sum of the 1000 pF internal capacitors and $C1$ or $C2$, also expressed in picofarads. If unequal external capacitors are used, then

$$R4 = R5 = 5.03 \times 10^7/f_o(C_A C_B)^{1/2}$$

where $C_A = C1 + 1000$ pF and $C_B = C2 + 1000$ pF.

The procedure given above results in a unity gain universal filter. If additional gain is required, then use the uncommitted operational amplifier ($A4$ is Fig. 7-4A) to provide the amplification desired. This stage obeys the standard rules for operational amplifiers.

Table 7-3. Datel FLT-O2 Noninverting Configuration Resistor Values.

Form	R_1	R_2	R_3
LP	—	316K/Q	100K/(3.16Q − 1)
HP	—	31.6K/Q	100K/(0.316Q − 1)
BP	—	100K	100K/(3.48Q − 1)

UNIVERSAL V/F-F/V CONVERTERS

The Datel VFV-10K and VFV-100K function modules are capable of operating as either voltage-to-frequency (V/F) or frequency-to-voltage (F/V) converters. The block diagram to this useful function module is shown in Fig. 7-5. The VFV-series modules will operate in V/F or F/V modes depending only on external connections.

The internal circuit is of the capacitor-charge integrator variety, but in this particular implementation, the inverting input of the operational amplifier used as the integrator can be accessed either directly or through a resistor to make current or voltage inputs, respectively. The name for this device, then, is a little bit of a misnomer because it will also function as a current-to-frequency converter.

Another feature of this module set is the uncommitted inverting amplifier preset with unity gain. This stage is used to accommodate negative inputs. The main V/F section, the integrator, always wants to see a positive voltage, so to accommodate negative voltages I must first invert them in this amplifier and connect the operational amplifier to the regular input.

The analog input ranges are 0 to + 10 volts if connected in the direct configuration and 0 to − 10 volts if the inverting amplifier is used. The full-scale range of the current input is 0 to 1 milliampere in direct and 0 to − 1 milliampere if through the inverting amplifier.

The principal difference between the two Datel modules in this series is in the output frequency range. A full-scale input current or voltage will produce a 10 kHz output in the VFV-10K and 100 kHz in the VFV-100K. For voltage inputs, the scale factors, then, are 1 hertz/millivolt and 10 hertz/millivolt, respectively.

These modules can be used to produce an output frequency from an analog voltage or current input, or conversely an output voltage from an input frequency. Such a converter may be used in computerized instruments either as part of intermediate processing, or in either digital-to-analog (i.e., F/V) or analog-to-digital (i.e., V/F) capacities. In the latter case, either an auxiliary frequency counter is used to measure the frequency, or a software frequency counter can be used. In general, the hardware counter is more accurate, but the software approach is more desirable if the error is tolerable.

When using the V/F converter in A/D service, one will have to give some thought to the coding of the counter's output. An off-the-shelf frequency counter will most likely produce a binary-coded decimal (BCD) output, as will most of the decade counter integrated circuits you might choose to implement a hardware counter of your own design. Similarly, the LSI CMOS counters now available that offer five or six decades on a single chip usually have a multiplexed BCD output. N-bit binary counters, of which several exist in both TTL and CMOS lines, output a binary word. These might be used directly (especially if an eight-bit computer input port is used), or you might use a set of cascaded four-bit binary counters (i.e., 7493) organized in hexadecimal form. Regardless of which coding scheme is selected, however, unless it is of a format that is already recognized by the computer, some sort of software code conversion is required.

Two V/F converters can be used to make *ratiometric* voltage measurements. These are especially useful where the ratio between two measured parameters is more accurately determined than absolute values, often the case in scientific and engineering instruments.

One converter is used to feed the input of a frequency counter, while the other converter is frequency divided by a divide-by-n counter chip, and is then used to control the frequency counter's main gate flip-flop. This results in an output count of

$$\text{Count} = 2n \ V1/V2 \qquad (7.4)$$

where $V1$ is the voltage applied to the V/F converter, $V2$ is the voltage applied to the divide-by-n stage, and n is the frequency division ratio set by the divide-by-n chips programming pins.

I am able to produce a pulse output equal to the difference between two input frequencies by using two F/V converters and one V/F converter in a suitable configuration.

If frequency $f1$ drives the input of one F/V module, and frequency $f2$ drives the input of the other F/V module, their respective outputs are directed to the $V+$ and $V-$ inputs of the V/F module. The output of the V/F module's internal inverting amplifier is connected to the current input of the integrator circuit. This connection scheme yields an output of

$$f_o = f1 - f2 \qquad (7.5)$$

The pulse output of this circuit can then be used as an analog-to-digital converter in the same manner as the frequency counter technique discussed earlier.

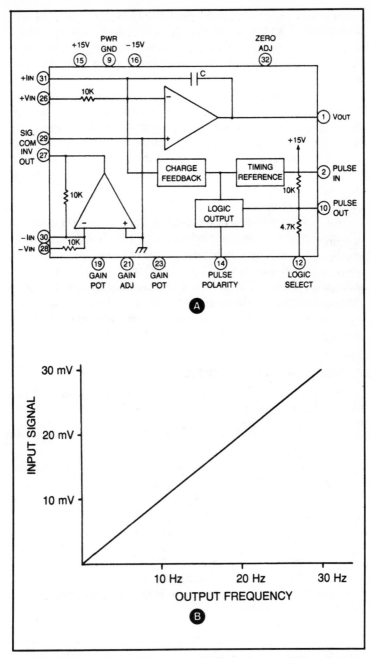

Fig. 7-5. Voltage-to-frequency converter. (A) Block diagram. (B) Transfer function. (Fig. 7-5A Courtesy of Datel.)

143

A primary class of applications for the V/F converter is in data transmission. An analog signal can be converted to a frequency, then transmitted over ordinary telephone or radio channels to or from a remote location to the computer or other processing instrument.

The data received from the telephone or radio receiver can be displayed in digital form, reconverted to analog form (in an F/V converter), or applied directly to the computer. In most cases, however, at least a Schmitt trigger or comparator will be required because the digital instrument receiving the data want to see TTL or CMOS compatible input pulses and the frequency response constrictions (plus other factors) will have converted the pulses from the V/F converter into near sine waves.

A related application or this function module is in the isolation of a signal source from the instrument. This is often done in medical electronic applications because of the relatively severe electrical safety standards enforced for patients who are somewhat more susceptible than healthy people. In industrial and scientific applications, isolation might be required because the transducer and preamplifiers (if any) must be located in an environment potentially hazardous to humans, or that is inaccessible most of the time.

A MULTIFUNCTION MODULE

The Burr-Brown 4301 and 4302 function modules obey the following transfer function:

$$E_{OUT} = V_Y (V_Z / V_X)^m \qquad (7.6)$$

where E_{OUT} is the output voltage, V_X, V_Y, and V_Z are input voltages and m is a constant such that $0.2 \leq m \leq 5.0$.

This analog function module is capable of, either singly or in combination with certain other components, the following output transfer functions: analog multiplication, analog division, squaring, square rooting, exponentiation, rooting, sine and cosine, arctangent (V_X / V_Y), and vector summation/subtraction.

Figure 7-6 shows the general circuit for using the multifunction converter. In the case where m is greater than unity (1), the constant m is given by:

$$m = (R1 + R2)/R2 \qquad (7.7)$$

In the case where m is less than unity,

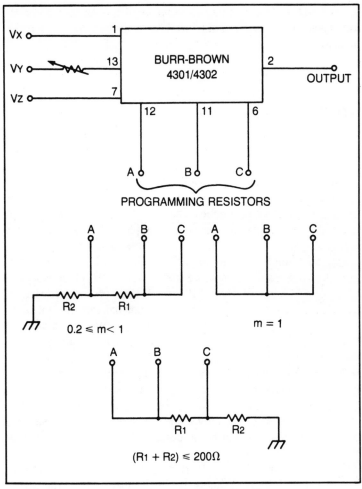

Fig. 7-6. Versatile signal-processing module by Burr-Brown.

$$m = R2/(R1 + R2) \qquad (7.8)$$

where $m = 1$, all three converter programming pins are connected together. In both Eqs. 7.7 and 7.8 the sum $(R1 + R2)$ must be equal to or less than 200 ohms.

The 4301 and 4302 modules can be used as a logarithmic amplifier by using pin 11 as the output, and connecting the circuit for m between unity and five. In that case only,

$$m = R2/(R1 + R2) \qquad (7.9)$$

for m greater than unity. In all other applications, use the equations given earlier for setting m. The transfer function for the logarithmic amplifier is

$$E_{OUT} = (kT/qm) [\ln(V)_X / V_Z)] \qquad (7.10)$$

OTHER FUNCTION MODULES

The range of possible function modules seems nearly limitless, and you are cautioned to check the catalogues of the function module manufacturers before undertaking a complex design project. The use of analog function modules is a little more costly than other approaches if you only consider the cost of the components, but becomes a lot more economic if the total design/construction time must also be considered. Low-cost, sample-and-hold circuits, for example, can be constructed from three integrated circuits (two operational amplifiers and an analog switch with TTL drive terminal), or one may be purchased from any of several function module sources for less than $10 as of this writing. Superfact, sample-and-hold circuits are tricky to design and actually build (layout becomes critical and circuits become a lot fussier), so very often the seemingly high cost of the equivalent analog function module is a justifiable expense.

Another often-encountered function module is the RMS-to-dc converter. Some converters are merely time averagers, while other purport to perform a calculation such as $E_{RMS} = 0.707E_{PEAK}$, but these methods are of use only if you can assure that only sinusoidal input waveforms will be processed. Otherwise, the accuracy suffers a tremendous amount.

Most of your better circuits that do the RMS-to-dc conversion, and that includes most function modules, operate from the definition of RMS voltage, namely,

$$E_{RMS} = \sqrt{\int_0^t (E_{IN})^2 2 \, dt} \qquad (7.11)$$

Most RMS-to-dc converter function modules are designed to be little dedicated analog computers, and are programmed to solve Eq. 7.11. They will generally operate to frequencies in excess of 1 megahertz, something that cannot always be said of the thermal and optical transducer types of RMS-to-dc converter.

146

Chapter 8

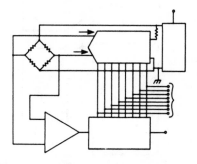

Controlling the World

WHILE MANY AMBITIOUS POLITICOS HAVE ATTEMPTED TO control the world, only to fail, I can control at least a small part of it using microcomputer I/O ports and some external circuitry. The approach taken in this chapter will be by example.

Example 8-1

Use one bit of a computer output port to turn on a 120-volt ac lamp.

Solution:

One possible solution to this problem is shown in Fig. 8-1. Here I use a magnetic relay to control the high ac potential from the power mains to the lamp.

Transistor $Q1$ is a relay driver and can be almost any npn silicone type that will do the job (i.e., handle the voltage and current levels present). Types 2N3906 and 2N2907 are often suggested for this service.

In this example I am assuming that TTL-compatible output ports are available in which logic 0 is 0 volts and logic 1 is $+5$ volts. Resistor $R1$ is selected for those levels. If a higher output voltage is used, then scale the value of $R1$ upwards proportionally.

When bit 1 is low, then transistor $Q1$ is cut off and no current flows in the coil of relay $K1$. Since $K1$ is deenergized, the circuit to the lamp is open and the bulb is turned off.

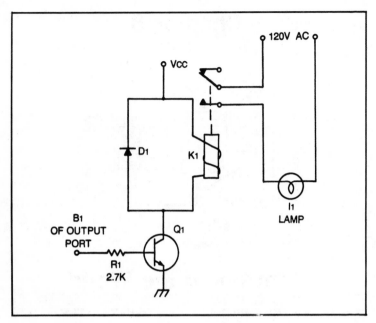

Fig. 8-1. Driving a relay-controlled 120V ac lamp circuit.

When bit 1 goes high, on the other hand, transistor $Q1$ is forward biased, so will conduct collector current. The collector of $Q1$, then, goes to ground, turning on $K1$.

If the energized position of $K1$ the contacts controlling the lamp are closed, so the lamp turns on.

Diode $D1$ is used to suppress voltage spikes generated by the inductive kick produced by the coil of $K1$. Diode $D1$ can be almost any rectifier diode in the 1N4000 series. Since the spike that is generated can be quite high, only 1N4004 through 1N4007 are recommended; 1N4001 through 1N4003 have too low a peak reverse voltage (PRV) or peak inverse voltage (PIV) rating.

If you doubt the need for diode $D1$, then examine the waveforms shown in Fig. 8-2. The waveform of Fig. 8-2A shows the situation when a step-function (i.e., switch turn-on) potential is applied to $K1$ with $D1$ present and doing its job. Notice that the waveform is essentially clean.

In Fig. 8-2B, however, diode $D1$ has been disconnected, and that has caused a high-amplitude negative-going voltage spike at turnoff. This spike, which had been suppressed by $D1$, is actually larger than shown here because the camera used to make the photograph could not write the record onto film fast enough to show the true height. Such spikes can, and frequently do, damage electronic circuitry (i.e., $Q1$ is vulnerable), and can also cause spurious pulses in digital circuitry. Even where the circuitry can absorb the pulse, there is the possibility that such a pulse will reset counters and flip-flops, or be propagated through gates, at exactly the wrong time. Experienced troubleshooters will recognize that as one of the most unnerving situations that can occur.

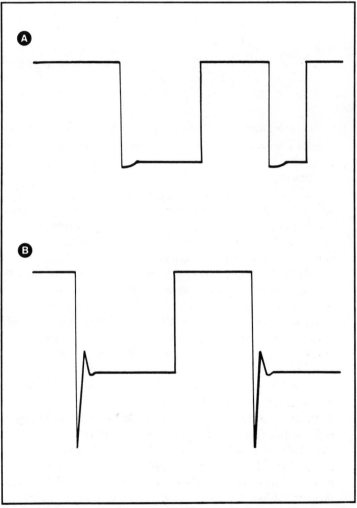

Fig. 8-2. Waveforms of circuit in Fig. 8-1. (A) Relay with diode D1. (B) Without D1.

Example 8-2

Turn on a relay, as in Example 8-1, using a logic IC instead of a driver transistor.

Solution:

Certain TTL and CMOS logic devices are designed with an open-collector bipolar output transistor, and in normal operation require a pull-up resistor between each output terminal and the +5-volt dc supply. This resistor will have a value in the 1K to 2K range. An example from the popular TTL series of devices is the 7406/7416 hex inverter. These devices are essentially the same type of IC, both being designed as relay or lamp drivers, and can connect to supplies of +15 volts and +30 volts, respectively.

To use either the 7406 or 7416, connect an appropriate relay coil between the output of one inverter section and a positive supply voltage. Otherwise, the circuit is as in Fig. 8-1.

Example 8-3

Do the same job as in the two previous examples, except eliminate $K1$.

Solution:

There are two approaches to this problem shown in Fig. 8-3; both use optoisolators and an SCR to control the lamp.

An opto-isolator is a special IC that contains a light-emitting diode (LED) and a phototransistor configured such that the transistor is on if the LED is lighted.

An SCR (i.e., silicon-controlled rectifier) is a rectifier diode that remains turned off unless a current is injected into the gate terminal.

In the circuit of Fig. 8-3A, two 7406 inverters keep the LED in the opto-isolator turned on (i.e., if the input to inverter 1 is low, then the output of inverter 2 is also low; a condition that grounds the cold side of the LED, thereby turning it on).

The LED, then, remains turned on, and that keeps the phototransistor conducting. The collector of $Q1$, therefore, remains at, or near, ground potential, and that keeps the SCR turned off.

Applying a high to the input of inverter 1 places a high on the output of inverter 2, turning off the LED; thereby turning off $Q1$ and turning on the SCR. When the collector voltage of $Q1$ goes high, a current is set up in the gate of the SCR that is sufficient to cause it to turn on.

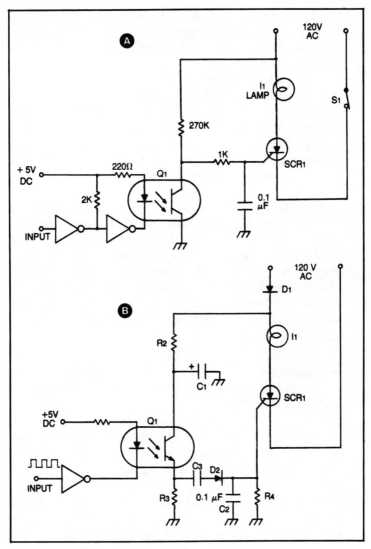

Fig. 8-3. SCR circuits. (A) Single pulse turn-on SCR circuit using opto-isolator. (B) Circuit that is input "turn-offable."

Once gated on, an SCR will remain on until the anode-cathode current drops below a certain critical threshold, or holding, current. In the circuit of Fig. 8-3A, commutation of the SCR is done manually, by opening switch S1. This arrangement allows a single pulse to start and hold the lamp in operation.

The circuit of Fig. 8-3B, on the other hand, must be continuously pulsed in order to keep the lamp turned on. As long as a pulse train is applied to the circuit, then the SCR remains in the on condition.

Diode $D1$ rectifies the 120-volt ac, so once every half-cycle the SCR anode-cathode current drops to zero. If the SCR gate is not excited when the next half-cycle begins, then the SCR remains turned off.

In some cases the SCR gate can be pulsed directly by the computer output port, but since 120 volts ac is involved, the isolation provided by the opto-isolator is very necessary.

The SCR gate circuit in Fig. 8-3B is controlled by RC network $R4$ $C2$ $C3$ and diode $D1$. If the LED is pulsed through the inverter, then the phototransistor is also pulsed, and these pulses appear across emitter resistor $R3$.

Pulses across $R3$ are coupled through $C3$ and $D2$, which charge capacitor $C2$. As long as the pulses continue, $C2$ remains charged, which keeps the SCR gate excited. But if the pulse train ceases, then $C2$ discharges. The SCR will remain turned off after the first zero-current half-wave cycle that the gate voltage is below the turn-on threshold.

Example 8-4

Turn on a motor if, and only if, the binary code on an output port is 01100111.

Solution:

All of the circuits discussed thus far can be used to turn on a motor, although be careful of SCR circuits (use a full-wave SCR, or *triac*); only selective decoding from the computer output port is needed.

Previously, a single bit was used to turn on the circuit, but in this example an entire output port is needed. Selective decoding can be done using an eight-input NAND gate and any needed inverters. Figure 8-4 shows how an eight-bit address decoder can be built from a 7430 eight-input NAND gate and an inverter.

A TTL NAND gate output will remain high if any of the eight input lines are low. To make a 7430 output high requires a code of 11111111.

But the desired code is 01100111, so we must invert bits 4, 5, and 8 (see Fig. 8-4). If this is done as shown, then the 7430

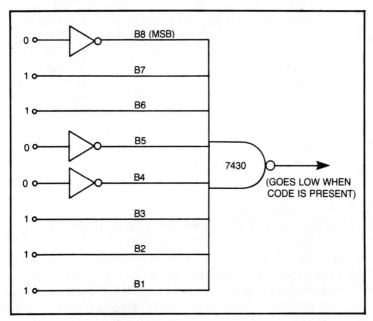

Fig. 8-4. Selective address decoder.

output will drop low when the code 01100111 appears on the input.

This technique is used whenever you want to use a single output port to control several devices, each of which is assigned its own unique code. The eight-bit parallel lines can be bussed together, and only the device addressed will respond.

Example 8-5

Perform the same job as in Example 8-4, except that now I require a means for the computer to verify that the desired action has taken place.

Solution:

A computer can verify the action using only one bit of an input port, provided that an appropriate transducer or other indicator is used.

The verification can be provided, with varying degrees of reliability, by any of the following: relay contact closures; detection of the output state of a gate, driver, or inverter; a tachometer ganged to the shaft of the motor (see Chapters 1 and 2).

153

The relay used to turn on a motor can be specified to have one more contact pair than is otherwise necessary. These can be connected to a + 5-volt dc source through a pull-up resistor and ground, such that a high is applied to an input port when the relay is energized. This scheme does not give an absolute indication of motor operation, but only that the relay has been energized and is telling the motor to operate. A defective motor would not be detected by this method. It is, however, very low cost.

Detection of the output state of one of the inverters or drivers can also yield the data but, again, it is possible for the command to be given, yet the motor be defective. This condition is not detected by this method.

The use of a tachometer on the motor shaft is the most reliable method for detection of motor operation. This device will produce a dc or ac (most common) output that can be detected by an appropriate electronic circuit and used to tell the computer that the motor has indeed started. Since the frequency of the ac tachometer, and amplitude of the dc type, is proportional to motor speed, then it is also possible to use these devices in a feedback control circuit.

Other related problems can be similarly solved, with the constraint that a suitable transducer be provided.

A furnace controller, for example, ignites the flame in a burner. A photoresistor or phototransistor looking into the spy-flame hole will detect whether or not the flame is actually turned on. The output of the detector could then be fed to a TTL output comparator such as the LM311 or the Precision Monolithics, Inc., CMP-01 and CMP-02. The comparator output will then serve to tell the computer that the flame is on.

Other transducers that provide a voltage output can be similarly applied to a comparator to supply verification.

Thus far, all of our applications have involved step-function control; that is, a device is either on or off. A continuously variable control circuit is possible if a digital-to-analog converter is used. For information on those circuits the reader is referred to Chapters 10, 11, and 13.

CONTROLLING SMALL DC MOTORS

Many control system projects require the use of a small, fractional horsepower dc motor as the prime mover. These can be used in several ways: on-off or continuously variable, in either closed-loop or open-loop configurations.

Turning a motor on and off is the most trivial problem and is an example of a simple open-loop system. The definition of an open-loop system is that there is no controlling negative feedback; the output causes no effects at the input.

Any of the methods used for ac loads earlier in this chapter will also work for dc motors, but I can also use a regular power transistor to control the motor (see Fig. 8-5).

Transistor $Q1$ in Fig. 8-5 serves as a driver for $Q2$. If $Q1$ is a 2N3053, and $Q2$ is a 2N3055, then motors drawing up to 8 or 10 amperes can be accommodated by this circuit. We do not want to drive the 2N3055 directly from the output port of a computer because the transistor may not have sufficient *beta* to drive the motor to full output with the current levels typically available from TTL output ports.

For the TTL compatible output port, i.e., those with +5 volts, the value of $R1$ shown will suffice. Scale $R2$ is proportional to the level of VCC. The operation of this circuit is as follows:

1. A low on the output port turns off both $Q1$ and $Q2$, so nothing happens.

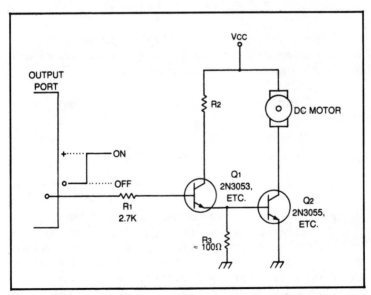

Fig. 8-5. Controlling dc motors through the use of power transistors in an open-loop system.

2. A high on the output port bit used to control this is circuit forward biases $Q1$ to saturation, making the voltage at its emitter high. This condition turns on $Q2$.
3. $Q2$ is now forward biased to saturation, so turns on the motor. (*Note:* Resistor $R3$ must have a value that limits the current flow in the base of $Q2$ to a safe value.)

In an open-loop control system the motor would simply turn on and off by command from the input, i.e., the computer. In a closed-loop system, on then other hand, the motor control commands can be modified by data received from the action of the motor. Consider the example of Fig. 8-6. Here we are using a dc motor to lift a load suspended from a rope from height $Y0$ to $Y1$. A precision multiturn potentiometer is ganged to the motor shaft, possibly through a gear train, so that it is, in effect, a position transducer (see Chapter 2).

Voltage E at the output of the potentiometer represents the height of the load (Y), and is a fraction of E_{REF} proportional to that height:

$$E = 0 \text{ at } Y0 \qquad (8\text{-}1)$$
$$E = E_{REF} \text{ at } Y1 \qquad (8\text{-}2)$$
$$E = E_{REF} (Y - Y0)/(Y1 - Y0) \qquad (8\text{-}3)$$

Both E and E_{REF} are connected to an operational amplifier voltage comparator that will produce an output level according to the following rules:

Condition	Comparator Output	
$E = E_{REF}$	Low	(8-4)
$E = E_{REF}$	High	(8-5)

The comparator output is connected to a single bit of the computer input port, while the motor is connected through a circuit such as Fig. 8-5 to a single bit of the computer output port.

The idea here is to write a program into the computer that will turn on the motor by setting the control bit of the output port high, then periodically test the $Y1$ position-indicating bit of the input port for a high condition. As long as that bit of the input port remains low, then the program keeps the motor running. But when a high is detected, indicating that the load is at position $Y1$, then the pro-

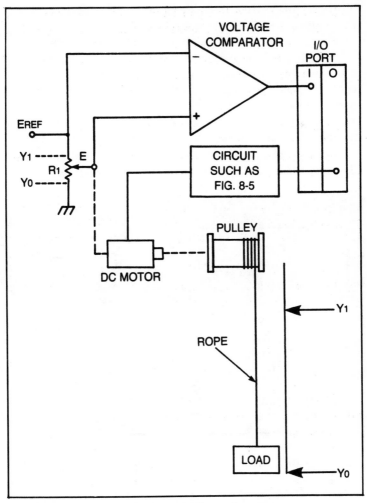

Fig. 8-6. A closed-loop system of controlling a dc motor.

gram resets the output port bit to low, thereby turning off the motor. This circuit is considered trivial for several reasons, not the least of which is the fact that the load cannot be lowered once it has reached the high position. Of course, another bit may be wired to a relay driver. If the relay is connected in the classic DPDT double-cross or X circuit, then the computer will be able to reverse the motor. In that case the reverse from the above protocol will indicate when the load has returned to $Y0$. The relay method is, however, too inelegant—tacky even. In a section that follows we will discuss an electronic motor reversal circuit that does not need archaic devices such as electromechanical relays.

The circuit in Fig. 8-6 could be used for such applications as automatic flag pole or window raiser. It suffers from the inability to recognize more than two states, i.e., $Y0$ and $Y1$. But what if you do not want your window open all of the way, or you want to mourn the passing of a national hero by running the flag to half-mast? In that case you will need a circuit that can be trained to recognize positions between the extremes.

Figure 8-7 shows a modification of the original circuit that will allow the computer to dictate height Y at which the motor stops

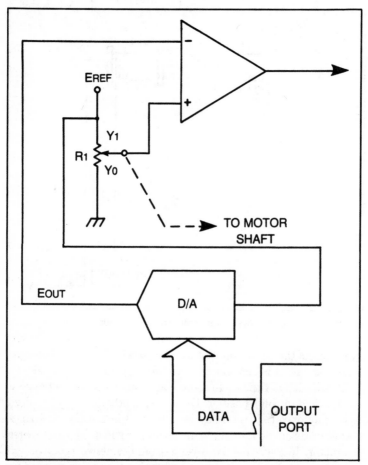

Fig. 8-7. A modified version of the circuit in Fig. 8-6 that recognizes more than two conditions.

158

turning. Voltage E_{REF} is still applied to the position potentiometer as before, but only a fraction of E_{REF} is applied to the comparator. You will learn in Chapter 10 that the output voltage from a D/A converter is

$$E_{OUT} = \frac{A}{2^n} \times E_{REF} \qquad (8\text{-}6)$$

where E_{OUT} is the D/A converter output potential, E_{REF} is the reference voltage, n is the number of bits at the D/A converter's digital inputs (usually eight in microcomputer systems), and A is the value of the digital word applied to the D/A converter input.

Example 8-6
Find the output voltage from an 8-bit D/A converter if E_{REF} is 10.00 volts and A is 10000000_2. (*Note:* $10000000_2 = 128_{10}$).
Solution:

$$E_{OUT} = \frac{(128)\,(10.00V)}{2^8} \qquad \text{(from Eq. 8-6)}$$

$$E_{OUT} = \frac{(128)\,(10.00V)}{(256)} = 5.00V$$

Since Y is proportional to E and a fraction of E_{REF}, and E_{OUT} is also a fraction of E_{REF}, I can set the height at which the comparator output goes low (telling the computer to halt the motor), by setting A in Eq. 8-6.

$$Y1 = 255/256 \qquad (8\text{-}7)$$
$$Y = A/256 \qquad (8\text{-}8)$$
$$Y0 = 0/256 \qquad (8\text{-}9)$$

Note that some D/A converters can give bipolar output voltages. They will produce a negative or positive output potential depending upon the digital code applied to the inputs. This feature opens up both the possibility of the controlling motor direction and speed.

Figure 8-8 shows the type of motor drive amplifier required to electronically reverse the motor. This circuit is based on the complementary symmetry circuit used in high-fidelity amplifiers. Note

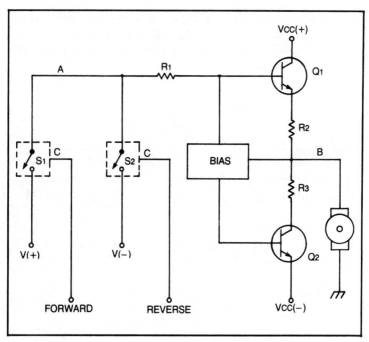

Fig. 8-8. Continuously variable motor control through the use of CMOS electronic switches.

that bipolar, V_{CC} (+) and V_{CC} (−), power supplies are required for this circuit to operate properly.

A positive potential applied to the input, i.e., point A in Fig. 8-8, will reverse bias transistor $Q2$ and forward bias transistor $Q1$, making the potential applied to the motor (i.e., point B) positive.

To reverse the motor's direction of rotation it is necessary to reverse the polarity at point B. This is done by applying a negative potential to point A. In that case, $Q1$ is reverse biased and $Q2$ is forward biased.

In the example of Fig. 8-8 we have used CMOS electronic switches such as the 4016 and 4066 to supply positive and negative inputs on command. The positive input is connected when the control line from $S1$ is high, while the negative potential is applied if the control input of $S2$ is made high.

Continuously variable motor speed control is available if the output of a D/A converter is applied to point A in Fig. 8-8 instead of a V (+) or V (−) source. If the D/A converter is a bipolar type, then both speed and direction of rotation can be dictated by the digital command from a computer output port.

160

Figure 8-9 shows analog and digital versions of a continuously controlled servomechanism using a dc motor. This circuit uses negative-feedback control techniques. Any such system can be described by the equation

$$\frac{E_{\text{OUT}}}{E_{\text{IN}}} = \frac{H}{1 + HB} \qquad (8\text{-}10)$$

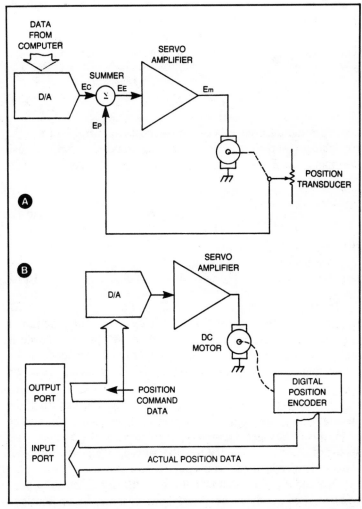

Fig. 8-9. Continuously controlled servomechanism. (A) Analog method. (B) Digital method.

where E_{OUT} is the output voltage, E_{IN} is the input in voltage, H is the gain of the forward path, and B is the gain of the feedback loop.

In the examples of Fig. 8-9 the value of H is the gain of the servo amplifier, while B is unity (there is neither gain nor attenuation in the feedback loop).

The analog version of the circuit is constructed of operational amplifiers and power transistors. The summer is merely a multiple input operational amplifier stage such as an inverting follower with more than one input source. If only two inputs are needed to produce a difference signal, as in Fig. 8-9A, then a simple dc differential amplifier (Fig. 3-5) will suffice.

The motor is driven by the output of the servo amplifier E_M. This voltage, E_M, is the product of the servo amplifier gain A_V and error voltage E_E,

$$E_M = A_V E_E \qquad (8\text{-}11)$$

and E_E is the difference between the position signal E_P from the transducer and the control signal E_C that indicates what the position should be. E_C can be supplied from a computer-controlled D/A converter.

$$E_E = E_C - E_P \qquad (8\text{-}12)$$

so,

$$E_M = A_V (E_C - E_P) \qquad (8\text{-}13)$$

If the position is correct, then $E_C = E_P$, so, by Eq. 8-13 E_M is zero—the motor is turned off. But if $E_C = E_P$, then the motor is turned on. Since the motor is a dc type, its operating speed is a function of E_M. Clearly, then, the greater the difference between the actual and correct positions, the greater the value of E_M, so the faster the motor speed.

The speed of response is set by the amplification or gain of the system, and by the frequency response. If the signal tends to be overcritically damped, i.e., too sluggish, increase the frequency response of the system. But if it is undercritically damped, i.e., it overshoots, then reduce the frequency response of the system.

A digital version is shown in Fig. 8-9B. In this case the position transducer is shown as a block because it may be a potentiometer and A/D converter, or it can be one of the digitially encoded transducers of Chapter 2.

In the digital version the comparison between the actual and correct position data is made in software. In an eight-bit system there are 256 different states to represent position points. The computer compares the eight-bit word indicating actual position with the eight-bit word indicating correct position by performing a binary subtraction that is analogous to Eq. 8-12. Both of these systems are essentially self-regulating.

Chapter 9

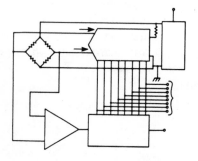

Digital Codes

DIGITAL ELECTRONIC CIRCUITS, AND THAT INCLUDES THE digital computer as well as less complex circuits, only recognize two different voltage levels, designated in most treatments as logic 1 and logic 0. But before the machine can be used to process intelligence, a code must be arranged that will represent numbers, characters, and other symbols.

STRAIGHT BINARY CODE

The two voltage levels recognized by computers can be called bits (short for binary digits), and can assume values of only 1 and 0. A two-state system can only represent two different things, which in the case of numbers are the quantities *one* and *zero*. To represent higher numbers we must resort to a weighted number system of binary numbers.

A weighted system is one in which a digit's position in a string of digits gives it added value. You are already familiar with the concept of weighted number systems in the form of our ordinary decimal (base-10) number system from everyday life. For example, the number 234 is actually a shorthand method of writing a quantity in a weighted form. In this type of notation each decimal digit can assume any of ten different values, 0 through 9, and its position gives it additional value. In any weighted system the digit is multiplied by the radix raised to an integer power. The radix of the decimal system is ten, so the positional weighting scheme is:

$$10^n + 10^{n-1} + \ldots + 10^3 + 10^2 + 10^1 + 10^0$$

So our example of the number 234 is actually a representation of

$$
\begin{aligned}
&= (2 \times 10^2) + (3 \times 10^1) + (4 \times 10^0) \\
&= 200 + 30 + 4 \\
&= 234_{10}
\end{aligned}
$$

Elementary school arithmetic teachers explain that the weighting system is:

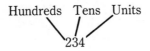

The binary number system deals with base 2, which is just like base 10 if you are missing eight fingers. The radix of the binary system is, then, two, so the weighting scheme is of the form:

$$2^n + 2^{n-1} + \ldots + 2^3 + 2^2 + 2^1 + 2^0$$

and each digit can only be 1 or 0. As an example consider the binary number 11001:

$$
\begin{aligned}
&= (1 \times 2^4) + (1 \times 2^3) + (0 \times 2^2) + (0 \times 2^1) + (1 \times 2^0) \\
&= 16 + 8 + 0 + 0 + 1 \\
&= 25_{10}
\end{aligned}
$$

(Note: When dealing with different number systems in the same text it is good practice to denote which is meant by a subscripted radix. This means that 25_{10} indicates that we are dealing with the decimal number twenty-five. This is 11001_2 in binary. The difference between binary and decimal is usually such that few errors would be made, but at a few critical junctures problems arise. For example, does $11_{20} = 11_2$? No, in base-10 *11* means eleven, while in base-2 *11* means three.)

In any positional numbers system with n digits, the maximum number of different things that can be represented is r^n, where r is the radix. The highest quantity that can be represented is r^{n-1}

becomes zero is a thing. In eight-digit decimal, for example, there are 10^8 different combinations (100,000,000), but the highest quantity that can be represented is the number obtained when all eight digits are 9s, in other words 99,999,999. In the binary system, again using eight digits, the number of different things that can be represented (including zero) is 2^8, or 256. The highest number, though, exists when all eight digits are 1s, which is 11111111_2 or 255_{10}.

TWO'S COMPLEMENT

A number system related to the binary system is the complement number system. The complement, also called ones complement, of any binary number is a number made up to have exactly the opposite binary digits. In other words, all the 1s are made 0s, and all 0s are made 1s.

The complement of binary 1 is 0, and the complement of binary 0 is 1.

Number	Complement
1	0
0	1

The one's complement of a binary number is formed by complementing each bit of the binary number. To form this, change all of the ones to zero, and all of the zeros to ones.

Example 9-1
   Form the one's complement of 11011_2.
Solution:

Binary number	1	1	0	0	1
One's complement	0	0	1	1	0

To find the two's complement of a number it is necessary to first find the one's complement, then add 1 to the least significant digit.

Example 9-2
Find the two's complement of 11001_2.

Solution:

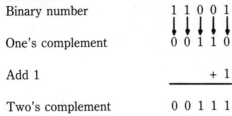

Binary number	1	1	0	0	1
One's complement	0	0	1	1	0
Add 1				+	1
Two's complement	0	0	1	1	1

Many people believe that binary addition is easier than decimal addition even though the decimal version is known by rote to most adults, and is exactly the same process conceptually. The rules for binary addition are:

$$0 + 0 = 0$$
$$1 + 0 = 1$$
$$0 + 1 = 1$$
$$1 + 1 = 0 \text{ plus carry } 1$$

Example 9-3

Add 01001_2 to 01110_2.

Solution:

```
      1
    0 1 0 0 1
  + 0 1 1 1 0
    1 0 1 1 1
```

Example 9-4

Add 00001_2 to 01001_2.

Solution:

```
            1
    0 1 0 0 1
  + 0 0 0 0 1
    0 1 0 1 0
```

It is possible to subtract in binary in the same manner as for decimal numbers, namely instead of carrying a 1 to the next higher order digit, borrow 1 from the next higher order position. The rules for binary subtraction are:

$$0 - 0 = 0$$
$$0 - 1 = 1 \text{ borrow } 1$$
$$1 - 0 = 0$$
$$1 - 1 = 0$$

Digital circuits that add are very easy to build, but subtractors are a little harder to achieve. To overcome this problem computers often perform subtraction in two's complement. If we add the two's complement of a number it is the same as if we had subtracted the number itself.

Example 9-5
Subtract 000110_2 from 001110_2. (6_{10} from 14_{10}).
Solution:
This problem is expressed by:

$$0\ 0\ 1\ 1\ 1\ 0_2 \qquad\qquad 14_{10}$$
$$\text{and}$$
$$\underline{-0\ 0\ 0\ 1\ 1\ 0_2} \qquad\qquad \underline{-6_{10}}$$
$$? \qquad\qquad\qquad 8_{10}$$

The procedure is to form the two's complement of the subtrahend and add it to the minuend.

1. Form the two's complement of the subtrahend.

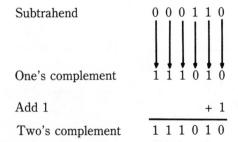

Subtrahend	0 0 0 1 1 0
One's complement	1 1 1 0 1 0
Add 1	+ 1
Two's complement	1 1 1 0 1 0

2. Add the two's complement of the subtrahend to the minuend.

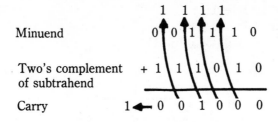

	1	1	1	1		
Minuend	0	0	1	1	1	0
Two's complement of subtrahend	+ 1	1	1	0	1	0
Carry	1 ← 0	0	1	0	0	0

Note that 001000_2 is the same as 8_{10}, so is the correct answer.

The circuitry required to form the two's complement consists of an inverter for each bit and a full adder to add the 1 to the complimented (i.e., inverted) binary number representing the subtrahend.

DIVISION & MULTIPLICATION

The rules for binary division are:

$$1/1 = 1$$
$$0/1 = 0$$
$$1/0 = \text{(illegal operation)}$$

The rules for binary multiplication are:

$$0 \times 0 = 0$$
$$1 \times 0 = 0$$
$$0 \times 1 = 0$$
$$1 \times 1 = 1$$

and the procedure is the same as for decimal multiplication.

Example 9-6
Multiply $0101_2 (5_{10})$ by $011_2 (3_{10})$.
Solution:

```
      0 1 0 1
    × 0 1 1
      0 1 0 1
    0 1 0 1
  0 0 0 0
  0 0 1 1 1 1₂
```

169

Note that $1111_2 = 15_{10}$.
Example 9-7
Divide 011110_2 (30_{10}) by 0110_2 (6_{10}).
Solution:

```
                    0 0 0 1 0 1
      0 1 1 0 ) 0 1 1 1 1 0
                    0 1 1 0
                    0 0 0 1 1 0
                    0 0 0 1 1 0
                    0 0 0 0 0 0
```

NUMBER SYSTEM CONVERSION

Numbers from any system can be converted to any other system if a suitable method is provided. The method that follows will convert numbers from any number to decimal form:

1. Write down the number being converted.
2. Multiply the most significant digit (left most) by the radix of the number.
3. Add the result of step 2 to the next most significant digit.
4. Repeat this process until no further digits are left. The result is the final answer.

Example 9-8
Convert 1101101_2 to decimal form.
Solution:

1101101	Multiply MSD by radix	$1 \times 2 = 2$
1101101	Add result to next digit	$2 + 1 = 3$
1101101	Add result to next digit	$2 = 1 = 3$
	Multiply result by radix	$3 \times 2 = 6$
1101101	Add result to next digit	$6 + 0 = 6$
	Multiply result by radix	$6 \times 2 = 12$
1101101	Add result to next digit	$12 + 1 = 13$
	Multiply result by radix	$13 \times 2 = 26$
1101101	Add result to next digit	$26 + 1 = 27$
	Multiply result by radix	$27 \times 2 = 54$
1101101	Add result to next digit	$54 + 0 = 54$
	Multiply result by radix	$54 \times 2 = 108$
1101101	Add result to next digit	$108 + 1 = 109$
	Final result	109

OCTAL & HEXADECIMAL NUMBERS

The octal number system has a radix of 8, and is a weighted system of the form:

$$8^n + 8^{n-1} + \ldots + 8^3 + 8^2 + 8^1 + 8^0$$

There are only eight permissible digits, namely 0, 1, 2, 3, 4, 5, 6, and 7. Formation of octal numbers greater than seven is through use of weighted notation in a manner much like decimal. In fact, a noted comedian once said, "base 8 is like base 10 . . . if you are missing two fingers."

Example 9-9

What is the decimal equivalent of 247_8?

Solution:

$$
\begin{aligned}
247^8 &= (2 \times 8^2) + (4 \times 8^1) + (7 \times 8^0) \\
&= (2 \times 64) + (4 \times 8) + (7 \times 1) \\
&\times 128_{10} + 32_{10} + 7_{10} \\
&= 167_{10}
\end{aligned}
$$

The same technique used in the binary conversion will also work to convert octal to decimal. In the example given:

$$
\begin{aligned}
247_8 \\
2 \times 8 &= 16 \\
16 + 4 &= 20 \\
20 \times 8 &= 160 \\
160 + 7 &= 167_{10}
\end{aligned}
$$

The hexadecimal (hex) number system has a radix of sixteen, and there are sixteen permissible digits. The first ten are the 0 through 9 of the decimal system, while the latter six are letters of the alphabet.

Decimal	Hex	Decimal	Hex
0	0	8	8
1	1	9	9
2	2	10	A
3	3	11	B
4	4	12	C
5	5	13	D
6	6	14	E
7	7	15	F

Example 9-10

Find the decimal equivalent value of $8F4_{16}$.

Solution:

$$8F4_{16} = (8 \times 16^2) + (F \times 16^1) + (4 \times 16^0)$$
$$= (8 \times 256) + (15 \times 16) + (4 \times 1)$$
$$= 2048_{10} + 240_{10} + 4_{10}$$
$$= 2292_{10}$$

And by the technique given earlier:
8F4

$8 \times 16 = 128$		$143 \times 16 = 2288$	
$128 + F = 143$		$2288 + 4 = 2292_{10}$	

Most machine language microcomputer programs are written in either hex or octal. Most of these machines have an eight-bit data format. Hexadecimal digits can be represented by four-bit binary words, namely:

Hex	4-Bit Binary	Hex	4-Bit Binary
0	0000	8	1000
1	0001	9	1001
2	0010	A	1010
3	0011	B	1011
4	0100	C	1100
5	0101	D	1101
6	0110	E	1110
7	0111	F	1111

The eight-bit binary word used by the computer is often represented by two hexadecimal digits. For example:

Binary Code	Hex Representation
11010001	D1
10011101	9D
11001001	C9

In the case of the first example, 11010001, the binary word is broken into two portions, 1101 and 0001. From the table above, 1101_2 is the same as D_{16}, and 0001_2 is the same as 1_{16}, so by combining terms, D1 represents 11010001.

BINARY-CODED DECIMAL

Binary-coded decimal (BCD) is a method for representing the ten decimal digits (0, 1, 2, 3, 4, 5, 6, 7, 8, and 9) in a four-bit binary format:

Decimal	BCD	Decimal	BCD
0	0000	5	0101
1	0001	6	0110
2	0010	7	0111
3	0011	8	1000
4	0100	9	1001

BCD words are grouped in the same manner as regular decimal digits, so a word's actual value is determined by its position in a power-of-ten system. For example:

Decimal	BCD
3	0011
32	0011 0010
324	0011 0010 0100

The BCD system is used most often in circuits requiring a numerical output display, such as in frequency counters, digit panel meters, clocks, and scientific instruments.

Several integrated circuits exist which convert BCD to a code used by certain display devices such as Nixie tubes and seven-segment readouts.

EXCESS-3 CODE

Excess-3 code is formed by adding binary 3_{10} (011_2) to BCD numbers. Namely;

Decimal	BCD	Excess-3 Code
0	0000	0011
1	0001	0100
2	0010	0101
3	0011	0110
4	0100	0111
5	0101	1000
6	0110	1001
7	0111	1010
8	1000	1011
9	1001	1100

Excess-3 code is used to make digital subtraction in BCD a little easier. Recall that you often use complement arithmetic with binary digits. But when the binary word is a BCD character this would result in a disallowed bit pattern on occasion that is not recognizable as a BCD character.

GRAY CODE

Shaft encoders and certain applications work best if the binary code generated changes only one bit at a time, for each change of state. An example of such a code is the Gray code.

Decimal	Binary	Gray Code
0	0000	0000
1	0001	0001
2	0010	0011
3	0011	0010
4	0100	0110
5	0101	0111
6	0110	0101
7	0111	0100
8	1000	1100
9	1001	1101
10	1010	1111

ALPHANUMERIC CODES

There are several different standard codes used to represent alphabetic or numeric characters. It is important to distinguish between representations of a numeric *quantity* and a coded expression of the *character* representing that quantity. Consider, for example, the number 8_{10} (1000_2). In one standard alphanumeric code the character *8* is represented by seven-bit binary word 0111000. The seven-bit binary word representing the quantity eight is 0001000. Clearly, if you were to attempt to use the *character* representation in an arithmetic operation, or the *quantity* representation in an attempt at creating a display output, the outcome would not be anything like what was expected. In this particular situation the character representation (0111000) mistakenly used to perform arithmetic would mean not 8_{10} but 56_{10}. Similarly, in the alphanumeric code being used in this example the quantity representation (0001000) is used as a machine control function (i.e., on a teletypewriter or CRT terminal) known as *horizontal tab*.

Now that I have hopefully cleared up the difference between binary numbers and the binary representation of characters let me proceed to describe some of the more common codes.

BAUDOT CODE

One of the earliest machine codes was the Baudot. It was used on most earlier teletype machines, although it has now been largely supplanted by another, more modern code. One will still find Baudot machines used, however, even in commercial service. Amateur radio operators who favor RTTY communications, and computer hobbyists looking for a low-cost hard-copy printer have kept them popular on the surplus market. (*Note:* This code is used with teletypewriters, so will have a regular keyboard and a keyboard with the shift key pressed in the same manner as a typewriter has lower-case letters in the regular keyboard and CAPITALS ON THE SHIFTED KEYBOARD.)

Baudot Code

B5	B4	B3	B2	B1	Regular	Shifted
0	0	0	0	0	Blank	Blank
0	0	0	0	1	E	3
0	0	0	1	0	linefeed	linefeed
0	0	0	1	1	A	-
0	0	1	0	0	space	space
0	0	1	0	1	S	Bell
0	0	1	1	0	I	8
0	0	1	1	1	U	7
0	1	0	0	0	Car. Ret.	Car. Ret.
0	1	0	0	1	D	$
0	1	0	1	0	R	4
0	1	0	1	1	J	'
0	1	1	0	0	N	,
0	1	1	0	1	F	!
0	1	1	1	0	C	:
0	1	1	1	1	K	(
1	0	0	0	0	T	5
1	0	0	0	1	Z	"
1	0	0	1	0	L	)
1	0	0	1	1	W	2
1	0	1	0	0	H	#
1	0	1	0	1	Y	6

Baudot Code

B5	B4	B3	B2	B1	Regular	Shifted
1	0	1	1	0	P	0
1	0	1	1	1	Q	1
1	1	0	0	0	O	9
1	1	0	0	1	B	?
1	1	0	1	0	G	&
1	1	0	1	1	(figures)	(figures)
1	1	1	0	0	M	.
1	1	1	0	1	X	/
1	1	1	1	0	V	;
1	1	1	1	1	(letters)	(letters)

The Baudot codes uses five bits, so is only capable of 32 (i.e., 2^5) different characters. A shift control, much like that on the typewriter, allows another 32 different characters, although not all are used on any given machine. In fact, one of the jobs facing the person who wishes to decode an unfamiliar teletypewriter is to find out which of numerous variations on the fundamental code are used in the particular machine.

ASCII

The American Standard Code for Information Interchange (AS-CII) is the code most often used on computer keyboards, especially those intended for hobbyist uses. It is a seven-bit code, plus an eight bit that is sometimes used as a parity indicator, but most often as a strobe to tell the computer that the data is stable and ready for use.

Since ASCII is a seven-bit code it is possible to represent a total of 128 (i.e., 2^7) different characters. In the ASCII listings to follow we use hexadecimal notation. The bit pattern for that character can be realized by converting from hex to binary. An & is given in hex as *26* so in binary would be 0010 0110, or simply 00100110.

ASCII

Hex Code	Meaning	Comments
00	NUL	null
01	SOH	start of heading
02	STX	start text
03	ETX	end text
04	EOT	end of transmission
05	ENQ	enquiry
06	ACK	acknowledgement
07	BEL	bell
08	BS	back space
09	HT	horizontal tab
0A	LF	line feed

ASCII

Hex Code	Meaning	Comments
0B	VT	vertical tab
0C	FF	form feed
0D	CR	carriage return
0E	SO	shift out
0F	SI	shift in
10	DLE	data link escape
11	DC1	direct control 1
12	DC2	direct control 2
13	DC3	direct control 3
14	DC4	direct control 4
15	NAK	negative acknowledgement
16	SYN	synchronous idle
17	ETB	end of transmission block
18	CAN	cancel
19	EM	end of medium
1A	SUB	substitute
1B	ESC	escape
1C	FS	form separator
1D	GS	group separator
1E	RS	record separator
1F	US	unit separator
20	(special)	—
21	!	—
22	"	—
23	#	—
24	$	—
25	%	—
26	&	—
27	'	—
28	(	—
29	)	—
2A	*	—
2B	+	—
2C	,	—
2D	-	—
2E	.	—
2F	/	—
30	0	—
31	1	—
32	2	—
33	3	—
34	4	—
35	5	—
36	6	—
37	7	—
38	8	—
39	9	—
3A	:	—
3B	;	—
3C	>	—
3D	=	—
3E	<	—
3F	?	—
40	@	—
41	A	—
42	B	—
43	C	—
44	D	—

177

ASCII

Hex Code	Meaning	Comments
45	E	—
46	F	—
47	G	—
48	H	—
49	I	—
4A	J	—
4B	K	—
4C	L	—
4D	M	—
4E	N	—
4F	O	—
50	P	—
51	Q	—
52	R	—
53	S	—
54	T	—
55	U	—
56	V	—
57	W	—
58	X	—
59	Y	—
5A	Z	—
5B	[	—
5C	\	—
5D	]	—
5E	^	—
5F			—
60	∧	—
61	a	—
62	b	—
63	c	—
64	d	—
65	e	—
66	f	—
67	g	—
68	h	—
69	i	—
6A	j	—
6B	k	—
6C	l	—
6D	m	—
6E	n	—
6F	o	—
70	p	—
71	q	—
72	r	—
73	s	—
74	t	—
75	u	—
76	v	—
77	w	—
78	x	—
79	y	—
7A	z	—
7B	{	—
7C			—
7D	}	—
7E	~	—
7F	DEL	—

EBCDIC

The code most often associated with IBM equipment is the Extended Binary Coded Decimal Interchange Code (EBCDIC). This code can possibly be explained best by referring to the standard Hollerith card shown in Fig. 9-1. This card was originally designed to automate the tabulation of the 1890 United States national census, but remains immensely popular even today, and is in fact one of the mainstays of the data-processing industry.

The card is divided into rows and columns. There are twelve horizontal rows and eighty vertical columns. There are also two areas on the card. The rows numbered 0 through 9 form one area, while rows 11 and 12 form the other. Punches made in rows 11 and 12 are called "zone punches."

Characters are represented by one-byte (eight-bit) binary words. For convenience in discussion I can divide the byte into two "nybbles" of four bits each. This allows me to use hexadecimal notation. The code for digits is:

Digit	Hex Code	Digit	Hex Code
0	F0	5	F5
1	F1	6	F6
2	F2	7	F7
3	F3	8	F8
4	F4	9	F9

Conversion to binary merely requires the substitution of the binary equivalents of the hexadecimal numbers. For example:

Decimal	EBCIDIC	Binary
6	F6	1111 0110

EBCDIC representation of numbers places a hexadecimal F_{16} in the most significant position, so the four most significant binary digits would be 1111_2. Those positions are used when representing alphabetic characters, and indicate which third of the alphabet the letter falls in. There are twenty-six letters in our alphabet, so I can divide the alphabet roughly into thirds.

Letter No.	1st Third	2nd Third	3rd Third
1	A	J	–
2	B	K	S
3	C	L	T
4	D	M	U
5	E	N	V
6	F	O	W
7	G	P	X
8	H	Q	Y
9	I	R	Z

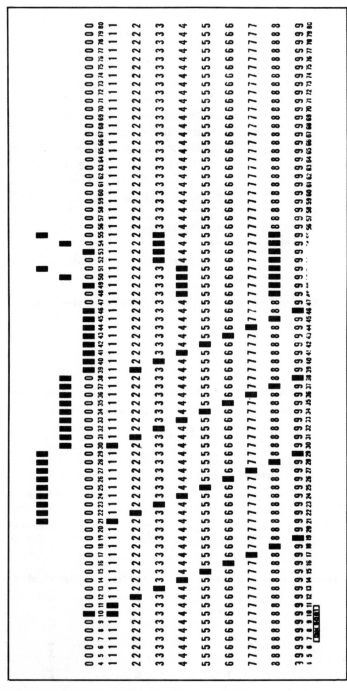

Fig. 9-1. Hollerith card character codes.

Two punches are used to represent a letter on an IBM card. A zone punch indicates the third of the alphabet in which that letter appears, and the numeric punch tells me the letter number. Letters in the first third of the alphabet have a row-12 zone punch, those in the second third of the alphabet are a row-11 zone punch, while those in the last third of the alphabet have a row-0 zone punch.

The letter *D* is the fourth letter in the first third of the alphabet, so it will be represented by a row-12 zone punch and a row-4 numeric punch, *both in the same column*. The letter *N*, on the other hand, is the fifth letter in the second third of the alphabet. It is represented by a row-11 zone punch and a row-5 numeric punch.

A slight departure from the rigor of the system is noted in the last third of the alphabet. There is no row-1 punch. The first letter in this group is *S* and it is represented by a row 0 zone and a row-2 numeric punch. The alphanumeric portion of the EBCDIC system follows.

EBCDIC

Character	Hex Code	Character	Hex Code
A	C1	K	D2
B	C2	L	D3
C	C3	M	D4
D	C4	N	D5
E	C5	O	D6
F	C6	P	D7
G	C7	Q	D8
H	C8	R	D9
I	C9	S	E2
J	D1	T	E3
U	E4	4	4
V	E5	5	5
W	E6	6	6
X	E7	7	7
Y	E8	8	8
Z	E9	9	9
0	0		
1	1		
2	2		
3	3		

CODE CONVERSION

There is, unfortunately, almost nothing resembling a universal standard in the microcomputer field, even in the limited application of the hobbyist field. So when you try to interface various devices, some of which are government or commercial surplus or even of dubious parentage, it may be necessary to build some sort of code conversion system or write a code conversion program. This is one of those areas in which hardware and software buffs are often at odds with each other, and both camps can offer arguments that support their particular pet solutions. Both, however, may use a table lookup procedure. In software, for example, the bit pattern for the desired code is stored in memory. When an input is received it is examined, and the location for the correct code is determined.

Electronic circuits can be constructed that do substantially the same thing. A read only (ROM) is programmed so that the desired code is stored in locations addressed by the input code. An example is an ASCII-to-Baudot converter, of which several dozen designs seemed to have been published in the hobby computer press. The binary representation for an input character is used to form an address in which the ROM stores the bit pattern in the other code for the same character.

Several manufacturers offer preprogrammed ROMs that convert from one code to another. National Semiconductor seems to have a particularly good selection of code converter chips.

A typical ASCII keyboard encoder is shown in Fig. 9-2. There are two basic methods for generating ASCII characters form key closures. One uses a special ROM, as shown in the figure, while the other is a timing method.

The ROM is entirely self-contained and can be used without the other chips shown (several low-cost keyboards do just that), but this design is a little more elegant.

The pushbuttons are arranged in a crosspoint X-Y matrix consisting of nine Y input lines and ten X lines. When a pushbutton is pressed it shorts out two lines, one from the X bank and one from the Y bank.

When the output data has settled and is ready for use, the two inputs of the NAND gate (IC_3) will go high, making the output drop low. This triggers the monostable multivibrator $(IC_2,$ a type 74121) to generate a 10-millisecond strobe pulse.

The computer will have a keyboard program that loops until it sees the strobe go high. At that time it will input the data on the keyboard output lines. It is standard practice to use the strobe pulse as bit 8 since ASCII requires only seven bits and most microcomputers have eight-bit input/output ports.

One must be forewarned to look closely at deals offering low-cost new or surplus keyboards in order to ascertain just what is being offered. Some cheapies are only a pushbutton matrix, and contain no electronics. Others, often offered as kits to the hobbyist market, are thrown together under the auspices of little or no decent engineering. Several are known that not only fail to deliver

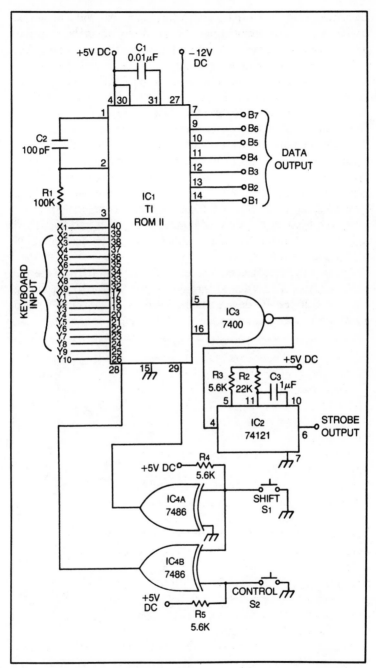

Fig. 9-2. Keyboard encoder circuit (ASCII).

183

the proper character when a key is depressed, but will deliver one character on the downstroke of the key and another on the upstroke. The result is a double entry, and there is no assurance that either will be the correct character (sigh).

The ROM type using a keyboard encoder IC is, in my view, superior, but here again there are pitfalls when buying on the surplus market. A large number of keyboards are available at seductively low cost, but are missing the ROM. Some suppliers offer the ROM at a nominal extra charge (thereby deteriorating the rosiness of the deal), but others leave you on your own.

Wringing out an undocumented keyboard can be quite a chore, especially, after you unwrap your new acquisition, when it is noticed that the printed-circuit edge connector is unlabeled. Don't panic yet. First, if you know who made it, or which company used it in their equipment, then apply to them for a circuit diagram. Failing that, do it the hard way.

Hopefully, the PC board will have the markings to indicate + 5 Vdc, – 5 Vdc, – 12 Vdc, B1 through B7, and so forth. But failing this, look at the IC devices being used. If they bear standard 74-series TTL markings, or are the standard pin-out keyboard en-

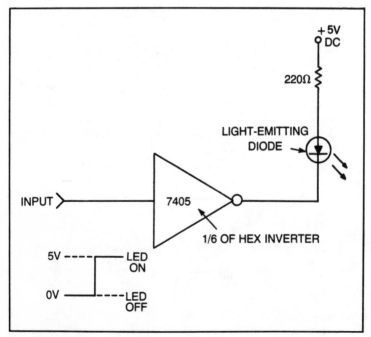

Fig. 9-3. Circuit for a simple logic probe.

Table 9-1. Common Number Systems.

Decimal	Binary	BCD		Octal	Excess-3 BCD		Hexadecimal
0	00000	0000		0	0011		0
1	00001	0001		1	0100		1
2	00010	0010		2	0101		2
3	00011	0011		3	0110		3
4	00100	0100		4	0111		4
5	00101	0101		5	1000		5
6	00110	0110		6	1001		6
7	00111	0111		7	1010		7
8	01000	1000		10	1011		8
9	01001	1001		11	1100		9
10	01010	0001	0000	12	0001	0011	A
11	01011	0001	0001	13	0001	0100	B
12	01100	0001	0010	14	0001	0101	C
13	01101	0001	0011	15	0001	0110	D
14	01110	0001	0100	16	0001	0111	E
15	01111	0001	0101	17	0001	1000	F
16	10000	0001	0110	20	0001	1100	10

coder, then the power terminals can be deciphered by looking for the corresponding pins on the IC then backtracking on the card edge connector. Similarly, if the encoder is standard, the − 12-volt pin can be located and backtracked. The outputs are best found by connecting an oscilloscope, but the LED circuit of Fig. 9-3 can be used if no oscilloscope is available (it should be). Note that the output lines will have a clock pulse, or some pulse trash on it until a key is pressed. When the key is closed, the correct code will appear stable in the B_1 through B_7 lines. Note that a device such as Fig. 9-3 will glow dimly until the key is pressed, but when the data is stable those bits that should be *zero* go out, and those that should be *one* go to full brilliance.

Refer to Table 9-1 for a comparison of some of the more common number systems.

Chapter 10

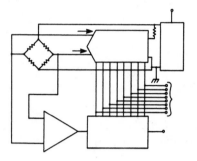

Basics of Data Conversion

DIGITAL COMPUTERS CANNOT DIRECTLY DEAL WITH ANALOG input signals, nor can they directly drive analog output devices. Some sort of interface must be provided that converts the analog input to a digital word that can be handled by the computer. Another type of circuit is required to convert a digital word at a computer output port to an analog current or voltage required to turn on a relay or drive a motor, or that represents some quantity to be displayed on a chart recorder or oscilloscope.

The job of converting analog signals to digital words is performed by an analog-to-digital (A/D) converter, while the inverse job is performed by another circuit called a digital-to-analog (D/A) converter.

Some people believe that the use of a computer, or any of several other digital circuits, is *de facto* evidence of higher accuracy and great precision. This belief is not necessarily justified. One is, perhaps, reminded of the person who will use a battered old wooden yardstick to measure the dimensions of a rectangle, then multiply the results together on a ten-digit calculator to find the area. This procedure results in a multidigit answer, but the wise person will know that only a few digits out of the ten will be accurate. The remaining digits are not significant at all. Only the fool will use and possibly act on information out to the last decimal point (sigh).

In digital computers, the maximum resolution is dependent upon the length of the basic computer word. A typical eight-bit microcomputer can only recognize 2^8 or 256 different states, of which one is zero (00000000). One implication of this is that an analog input signal can only be resolved on levels of $E/256$. If better resolution is needed at any given application, then a longer word length (i.e., ten, twelve, or sixteen bits) will be needed. One alternative is to use double-length computer words to increase precision. Two successive bytes of eight bits each can be used to represent the output of a sixteen-bit A/D converter.

SAMPLED DATA

Analog signals can be plotted as in Fig. 10-1A to have continuous range and continuous domain. This is the type of signal ordinar-

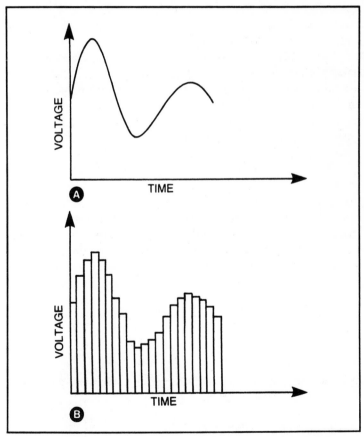

Fig. 10-1. Data conversion. (A) Continuous analog function (continuous range and domain). (B) Sampled analog signal (continuous range, discrete domain).

ily available from analog electronic circuits and many scientific instruments.

The same signal can also be represented in *sampled* form as in Fig. 10-1B. In this case the range (i.e., volts) is continuous but the domain is discrete. That is to say only specific values of time are allowed.

There is an error inherent in the representation of Fig. 10-1B, but this error can be reduced to a negligible point by taking a sufficiently large number of samples (about which, more later).

Figure 10-2 shows an example of a *digitized* analog data signal. In this case, I am dealing with a three-bit digital word, so there are 2^3 discrete output levels, or eight levels. The range values (vertical axis) are only allowed to assume certain specific, discrete, levels; namely, the binary digits representing decimal numbers 0 through 7. These are used to represent input voltage from 0 to 7 volts.

The steps between discrete levels are all equal, and are symbolized in Fig. 10-2 by the letter Q. Midway between each analog level is a decision point above which the next higher code is used, and below which the next lower code is appropriate. This means, for example, that all analog voltages between 3.5 volts and 4.5 volts are represented by the digital word 100, the code for 4.0 volts. We have an error that varies between limits of $-1/2Q$ and $+1/2Q$, with the error being zero only where the code is absolutely correct. This situation is obtained only when the analog input voltage is exactly the same as the coded representation.

Again I have a situation where the encoding error can be quite large unless an adequate number of samples are taken. For the scheme of Fig. 10-2 I have only three data bits, so by the $2^n - 1$ rule I have only seven decision levels. Since the full-scale analog voltage (E_{FS}) is 7.0 volts, I have a resolution of

$$\text{res} = E_{FS}/(2^n - 1) \qquad (10.1)$$
$$\text{res} = 7 \text{ volts}/7 \qquad (10.2)$$
$$\text{res} = 1 \text{ volt} \qquad (10.3)$$

If an eight-bit word, common in microcomputers, were used to represent 0 volts, then the resolution would be:

$$\text{res} = E_{FS}/(2^n - 1) \qquad (10.4)$$
$$\text{res} = 7 \text{ volts}/(2^8 - 1) \qquad (10.5)$$
$$\text{res} = 7 \text{ volts}/255 \qquad (10.6)$$

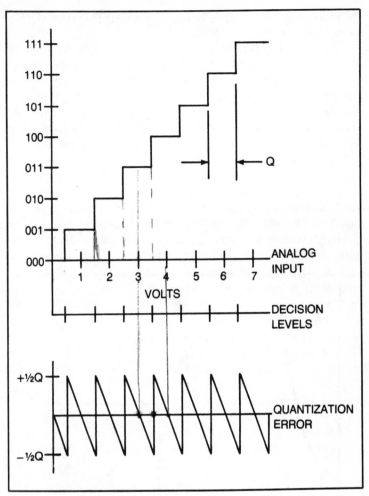

Fig. 10-2. Digitized analog signal (discrete range and domain).

$$\text{res} = 0.027 \text{ volts} \qquad (10.7)$$

Clearly, the more bits used in any given situation, the better the resolution that is possible. There is, of course, a limit to the practical resolution, but in general the more the better.

When I digitize an analog signal so that it can be processed in a computer it is quantized in both range and domain. The range of permissible digital values, as shown in Fig. 10-2, is quantized, but so is time. I only sample the signal at specific times. Accurate representation of the input waveform is dependent upon the number of bits making up the digital word and the sampling rate (how many times per second a sample is taken).

Nyquist's theorem requires the sampling rate to be not less than *twice* the highest frequency component of the input waveform. Recall that all nonsinusoidal waveforms can be represented by a Fourier series of sines, and cosines, and their respective coefficients that give information as to phase and amplitude. The frequency used to set the sampling rate of a given waveform is the frequency of the highest significant Fourier component. A waveform might have a fundamental frequency of, say, 10 hertz. But if there are steep leading and trailing edges there might easily be Fourier components out to some frequency in the 200- to 1000-hertz range. The sampling rate must be *twice* the frequency of the highest significant Fourier frequency components.

Every reading from an A/D converter is subject to another type of error resulting from the time required to make the conversion.

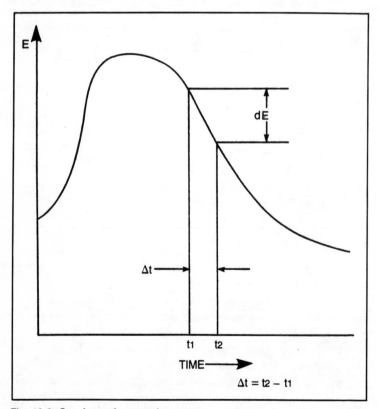

Fig. 10-3. One form of conversion error.

Although different A/D converters require different conversion times, none will do the job in zero time. If the input signal changes any significant amount during this period an error will result (see Fig. 10-3). If dT represents the conversion time of the A/D converter, then there is an uncertainty in the final measurement of

$$\Delta E = dE \, \Delta t / dt \qquad (10.8)$$

D/A BASICS

A digital-to-analog (D/A) converter is a device that converts a digital word at its input to an analog voltage or current at its output that is proportional to that word. The D/A converter is actually a form of multiplier circuit, so has the following transfer function:

$$F = A \times B \qquad (10.9)$$

where F is the output voltage or current function, A is the digital input value, and B is the analog reference value. In most cases the output function, F, is proportional to a fractional part of the reference because digital function A is itself fractional.

Take as an example the Precision Monolithics DAC-08. It is an eight-bit device that has an external reference current, and produces an analog output current. The transfer function is:

$$I_{OUT} = (A/256) \times I_{REF} \qquad (10.10)$$

where I_{OUT} is the output current, I_{REF} is the reference current in the same units as I_{OUT}, and A is the value of the digital eight-bit word applied to the input. Notice that in this case the digital word function is a fraction equal to $A/2^n$, or $A/256$. A half-scale word (10000000) would, then, produce an output current of 128/256 times the reference.

Example 10-1

Find the output current at full-scale output when I_{REF} is 2 milliamperes. (*Note:* full-scale output is $2^n - 1$.)
Solution:

$$I_{OUT} = (2^n - 1)/256 \times 2 \text{ mA} \qquad (10.11)$$
$$I_{OUT} = (255/256) \times 2 \text{ mA} \qquad (10.12)$$
$$I_{OUT} = 1.992 \text{ mA} \qquad (10.13)$$

Most D/A converters use a current summation technique to perform the conversion. The two most common methods are the binary resistor ladder and the R-$2R$ resistor ladder.

Figure 10-4 shows an example of a binary resistor ladder type of D/A converter. Since amplifier $A1$ is an operational amplifier, I know that

$$-I_F = I_O \qquad (10.14)$$

and that,

$$E_{OUT} = -I_O R_F \qquad (10.15)$$

but,

$$I0 = I1 + I2 + I3 + \ldots + I_n \qquad (10.16)$$

$$I0 = \sum_{i=1}^{n} \frac{E_{REF}}{R_i} \qquad (10.17)$$

I make the circuit of Fig. 10-4 into a genuine D/A converter by giving the resistors $R1$ through R_n binary weights, namely, (letting $R_F = R1 = R$)

$$R1 = R$$
$$R2 = 2R$$
$$R3 = 4R$$
$$\bullet$$
$$\bullet$$
$$\bullet$$
$$R_n = 2^{n-1}R$$

In the converter of Fig. 10-4, SPDT switches $S1$ through S_n are connected to the reference voltage, and represent the bits of the input word. A logic level 1 exists when a switch is connected to E_{REF}, and a logic level 0 exists when the switch is connected to ground. I will assume a four-bit circuit at full-scale output:

$$E_{OUT} = RE_{REF} (1/R1 + 1/R2 + 1/R3 + 1/R4) \qquad (10.18)$$
$$E_{OUT} = RE_{REF} (1/R + 1/2R + 1/4R + 1/8R) \qquad (10.19)$$

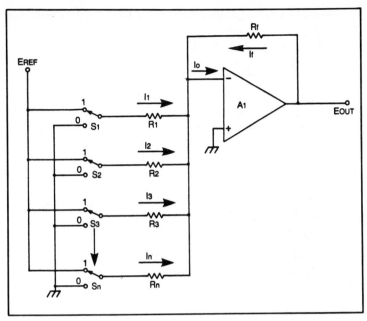

Fig. 10-4. Resistor ladder D/A converter (binary weighted resistor values).

$$E_{OUT} = E_{REF} (1 + 1/2 + 1/4 + 1/8) \qquad (10.20)$$
$$E_{OUT} = 1.875\ E_{REF} \qquad (10.21)$$

If n is greater than about five we can claim that

$$R_{IN} = 1/2R \qquad (10.22)$$

From the basic operational amplifier inverting follower transfer function I know that

$$E_{OUT}/E_{REF} = -R_F/R_{IN} \qquad (10.23)$$
$$E_{OUT}/E_{REF} = -R_F/(R/2) \qquad (10.24A)$$
$$E_{OUT}/E_{REF} = -R_F/R \qquad (10.24B)$$

To find practical values for R_F we set E_{OUT} to its full-scale (FS) value, and solve Eq. 10.24B for R_F, which results in:

$$R_F = (E_{OUT(FS)}/E_{REF}) \times R/2 \qquad (10.25)$$

Example 10-2

Assume that the reference voltage is precisely 10.00 volts,

193

and that I want a full-scale output of 2.56 volts. Find the value of R_F if R = 10K.

Solution:
By Eq. 10.25:

$$R_F = (2.56 \text{ volts}/10 \text{ volts}) \times (1/2 \times 10^4 \text{ ohms}) \qquad (10.26)$$
$$R_F = 1.28 \times 10^3 \text{ ohms} \qquad (10.27)$$
$$R_F = 1280 \text{ ohms} \qquad (10.28)$$

Example 10-3

An eight-bit D/A converter has the word 11010011 at its input lines. If the reference is 10 volts, and the full-scale output voltage is 2.56 volts, what is the actual output voltage? Assume that R_F is 1.28K and that R 10K (as in the previous example).

Solution:
The output is given by the transfer equation:

$$E_{OUT} = \frac{R_F E_{REF}}{R} \times \left(\frac{B1}{1} + \frac{B2}{2} + \frac{B3}{4} \right.$$
$$\left. + \frac{B4}{8} + \frac{B5}{16} + \frac{B6}{32} + \frac{B7}{64} + \frac{B8}{128} \right) \qquad (10.29)$$

where $B1$ through $B8$ are either 1 or 0 depending upon the bit value. In this particular example the binary word is 11010011, so

B1	B2	B3	B4	B5	B6	B7	B8
1	1	0	1	0	0	1	1

Eq. 10.29 then, will become:

$$E_{OUT} = \frac{(1.28 \times 10^3)(10)}{10^4} \times$$

$$(1 + 1/2 + 0 + 1/8 + 0 + 0 + 1/64 + 1/128)$$

$$E_{OUT} = 1.28(1.648) \qquad (10.31)$$
$$E_{OUT} = 2.11 \text{ volts} \qquad (10.32)$$

Another way to write Eq. 10.29 is

$$E_{\text{OUT}} = \frac{R_F E_{\text{REF}}}{2^{n-1}R} \sum_{i=1}^{n} 2^{n-i}B_i \qquad (10.33)$$

or for the eight-bit case,

$$E_{\text{OUT}} = \frac{R_F E_{\text{REF}}}{128R} \sum_{i=1}^{8} 2^{8-i}B_i \qquad (10.34)$$

and if $R_F = R/2$,

$$E_{\text{OUT}} = \frac{R_F E_{\text{REF}}}{256R} \sum_{i=1}^{8} 2^{8-i}B_i \qquad (10.35)$$

$$E_{\text{OUT}} = \frac{E_{\text{REF}}}{256} \sum_{i=1}^{8} 2^{8-i}B_i \qquad (10.36)$$

Two problems present themselves when actually trying to build a binary resistor ladder D/A converter. First, it is difficult to obtain resistors in precise power of two multiples of the basic R value. Second, when more than eight or ten bits are used the currents in the summing network for the low-order bits become very low. For example, if the reference voltage is 10 volts dc and $r = 10K$, then $I1$ will be 0.001 ampere, or 1 milliampere. The current produced by $B8$ is then only 7.8 microamperes, $B9$ is 3.9 microamperes, $B10$ is 1.9 microamperes, and so forth. These low-level currents cannot be handled in a low-to-moderate, cost operational amplifier. The need for premium-grade devices make the cost higher.

Figure 10-5 shows a better method for implementing a D/A converter using an R-$2R$ resistor ladder method. The output voltage is given by

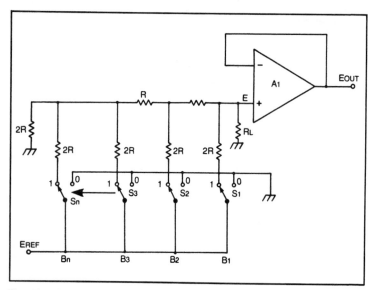

Fig. 10-5. R-2R resistor ladder D/A converter.

$$E_{\text{OUT}} = \frac{R_L}{2^n(R_L + R)} \sum_{i=1}^{n} E_{B(n-i)} 2^{i-1} \qquad (10.37)$$

The term $R_L / [2^n(R_L + R)]$ represents a scale factor determined by the number of bits and the load resistance. The summation term represents the value of the digital word applied to the D/A converter input lines.

If amplifier $A1$ is a high-grade operational amplifier with a very high input resistance relative to the value of R, I can rewrite in the form:

$$E_{\text{OUT}} = \frac{1}{2^n} \sum_{i=1}^{n} E_{B(n-i)} 2^{i-1} \qquad (10.38)$$

Example 10-4

Find E_{OUT} if a four-bit converter has the word 0110 at the input, and E_{REF} is 10.00 volts dc.

196

Solution:

$$E_{OUT} = \frac{1}{2^4} \sum_{i=1}^{4} E_{B(4-i)} 2^{i-1} \qquad (10.39)$$

$$(1/16)[E_{B3}(2^0) + E_{B2}(2^1) + E_{B1}(2^2) \times E_{B0}(2^3)] \qquad (10.40)$$

E_{Bi} will be E_{REF} when the bit is *one* and 0 volts when the bit value is *zero*, so $B3 = 0$, $B2 = 1$, $B1 = 1$, and $B0 = 0$.

$$E_{OUT} = (1/16)(0 + 2^1 E_{REF} + 2^2 E_{REF} + 0) \qquad (10.41)$$
$$E_{OUT} = (1/16)(2E_{REF} + 4E_{REF}) \qquad (10.42)$$
$$E_{OUT} = (6/16)E_{REF} \qquad (10.43)$$
$$E_{OUT} = (6/16)(10.00 \text{ volts}) \qquad (10.44)$$
$$E_{OUT} = 3.75 \text{ volts} \qquad (10.45)$$

Equation 10.46 gives the equation ordinarily given as a D/A converter transfer function, assuming that amplifier $A1$ has unity gain. If an inverting follower operational amplifier circuit is used, then assume that $R_{IN} = R_F = R$.

$$E_{OUT} = E_{REF} \frac{C1}{2^1} + \frac{C2}{2^2} + \frac{C3}{2^3} + \ldots + \frac{C_n}{2^n} \qquad (10.46)$$

where $C1$ through C_n will be 0 when the corresponding bit is *zero* and 1 when the corresponding bit is *one*.

Example 10-5

E_{REF} is 10.0 volts, and the eight-bit input word is 01101010. Find E_{OUT}.

Solution:

$$E_{OUT} = (10)$$

$$\left(\frac{0}{2^1} + \frac{1}{2^2} + \frac{1}{2^3} + \frac{0}{2^4} + \frac{1}{2^5} + \frac{0}{2^6} + \frac{1}{2^7} + \frac{0}{2^8} \right) \qquad (10.47)$$

$$E_{OUT} = (10) \left(\frac{1}{4} + \frac{1}{8} + \frac{1}{32} + \frac{1}{128} + \frac{1}{256} \right) \qquad (10.48)$$

$$E_{OUT} = 10/4 + 10/8 + 10/32 + 10/128 + 10/256 \quad (10.49)$$

$$E_{OUT} = 4.18 \text{ volts} \quad (10.50)$$

A/D BASICS

There are a number of methods for performing an analog-to-digital conversion, but most of them actually fall into only a few different classes. We will consider integration, voltage-to-frequency, parallel, binary counter (also called servo or ramp), and the successive-approximation methods.

Integration Methods

A block diagram to a dual-slope integrator A/D converter is shown in Fig. 10-6A. It consists of an input control switch, a reference source, an operational amplifier integrator, a comparator, a clock driven binary counter, and a control logic section that ties the whole thing together.

Switch $S1$ is shown here as an ordinary switch, but in actual dual slope integrators it would be a CMOS electronic switch.

A pulse applied to the START line will clear the binary counter, reduce the charge on the integrator capacitor to zero (it closes $S2$ briefly), and insures that switch $S1$ is connected to the analog input.

The voltage at the output of the integrator will begin to rise as soon as the analog input is connected. Since the comparator is ground referenced its output will snap high as soon as the integrator output is greater than a few millivolts, the normal comparator uncertainty (i.e., hysteresis).

When the comparator output is high the control section will gate clock pulses into the cleared binary counter. These are allowed to increment the counter until overflow occurs. All the while the integrator output (E_A) is rising to some final value. When the counter overflows a pulse is sent back to the control logic section and this switches $S1$ to the reference voltage input. The reference voltage is constant, and is of a polarity opposite that of the analog input voltage, so it will *discharge* the accumulated integrator output voltage at a constant rate.

The counter outputs will all be zero (0000) when overflow occurs. This means that the counter will be incrementing from zero, once each time a clock pulse is received, while the reference voltage is discharging the integrator capacitor.

When the integrator output has discharged all the way back

198

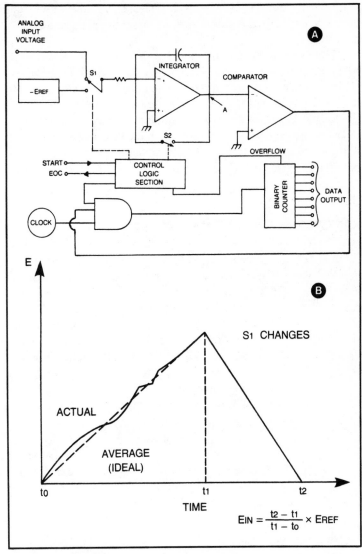

Fig. 10-6. Dual-slope integrator A/D converter. (A) Circuit. (B) Waveform.

$$E_{IN} = \frac{t_2 - t_1}{t_1 - t_0} \times E_{REF}$$

to zero the comparator output will snap low. This condition causes the control section to stop the counter, but does not reset the counter (its final value is held). Also at this time an end of conversion pulse is generated. The output data word will be proportional to the analog input voltage applied to the input.

Conversion time for the dual-slope integrator form of A/D converter is on the order of 2^{n+1} clock cycles, so for an eight-bit converter a total of 512 pulses are required to make a full-scale conversion.

The integrating type of converter is too slow to follow fast-breaking signals, but is considered very good where long conversion times can be tolerated. One reason for the popularity of the integrating A/D converter is the inherently good noise rejection of the circuit. Recall, if you please, that an integrator is also a form of low-pass filter, so noise artifacts are reduced considerably in amplitude. Most digital voltmeters take advantage of this facet of the converter, so several IC devices are made that will perform such conversions cheaply.

Voltage-to-Frequency Converter

The voltage-to-frequency (V/F) converter such as those discussed in Chapter 7 can be used to make an analog-to-digital conversion. The V/F converter will generate an output pulse train of a frequency that is proportional to a voltage applied to the input. A frequency counter circuit, or a software program that makes the computer think that it's a frequency counter, is then used to derive binary words that represent the frequency. In the counter approach, a binary counter (or BCD output decade counter if binary code conversion is also provided) is gated to receive pulses for a specified length of time. The accumulated count at the end of this time period is a measure of events per unit of time, or pulses per second—frequency.

The V/F converter is capable of respectable dynamic range, and moderate-to-slow conversion times. Its main use is where data may have to be transmitted over a distance, in which case the frequency components can be transmitted over a telephone line, or through a radio-telemetry channel.

Parallel Converters

The fastest type of A/D converter is the parallel circuit of Fig. 10-7. This circuit uses 2^{n-1} voltage comparators ($A1$ through A_n) to make the conversion. One input on each comparator is tied to the analog voltage input, while the alternate inputs are tied to specific reference voltage sources. The bias levels applied to these comparators differ from stage to stage by the amount of the least significant bit.

When an analog input signal is applied to this converter all of those comparators biased above its value will be off (i.e., output low), while those biased below the level of the analog signal are turned on (output high).

The outputs of the comparators *do not* represent a binary word, so the output states must be decoded in logic to follow the parallel output lines. Such logic will have to be set up to find the highest order bit that is high. The output states for a hypothetical A/D converter are shown in Fig. 10-7B.

A more elegant approach to logic conversion of the output data might be software decoding. Connect the parallel output lines to an input port of a computer. Note in Fig. 10-7C we have the possible bit patterns that will represent 0 to 7 volts. A program could be written to examine the input and create the proper binary code for the word appearing at the input. Either a table lookup or a method for testing successively higher order bits should work. Incidentally, this will slow down the conversion process, so one must be willing to tolerate the slower speed, or provide a buffer memory to hold words until conversion in software is completed.

Binary Counter A/D Converters

The binary counter type of A/D converter, also called the servo or ramp type in some texts, is shown in Fig. 10-8A. The conversion sequence is shown in Fig. 10-8B. The components of this type of converter are a reference voltage source, comparator, D/A converter, binary counter, clock, and control logic section.

You may have wondered why I covered the D/A converter first, since to many it seems almost contradictory to do so. The reason is that this type of A/D converter, as well as several others, use a D/A converter in a feedback method to perform an A/D conversion. They are part of a larger class of A/D converters known generically as feedback converters.

When a start pulse is received in the control logic section it clears the binary counter and gates clock pulses into the counter input. The binary counter is incremented by each successive clock pulse.

The output lines from the binary counter are connected to the digital input terminals of the D/A converter. So when the binary counter output changes I note that the D/A converter output also changes. The D/A converter output voltage is applied to one input of a comparator. The other input to the comparator is connected to the analog input voltage. If the analog input voltage is greater than the D/A converter output voltage, then the comparator out-

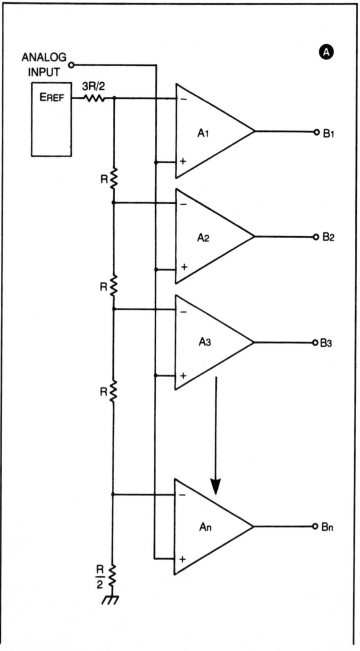

Fig. 10-7. Parallel converter. (A) Circuit. (B) Output state graph. (C) Output state table.

202

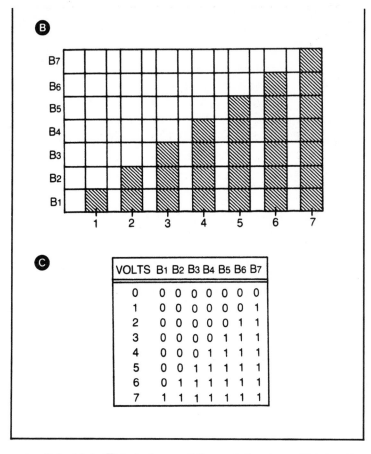

B

C

VOLTS	B1	B2	B3	B4	B5	B6	B7
0	0	0	0	0	0	0	0
1	0	0	0	0	0	0	1
2	0	0	0	0	0	1	1
3	0	0	0	0	1	1	1
4	0	0	0	1	1	1	1
5	0	0	1	1	1	1	1
6	0	1	1	1	1	1	1
7	1	1	1	1	1	1	1

put will be high. This is the condition existing immediately after the start pulse is received.

When the D/A converter output has been incremented to a point where it is equal to the analog input voltage the comparator output will snap low, stopping the conversion action. The binary counter output will remain latched, holding the data word, and an end-of conversion pulse is generated.

The binary counter type of A/D converter is relatively easy to build using low-cost components. The output of the D/A converter, hence the maximum analog input, is equal to the product of the reference voltage and a fraction representing the digital word. The full-scale output is:

$$E_{IN(FS)} = (E_{REF})[(2^n - 1)/2^n] \qquad (10.51)$$

203

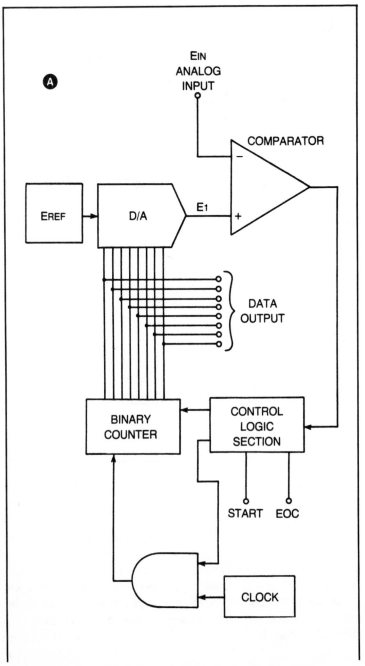

Fig. 10-8. Binary counter A/D converter. (A) Circuit. (B) Waveform.

204

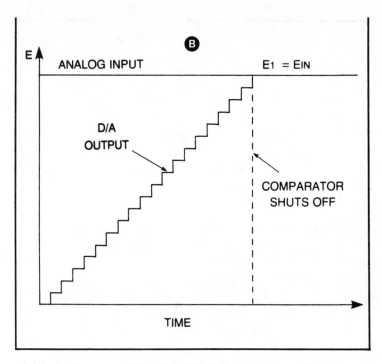

which, in the case of the eight-bit A/D converter, will be

$$E_{IN(FS)} = (E_{REF})(255/256) \qquad (10.52)$$

The conversion time will be on the order of 2^n clock pulses for a full-scale conversion. For our ubiquitous eight-bit A/D converter this works out to 256 clock pulses. If a D/A converter capable of tracking a 2.5 MHz clock (most binary counters will track a 2.5 MHz rate, the limiting factor in most cases is the D/A converter) I will receive a clock pulse every 400 nanoseconds (4×10^{-7} seconds). A full-scale conversion is possible, therefore, every 102 microseconds.

Successive-Approximation Converters

One of the fastest A/D conversion techniques, although slower than the parallel method, is the successive-approximation circuit of Fig. 10-9A. Although not as fast as the parallel conversion it is considered superior because its output is a readable binary code that requires no additional processing before being sent to the computer input port.

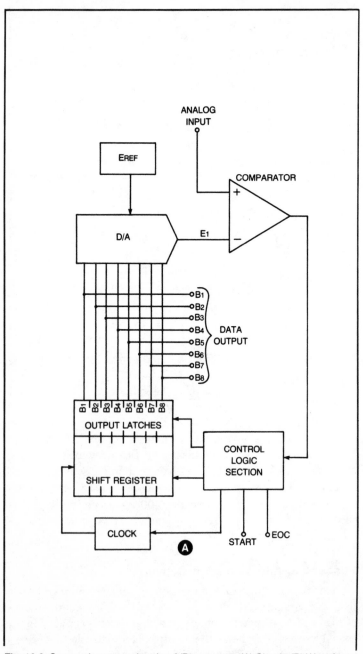

Fig. 10-9. Successive-approximation A/D converter. (A) Circuit. (B) Waveforms for E_{IN} above half-scale. (C) For E_{IN} below half-scale.

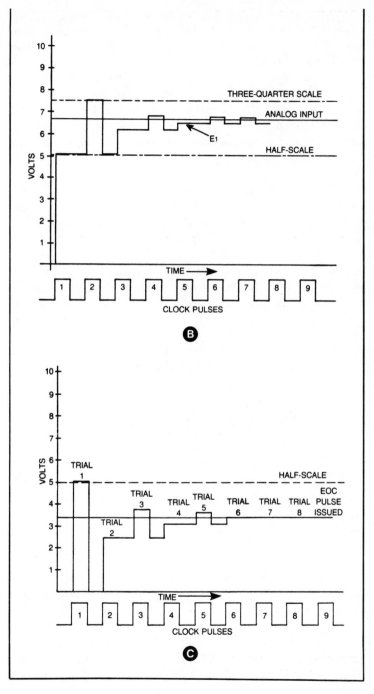

207

The constituent parts of a successive-approximation A/D converter include a D/A converter (again a feedback-class converter), a comparator, a shift register, register output latches, a clock, and a control logic section.

The digital input lines of the D/A converter are connected to the parallel output lines from the shift register. The analog output of the D/A converter is connected to one input of the comparator, while the remaining comparator input is connected to the analog input voltage. The comparator output will remain high whenever the analog input voltage is greater than the D/A converter output voltage.

The actual operation of the successive-approximation A/D converter is slightly different for two cases: E_{IN} less than half-scale and E_{IN} greater than half-scale. I will consider both cases separately even though they are similar.

Let me consider first a conversion in which the analog voltage is between one-half scale and three-quarter scale, say 6.6 volts. An eight-bit converter is used, and $B1$ is the most significant bit.

When a start pulse is applied to the control section the shift register is cleared and the clock is gated on. This type of converter operates in a synchronous mode, which means all action is keyed by the clock pulses. On the first clock pulse $B1$ is set to *one* and all other bits are *zero*. The binary word applied to the D/A converter, then, is 10000000. This word places the D/A converter output at half-scale, or (for this case) 5 volts.

In this instance the analog input voltage is still greater than the D/A converter output, so the comparator output remains high. Bit $B1$ is latched to a *one* and is held.

On the next clock pulse bit $B2$ is set high making the word applied to the D/A converter inputs 11000000, and this sets the D/A converter output voltage to three-quarter scale ($E1 = 7.5$ volts). Here I find that $E1$ is greater than the analog input voltage, so $B2$ is reset to *zero*. The $B2$ register is then latched with a *zero*, so the output word goes back to 10000000.

On the third clock pulse bit $B3$ is set high, making the D/A converter see an input word of 10100000. Under this input $E1$ is equal to 6.25 volts, which is below the analog input voltage, so the comparator output is high. This condition tells the control logic section to latch $B3$ to *one* leaving the output word at 10100000.

Bit $B4$ is set high on the fourth clock pulse. The D/A converter sees an input word of 10110000, making $E1$ equal to 6.875 volts. This potential is greater than the input voltage, so the comparator output goes low, signaling the control logic to reset $B4$ to *zero*. The register contents at the conclusion of the fourth clock period will remain at 10100000.

At the beginning of the fifth clock pulse bit $B5$ will be set to *one*, so the D/A converter will see an input word of 10101000, and $E1$ is equal to 6.563 volts. Again $E1$ is less than the analog input voltage so $B5$ is latched at *one*.

On the next trial bit $B6$ is set high, making the D/A converter input 10101100. The value of $E1$ that is produced in this case is 6.72 volts, so $B6$ is reset to *zero*. Trial 6 generates a code of 10101000. On trial 7 the D/A converter sees 10101010 at its input, so $E1$ is 6.64 volts. Again $E1$ is too high, so $B7$ is reset to *zero*. The register output is 10101000.

Bit $B8$ is set high on the last trial. This lets the D/A converter see an input code of 10101001, so $E1$ is 6.6 volts. The comparator output drops low, latching $B8$ at a *one*. On the next clock pulse (9) the register will overflow, and this becomes my end-of-conversion (EOC) pulse. The successive-approximation method requires only $n + 1$ clock pulses, so is faster than other methods.

The conversion shown in Fig. 10-9C is the sequence of events that will occur when the analog signal voltage is less than half-scale. Again, each trail is synchronized by clock pulses. In this example the analog input voltage is $+3.4$ volts.

When the first clock pulses after the start command arrives, the shift register is cleared and $B1$ (the MSB) is set to *one*. The digital word applied to the D/A converter input is 10000000, and the D/A converter output voltage ($E1$) is $+5$ volts. $E1$ is greater than the input voltage, so $B1$ is reset to *zero*.

When the second clock pulse arrives bit $B2$ is set high, making the digital word at the D/A converter input 01000000. The value of $E1$ is quarter-scale, or $+2.5$ volts. The comparator tells the control section that this voltage is less than the input voltage, so bit $B2$ is latched high. The D/A converter output remains at 01000000.

On the third trial bit $B3$ is set high making the digital word at the input of the D/A converter 01100000, and voltage $E1$ is $+3.75$ volts. This voltage is greater than the input voltage, so $B3$ is reset to *zero*. The D/A converter output remains 01000000.

The fourth trial sees bit $B4$ set high to create a D/A converter input code of 01010000. The D/A converter output voltage ($E1$) is $+3.13$ volts. $E1$ is lower than the input voltage, so $B4$ is latched high. The DAC code remains 01010000.

On the fifth trial bit $B5$ is set high, making the D/A converter input code 01011000. Voltage $E1$ is $+3.44$ volts. Again too high, so $B5$ is reset to *zero*. This process continues through $B6$, $B7$, and $B8$ until the final code existing is 01010111, and $E1$ is $+3.4$ volts. The ninth clock pulse overflows the register, and is used as the EOC pulse. One must take care that the analog input voltage is within the range of the converter, or an overflow condition will tell the computer that EOC has occurred, but the code will be erroneous (11111111).

SOFTWARE A/D CONVERTERS

A minimum hardware A/D converter can be built along the lines of either binary counter or successive-approximation methods by connecting an eight-bit D/A converter directly to the microprocessor. In one approach actual I/O ports are used, while in another the address lines are used. In both cases a comparator examines the output of the D/A converter and the analog input voltage.

In the counter method a program in the microcomputer generates a ramp at the input of the D/A converter. Figure 10-10 shows a case where the circuit uses computer I/O ports. One complete output port is used to drive the digital inputs of the D/A converter,

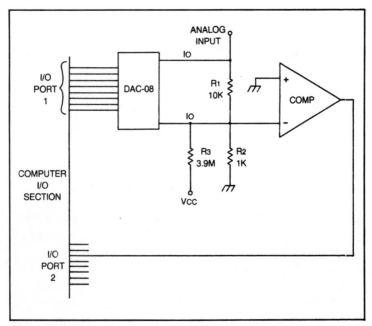

Fig. 10-10. Hardware for a software A/D converter using I/O ports.

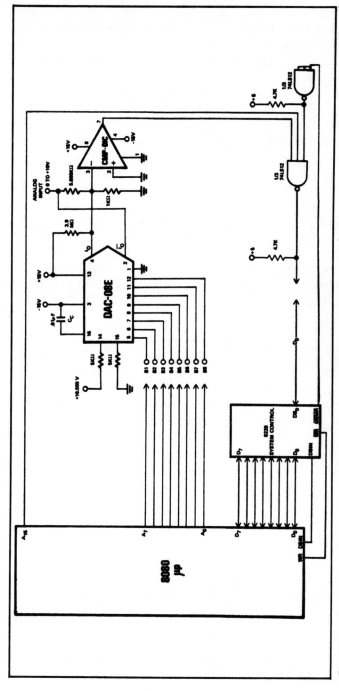

Fig. 10-11. Circuit for a software A/D converter using memory mapping. The DAC-08 is treated as memory with an address above 32K. (Courtesy of Precision Monolithics, Inc.)

while one bit of another is used to monitor the status of the comparator output. This is not quite as wasteful as it may seem because many microcomputers have a portion of one port already dedicated to some function such as serial data transmission or peripheral control, etc.

A program must be written that will generate a ramp at I/O port 1. This can be done by creating a loop in which the accumulator is incremented and outputted once during each pass through the loop. The elements required in the program will be (1) clear the accumulator, (2) increment the accumulator by one, (3) output accumulator, (4) test the appropriate bit of input port 2 to find out the status of the comparator output, and (5) if the comparator output is high, then loop back to (2) and continue, but if the comparator output is low, then break out of the loop and use the word in the accumulator as your input data.

Figure 10-11 shows the use of a microprocessor's address lines in a successive-approximation A/D converter operated under software control. This particular circuit is based on the Intel 8080A microprocessor and the Precision Monolithics, Inc. DAC-08 (of which more is given in Chapter 11).

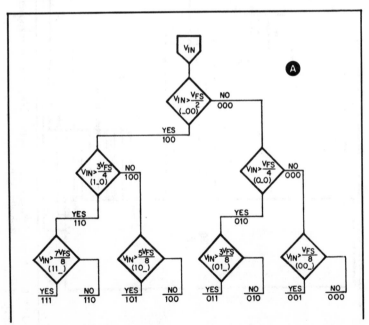

Fig. 10-12. Successive-approximation algorithm. (A) Concept. (B) Program flow chart. (Courtesy of Precision Monolithics, Inc.) (Continued on next page.)

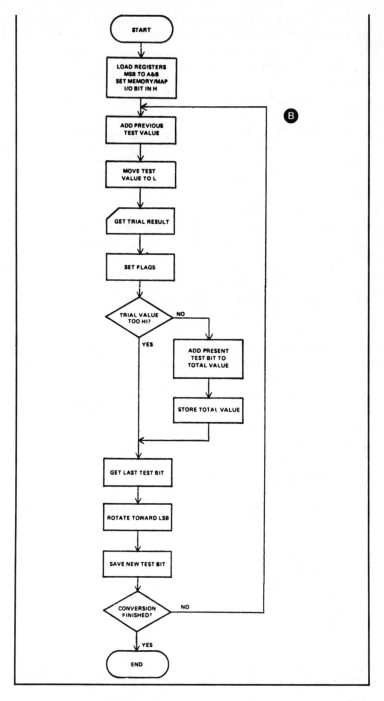

213

Table 10-1. DAC-08 A/D Conversion Routine.

Start:	LXI B, 08000H	;LOAD MSB IN B, CLEAR C
	MOV A,B	;MSB TO ACC
	MOV H,A	;SET MEM/MAP I/O
TEST:	ORA C	;ADD LAST TEST VALUE
	MOV L,A	;MOVE PRESENT TEST TO L
	MOV A,M	;GET COMP OUTPUT
	ANA A	;SET FLAGS
	JPO T0OHI	;DISCARD PRESENT TEST BIT
	MOV A,B	;GET PRESENT TEST BIT
	ORA C	;ADD TOTAL SO FAR
	MOV C,A	;SAVE TOTAL
T0OHI:	MOV A,B	;GET LAST TEST BIT
	RAR	;ROTATE TOWARD LSB
	MOV A,B	;SAVE NEW TEST BIT
	JNC TEST	;JUMP IF NOT FINISH
	END	;FINAL VALUE IS IN C

This circuit uses a technique called memory-mapped I/O. Most microcomputers and microprocessor instrumentation applications do not require anywhere near the total allotted addressable memory. The 8080A, Z-80 and most other eight-bit microprocessor chips have a sixteen-bit address bus, so can address up to 65K of different memory locations. Most applications, however, require considerably less in the way of active memory, with most being in the under 32K class. In Fig. 10-11 the D/A converter (DAC-08E) is connected to address lines $A0$ through $A7$. Address line $A15$ is used as an I/O control flag. If $A15$ is high, then the I/O function is enabled, but if $A15$ is low then the memory is active (note that $A15$ divides the upper and lower 32K of possible memory addresses so would never be used to address memory in a system of less than 32K memory).

The use of memory mapping, as in Fig. 10-11, requires that the peripheral (i.e., the DAC-08 in this example) recognize and conform to the memory address bus timing and control signals. Table 10-1 gives the 8080A assembler code for using a circuit to implement a successive-approximation algorithm (an example of which is shown in Figs. 10-12A and 10-12B).

Chapter 11

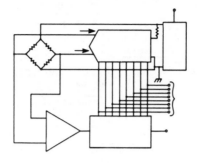

D/A Converters:
Some Real Products

IT HAS BEEN POINTED OUT THAT BOOKS AND ARTICLES FOR
amateur computerists are often written proclaiming this or that
feature of some exotic software program, but without much regard
for how the data will be obtained that feeds the program. In this
chapter, and that to follow, I will discuss some of the actual data
conversion products on the market, but will use a practical approach
rather then the theoretical discussion of the previous chapter.

In the case where the information is contained in an analog current or voltage used to represent some physical parameter, the
proper interface unit contains an analog-to-digital converter. Where
the computer must control some external analog device, a digital-
to-analog converter is used. Although the order might seem a little backwards, I will consider first the D/A products, then proceed
in Chapter 12 to consider some practical A/D circuits. This is done
because many A/D circuits use a D/A converter in a negative-
feedback loop.

THE FERRANTI ZN425E

Figure 11-1 shows the block diagram for the Ferranti ZN-425E
multimode data converter integrated circuit (Ferranti Semiconductors, East Bethpage Road, Plainview, NY 11803). This low-cost chip
uses a binary counter to drive the electronic switches of the standard *R-2R* ladder, *if* the *logic select* pin is turned on. In this mode,

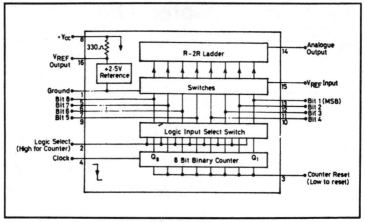

Fig. 11-1. Block diagram to the Ferranti ZN-425E building block. (Courtesy of Ferranti Semiconductors, Inc.)

it operates not as a straight D/A converter, but as a 255-step ramp generator with *current* output.

The ZN-425E will also operate as a regular D/A converter, or as an eight-bit A/D converter, depending upon the external circuit configuration and the logic level applied to the logic select terminal. The output of the Ferranti ZN-425E is a current that is proportional to the product of the reference and the digital word applied to the binary inputs. The use of suitable external circuitry, and a switching method, will allow the ZN-425E to operate in any of the modes under external command.

THE PRECISION MONOLITHICS DAC-08

One of the more popular integrated circuit D/A converters on the market is the DAC-08 (Fig. 11-2) by Precision Monolithics, Inc. (PMI, 1500 Space Park Drive, Santa Clara, CA 95050). This device contains the standard *R-2R* ladder, electronic switches, and reference amplifier, but requires an external reference voltage source. This makes the DAC-08 a *multiplying D/A converter*. The need for an external reference makes this IC less useful for some applications, but more useful for certain others in which the multiplication feature is needed.

The DAC-08 produces two output currents that are said to be complementary because one is decreasing while the other increases, but their sum is a constant. The sum value, full-scale current, is

216

shown to be

$$I_{FS} = (V_{REF}/R_{REF})(255/256) \qquad (11.1)$$

where I_{FS} is the full-scale current and V_{REF} and R_{REF} are the reference voltage and resistance.

Example 11-1

What is the full-scale current of a DAC-08 in which the reference voltage is $+7.5$ volts dc and the reference resistor (R_{REF}) is 9.1K?

Solution:

$$\begin{aligned} I_{FS} &= (V_{REF}/R_{REF})(255/256) \qquad (11.1)\\ &= (7.5/9100)(255/256)\\ &= 0.0008 \text{ amperes}\\ &= 0.8 \text{ milliamperes} \end{aligned}$$

and the output currents are,

$$I_{FS} = I_{OUT} + I_{OUT} \qquad (11.2)$$

where I_{FS} is the full-scale current as set by Eq. 11.1, I_{OUT} is the current flowing into pin 4, and I_{OUT} is the current flowing into pin 2.

Example 11-2

The full-scale current is 2 milliamperes, and the current flowing into pin 2 is 0.35 milliamperes. What current is flowing into pin 4?

Solution:

$$\begin{aligned} I_{FS} &= I_{OUT} + I_{OUT} \qquad (11.2)\\ (2) &= I_0 + (0.35)\\ I_{OUT} &= 2 - 0.35 \text{ milliamperes}\\ &= 1.65 \text{ milliamperes} \end{aligned}$$

The DAC-08 is capable of monotonic multiplying operation over a range of approximately 32 dB of reference current. It will achieve as fast as 85-nanosecond settling times if good layout and design practices are followed. It must be noted that many amateurs fail to achieve most rapid performance, as promised by the manufacturer's specification sheet, solely because of poor layout practices and sloppy construction technique; both of which conspire to slow down digital circuitry.

The DAC-08 is available in various models that have different quality levels and temperature ranges (at different prices, of course). The suffix of the part number denotes the quality level, namely:

Model	Temp. Range °(C)	Non-linearity
DAC-08AQ	– 55/ + 125	± 0.1 %
DAC-08Q	– 55/ + 125	± 0.19%
DAC-08NQ	0/ + 70	± 0.1 %
DAC-08EQ	0/ + 70	± 0.19%
DAC-08CQ	0/ + 70	± 0.39%

A feature of the DAC-08 series that proves especially useful in some applications is the fact that there are two complementary current outputs, I_{OUT} and I_{OUT}. The relationship between these currents is shown in Fig. 11-2B.

At the low end of the scale, when the binary number at the input is 00000000_2 I find that I_{OUT} is zero and I_{OUT} is maximum, in this case 2 milliamperes. At half-scale operation (input word set to 10000000_2) the two currents are equal, which in the example puts them at 1 milliampere. At full-scale operation (input word 11111111_2) I_{OUT} will be 2 milliamperes, and I_{OUT} is zero. The complementary outputs makes the DAC-08 especially suited to applications where a push-pull or differential load is to be accommodated.

Examples of applications where the complementary outputs are particularly useful are line drivers for data communications, driving the vertical or horizontal deflection plates of the cathode-ray tube, or in driving an X-Y recorder. In a circuit using the DAC-08 as an A/D converter (Chapter 12), the complementary output is arranged so that it keeps the input impedance seen by the analog voltage source constant, even though the impedance would normally vary over a ten-to-one range.

Figure 11-3A shows the basic connections required to make the DAC-08 work in a real circuit. In later circuits, I will delete the power connections for purposes of simplicity, so be prepared to look back at this figure should you want to actually try building any subsequent circuits.

The terminals marked $V(+)$ and $V(-)$ are the positive and negative power supply connections, respectively. Note that these are not the positive and negative points in a single supply, but re-

218

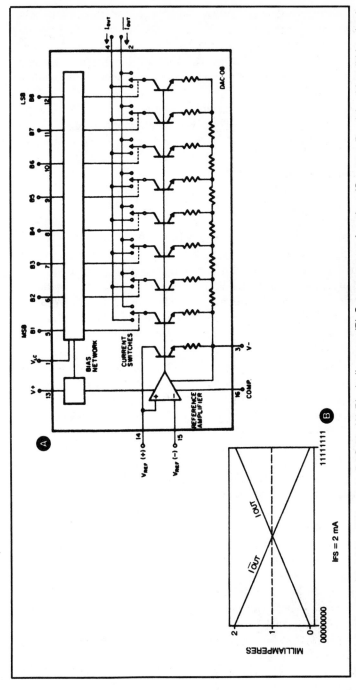

Fig. 11-2. The Precision Monolithics, Inc. DAC-08. (A) Block diagram. (B) Output current function. (Courtesy Precision Monolithics, Inc.)

219

fer to supplies that are positive with respect to ground and negative with respect to ground. The DAC-08 requires a bipolar power supply like an operational amplifier.

The potential applied to the $V(+)$ and $V(-)$ terminals can be anything between 4.5 volts and 18 volts dc, but most of the manufacturer's examples give 15 volts as the power supply levels. Note, however, that a very low power consumption of approximately 33 milliwatts is possible if the power supply voltages are 5 volts.

The terminal marked V_{LC} is a level, or threshold control, and should be fixed at some potential between $V(-)$ and $V(+)$. In most circuits, where TTL logic levels are used, V_{LC} will be grounded.

The threshold voltage (V_{LC}) controls the voltages that will be recognized by the input terminals as digital logic levels. Normally, when the power supply potentials are 15 volts dc, the digital inputs can swing over the range -10 volts to $+18$ volts. But logic levels in any given circuit tend to be fixed. TTL, for example, uses zero volts for logic 0, and $+5$ volts as logic 1. Depending upon whose expertise I trust, the transition point between *zero* and *one* occurs someplace between $+1.4$ volts and $+2.2$ volts. Whatever the "magic number" happens to be in any given chip, voltages below that level will be recognized as *zero* and those voltages above that level are seen as *one* states.

The V_{LC} terminal on the DAC-08 allows it to interface with a large range of different circuitry, without the need for level translators or other peripheral circuitry. For example, the DAC-08 can be interfaced with CMOS digital integrated circuits that use a *one* level in the $+5$ to $+15$-volt region, even though the power supply terminals of the DAC-08 are connected to 5-volt dc sources.

Another case that is often a real bear to solve is interfacing new circuitry, to perform a new job, with older instruments. Almost all modern instruments will use either TTL or CMOS logic elements, so the specifications for *zero* and *one* are fixed to a standard. But if you want to connect older instruments into a circuit with a D/A converter, it might be necessary to accommodate obsolete logic levels. When digital instruments were manufactured using discrete transistors to make the logic circuits, the designer was almost totally free to select logic level voltages. I have seen examples of diode logic AND gates in which -9 volts (or -12 volts) represented the logic 0 condition, and $+5$ (or $+12$) volts represented logic 1. Most D/A converter integrated circuits would not be able to handle nonstandard logic levels unless transistor level translators were used between the output of the instrument and the input

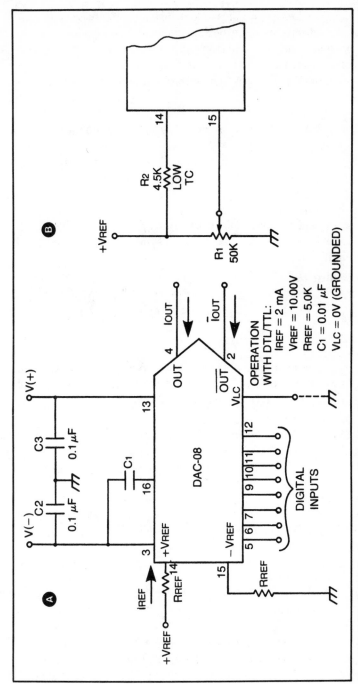

Fig. 11-3. DAC-08. (A) Basic connections. (B) Output trimming.

221

of the D/A converter. These older instruments, incidentally, are not all that uncommon, especially in smaller companies, amateur or hobbyist markets, or in many university laboratories where the pinch in basic research money is painfully evident.

The minimum input logic swing and logic threshold voltage (V_{TH}) are given by:

$$V_{TH} = V(-) + 2.5 \text{ volts} + (I_{REF} \times 1000 \text{ ohms}) \quad (11.3)$$

where V_{TH} is the threshold voltage, $V(-)$ is the negative power supply voltage, and I_{REF} is the reference current in amperes.

Example 11-3
Find the threshold voltage, V_{TH}, if the negative power supply potential is -12 volts, and the reference current is 2 milliamperes.

Solution:

$$
\begin{aligned}
V_{TH} &= V(-) + 2.5 + (I_{REF} \times 1000) \quad (11.3) \\
&= (-12) + 2.5 + (0.002 \times 1000) \\
&= (-12) + 2.5 + 2 \\
&= -7.5 \text{ volts}
\end{aligned}
$$

The logic threshold can be varied externally by applying a voltage to pin 1 of the DAC-08. The threshold voltage will be approximately 1.4 volts higher than V_{LC}.

$$V_{TH} = V_{LC} + 1.4 \text{ volts} \quad (11.4)$$

where all terms are as previously defined.

Example 11-4
Find the threshold voltage (a) if -6 volts dc is applied to the V_{LC} terminal, and (b) for TTL compatibility, in which case the V_{LC} terminal is grounded ($V_{LC} = 0$).

Solution:

$$
\begin{aligned}
\text{(a)} \quad V_{TH} &= V_{LC} + 1.4 \\
&= (-6) + 1.4 \\
&= 4.6 \text{ volts} \\
\text{(b)} \quad V_{TH} &= V_{LC} + 1.4 \text{ volts} \\
&= (0) + 1.4 \text{ volts} \\
&= +1.4 \text{ volts}
\end{aligned}
$$

Note that logic signal in (a) must be lower than −4.6 volts for logic 0 and greater than −4.6 volts for logic 1, allowing the chip to accommodate older, obsolete but still useful equipment. In (b) the standard +1.4-volt transition point for TTL logic is also the level that is used by the DAC-08, so the DAC-08 will directly interface with TTL and DTL circuitry if the V_{LC} terminal (pin 1) is grounded.

The threshold voltage has been defined as the swing between voltages representing logic 0 and logic 1 conditions at the input and the trip point at which the DAC-08 decides that the input logic level has changed. In TTL and DTL circuits, this level is approximately +1.4 volts, so V_{LC} is set to zero (i.e., it is grounded). In CMOS, the trip point will be half the positive supply potential, so if $V(+)$ is, say, +12 volts, the logic trip point will be +6 volts. The voltage applied to V_{LC} should be set to represent this difference (use Eq. 11.4). In that case, rearranging the equation gives us:

$$
\begin{aligned}
V_{LC} &= V_{TH} - 1.4 \\
&= (+6) - 1.4 \\
&= +4.6 \text{ volts}
\end{aligned}
$$

The reference current can be at any level between 0.2 milliamperes and 4 milliamperes, but the manufacturer recommends 2 milliamperes for TTL and DTL circuits and 1 milliamperes for ECL logic circuits. The input current requirements of the DAC-08, incidentally, is only 2 microamperes (μA), so it will load the digital output of the driving circuitry only very lightly.

The reference current is set by the reference voltage and the resistance (R_{REF}) of the reference input resistor. By Ohm's law:

$$
I_{REF} = V_{REF}/R_{REF} \tag{11.5}
$$

where I_{REF} is the reference current in amperes, V_{REF} is the reference voltage in volts, and R_{REF} is the reference input resistance in ohms.

Example 11-5
Find the value of the reference input resistor that will create a 2-milliampere reference current from a +10.00-volt reference voltage.

Solution:

$$
I_{REF} = V_{REF}/R_{REF}
$$

So,

$$R_{REF} = V_{REF}/I_{REF}$$
$$R_{REF} = (10.00)/(0.002)$$
$$R_{REF} = 5000 \text{ ohms}$$

Note that R_{REF} is connected to the $+V_{REF}$ terminal of the DAC-08 (pin 14), and another, identical resistor is connected between the $-V_{REF}$ terminal (pin 15) and ground. This resistor is a bias compensation measure that can be deleted if a slight error is tolerable in the particular application where the DAC-08 is selected.

Operation from a negative reference source can be accommodated by simply switching the roles of the two input resistors. The resistor from pin 14 will be grounded, while the resistor from pin 15 will be connected to the negative reference voltage instead of ground.

Regardless of the reference polarity, however, large errors could result if ordinary carbon composition resistors are used for establishing the reference current. The use of precision, low-temperature-coefficient resistors are highly recommended.

An alternate circuit for the input reference current is shown in Fig. 11-3B. Normally, the reference current will track the full-scale current very closely, so trimming of the reference current at the full-scale point often proves unnecessary. But in those cases where trimming is required, the circuit of Fig. 11-3B can be substituted for the original input circuit using just the resistors. Both the fixed resistor and the potentiometer should be low-temperature-coefficient devices, and, additionally, the potentiometer should be a ten-turn or more trimmer pot.

Although it is not always necessary, a bypass capacitor is recommended whenever dc reference sources are used on the DAC-08. The proper procedure is to split the reference resistance between two resistors in series, then bypass the junction of the two resistors to ground through a 0.1 μF capacitor.

The main power supply connections to the DAC-08 are straightforward, so require little comment beyond the usual admonition to keep the 0.1 μF bypass capacitors (C2 and C3 in Fig. 11-3A) as close to the body of the DAC-08 as possible. Additionally, a 0.01 μF frequency-compensation capacitor (C1 in Fig. 11-3A) is connected between $V(-)$ and the compensation terminal (pin 16). Make the leads of C1 as short as possible also.

The reference voltage is critical to the accuracy of any D/A converter. Nonmultiplying D/A converters use an internal reference source that may or may not be externally trimmable by a potentiometer. Multiplying D/A converters such as the DAC-08, on the other hand, require that an external reference voltage be provided. Errors in the reference supply are directly reflected as errors in the output current, so precision is required, unless (of course) accuracy is unimportant.

Never use the $V(+)$ or $V(-)$ power supply voltages as the reference supply. If they change, and they will change, then an error in the output current will result. The only proper way to use the power supply as the reference is to use a voltage regulator between the supply and the reference input on the DAC-08. There are no circumstances where it is proper to use a +5-volt supply that also serves TTL or DTL circuitry as the reference.

In low-precision applications, a simple zener diode and series resistor combination may suffice for the reference voltage supply. But this type of regulator is subject to thermal drift and large, non-trimmable errors in the actual voltage (the rated voltage on a zener diode is a nominal value unless special reference diodes are obtained, in which case a premium price is paid to have someone sit down at a test console at the plant and hand select those that are very close to the rated zener potential). Two alternative reference power supplies are shown in Fig. 11-4.

Figure 11-4A shows the use of a Precision Monolithics REF-01 (or REF-02 if a 5.00-volt reference is required) precision reference voltage source. The REF-01 is an integrated circuit regulator in an eight-pin metal can; it will provide an output that is trimmable to exactly +10.00 volts. The input voltage can be the $V(+)$ power supply, provided that $V(+)$ is greater than about +12 volts dc. The REF-01 and its 5-volt cousin, the REF-02, are highly recommended to create +10- and +5-volt reference supplies.

A slightly inferior, but still quite useful, method for obtaining the reference potential is shown in Fig. 11-4B. This circuit uses an operational amplifier and a reference-grade zener diode to produce an output voltage that can be used as V_{REF}. The operational amplifier should be a moderate to high-grade (preferably frequency compensated) type such as the 101/201/301 series, or the RCA CA3060. The zener diode should be a reference-grade diode to insure that V_Z will remain stable over moderate temperature changes. Be aware that National Semiconductor makes a four-terminal zener in which the reference diode is across two terminals and a heater element that keeps the internal temperature constant is across the others. All resistors used in this circuit must be precision, low-temperature-coefficient types in order that thermal drift is minimized.

The output voltage can be found from the ordinary rules for operational amplifiers, namely:

$$I2 = I3 \tag{11.6}$$
$$I2 = V_z/R2 \tag{11.7}$$
$$I3 = (V_{REF} - V_z)/R3 \tag{11.8}$$

So, by substituting Eqs. 11-7 and 11.8 into Eq. 11.6:

$$\frac{V_z}{R2} = \frac{V_{REF} - V_z}{R3} \tag{11.9}$$

$$\frac{V_z}{R2} = \frac{V_{REF}}{R3} - \frac{V_z}{R3} \tag{11.10}$$

$$\frac{V_z}{R2} + \frac{V_z}{R3} = \frac{V_{ref}}{R3} \tag{11.11}$$

$$V_z \times \left(\frac{R3}{R2} + \frac{R3}{R3} \right) = V_{REF} \tag{11.12}$$

$$V_z \times \left(\frac{R3}{R2} + 1 \right) = V_{REF} \tag{11.13}$$

Example 11-6
Find the reference voltage created by the circuit in Fig. 11-4B if the zener potential is + 2.5 volts, and the resistor values are $R3 = 5.6K$, and $R2 = 1K$.

Solution:
$$
\begin{aligned}
V_{REF} &= V_z [(R3/R2) + 1] \quad \text{(from 11.13)} \\
&= (2.5)[(5.6/1) + 1] \\
&= (2.5)(6.6) \\
&= 16.5 \text{ volts}
\end{aligned}
$$

I may wish to combine these two methods for creating refer-

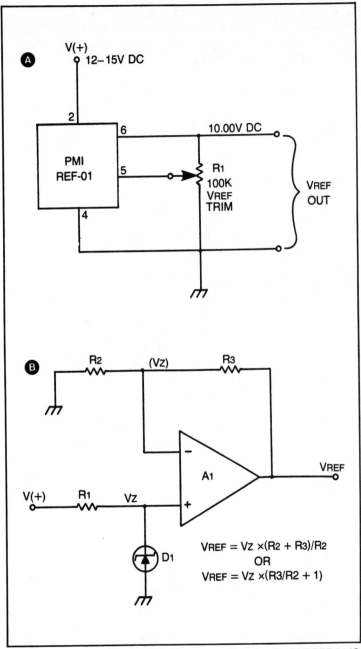

Fig. 11-4. Precision reference voltage sources. (A) Using the PMI REF-01 IC regulator. (B) Operational amplifier circuit.

227

ence voltages by substituting either the REF-01 or REF-02 for the zener diode in Fig. 11-4B. In later circuit examples, I will require a +2.56-volt power supply for a reference source, and will require stability greater than is normally obtainable using zener diodes.

Example 11-7

Find a resistor ratio $R3/R2$ that will produce a +2.56-volt output from a circuit such as Fig. 11-4B if a REF-01 reference supply is adjusted to give an output of +10.00 volts.

Solution:

Use Eq. 11.13 and solve for the case where V_Z is 10.00 volts, V_{REF} is +2.56 volts, and the expression inside of the brackets evaluates to 2.56/10, or 0.256.

$$[(R3/R2) + 1] = 0.256$$
$$R3/R2 = 0.256 - 1$$
$$R3/R2 = 0.744$$

If $R2$ is set to 2.7K, then $R3$ will be 2.0088K, which can be trimmed by using a 2K fixed resistor and a potentiometer in series. Note that there is only 8.8 ohms, or 0.4% error if the 2K precision resistor is used without the potentiometer. In fact, a 0.05% tolerance precision resistor has a range of possible values that is greater than the 8.8 ohm difference.

Thus far we have been assuming a current output for the DAC-08, but voltage outputs can also be accommodated if I take advantage of Georg Simon Ohm's magnificent discovery: the simplest technique for generating a voltage output is to connect a resistor between the I_{OUT} output and ground (Fig. 11-5A), and another, identical resistor between $\overline{I_{OUT}}$ and ground. The output voltages that will be generated if a 2-milliampere reference is used are as follows:

Condition	Binary Input	I_o*	E_o*	I_o*	E_o*
full-scale	11111111	1.992	−9.960	0.000	0.000
half-scale	10000000	1.000	−5.000	0.992	−4.960
zero scale	00000000	0.000	0.000	1.9924	−9.960

*—All currents in milliamperes, voltages in volts.

The circuit in Fig. 11-5A suffers from a high output impedance, which could prove troublesome in many applications. A perfect voltage source, after all, has a zero-ohms output impedance, and a good

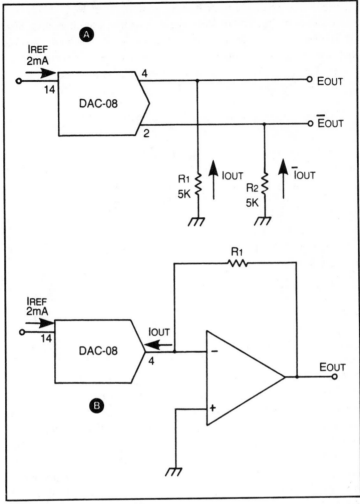

Fig. 11-5. Voltage output circuits (A) Simple type. (B) Low-impedance type.

voltage source will have an output impedance that is very low, less than 100 ohms, preferably.

A low output impedance is obtained by using an operational amplifier current-to-voltage converter following the DAC-08. For negative operation use a noninverting follower circuit and for positive output use the inverting follower.

The noninverting case is not shown, but is simple enough. Just connect a noninverting follower (unity-gain version unless you want the ability to trim the full-scale output without upsetting the reference current network) to either E_{OUT} or $\overline{E_{OUT}}$ in Fig. 11-5A.

Positive output voltages are attained using a circuit such as Fig. 11-5B. Here we use an operational amplifier in the inverting follower mode as a current-to-voltage converter. Since the output of the DAC-08 is a current, no input resistance is necessary for the operational amplifier. Output voltage E_{OUT} is given by

$$E_{OUT} = I_{OUT} \times R1 \qquad (11.14)$$

where E_{OUT} is the output voltage in volts, I_{OUT} is the DAC-08 output current in amperes, and $R1$ is given in ohms. (*Note:* Most DAC-08 literature gives the values of I_{OUT} in milliamperes. In that case, $R1$ would be given in kilohms).

Example 11-8
Find the output voltage E_{OUT} if $R1$ is 2200 ohms, and I_{OUT} is 1.5 mA.

Solution:

$$
\begin{aligned}
E_{OUT} &= I_{OUT} \times R1 \qquad \text{(from 11.14)} \\
&= (0.0015)\,(2200) \\
&= 3.3 \text{ volts}
\end{aligned}
$$

Note that $R1$ in Fig. 11-5B can be used as a full-scale trimmer, if desired. Simply make $R1$ a potentiometer, or better yet, a series combination of a potentiometer and a low-temperature-coefficient fixed resistor. For 10-volt full-scale operation with a reference current of 2 milliamperes, $R1$ should be 5K.

All of our examples up to this point have been unipolar, the output voltage could only be negative or positive. Bipolar operation of the DAC-08 allows both positive and negative output voltages. Of course, this is not exactly free; a few trade-offs are involved. Either the maximum output voltage, or the resolution suffers in bipolar (as opposed to the levels obtained in unipolar) operation.

The *span* is the difference between the maximum and minimum output voltages. In a unipolar voltage output circuit the span is the maximum output voltage (the minimum being zero), and the resolution is the maximum output voltage divided by 2^n, which is 256 for eight-bit circuits. In our eight-bit example with a nominal full-scale voltage of 10 volts, the span is 10 volts, and the resolution (volt/step) is 10.00/256, or 39 millivolts (nominally 40) per step.

In bipolar operation, I lose either the maximum output voltage or the resolution in most cases. If, for example, I want to keep the 40 millivolt/step revolution, then I must limit the maximum output voltages to ±5 volts, rather than ±10 volts. If, on the other hand, I wish to maintain the maximum output voltage (i.e., ±10 volts), I will have a span of (10) − (−10), or +20 volts. In that case the resolution deteriorates to 20/256, or about 80 millivolts per step. The decision that must be made, therefore, is whether the important property of the converter is a higher maximum output voltage or tighter resolution.

The simplest bipolar output voltage circuit is shown in Fig. 11-6A. Resistors $R1$ and $R2$ are connected back to the positive reference voltage. The output levels for this circuit are as follows:

Condition	Binary Input	E_{OUT}	$\overline{E}_{OUT}$
Full-scale (+)	11111111	− 9.920	+ 10.000
Zero-scale	10000000	0.000	+ 0.080
Full-scale (−)	00000000	+ 10.000	− 9.920

Note that the output voltage (i.e., E_{OUT}) is *asymmetrical* about zero because the full-scale positive and negative voltages are *not equal*. This is the result of having an even number of states (i.e., 256) evenly distributed about zero, there is no unique state that can represent true zero because zero is neither positive nor negative. The circuit in Fig. 11-6B shows a low output impedance circuit that will create a *symmetrical* output voltage function, which is given as follows:

Condition	Binary Input	E_{OUT}
Full-scale (+)	11111111	+ 9.920
(+) zero	10000000	+ 0.040
(−) zero	01111111	− 0.040
Full-scale (−)	00000000	− 9.920

In the unipolar and asymmetrical bipolar circuits there is no ambiguity over zero. In the case of the unipolar output, zero is 0 volts output, while in the asymmetrical bipolar zero is assigned a voltage of +80 millivolts. But in symmetrical bipolar operation, the nearest I can get to 0 volts output is ±1/2 LSB voltage (i.e., the resolution voltage, in this case 40 millivolts). In the circuit shown, the resolution voltage is that voltage that will exist when the input

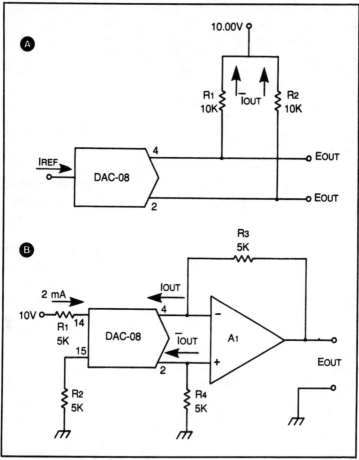

Fig. 11-6. Bipolar voltage output circuits. (A) Simple type. (B) Low-impedance type.

is LSB (00000001) is 80 millivolts), so 1/2 LSB will be 1/2(80), or 40 millivolts. The zero point will be ±40 millivolts. I must, in any given application, define true zero as either +40 millivolts or −40 millivolts, a concept often referred to as "± zero". I can obtain a true zero output in which the code 10000000 produces 0.00 volts output by offsetting the reference current, provided that my range requirements are such that I can live with shifting the codes up or down, and the loss of the maximum output voltage at one limit.

The DAC-08 can be used with both ac and pulsed reference supplies, in addition to the dc supplies discussed thus far. This feature allows the DAC-08 to be used in a wide variety of applications not open to nonmultiplying D/A devices. Figure 11-7 shows the proper circuit configurations appropriate for this type of operation.

Operation using ac reference is shown in Fig. 11-7A and 11-7B. In the example of Fig. 11-7A, the reference dc and ac signals are combined at pin 14 of the DAC-08. A constraint imposed on this type of operation is that the dc reference current must be greater than or equal to the peak negative swing of the ac reference sine wave.

An alternate circuit is shown in Fig. 11-7B, and it provides a high impedance input for the ac reference voltage (the reference voltage inputs of the DAC-08 are operational amplifier inputs). In this circuit, both of the DAC-08 reference pins must be connected to their own reference resistor. A constraint imposed on this circuit is that the positive dc reference voltage must be greater than the peak positive swing of the ac reference signal.

The reference amplifier inside of the DAC-08 must be frequency compensated for ac operation. A capacitor from pin 16 of the DAC-08 to the $V(-)$ power supply terminal must be provided. The material to follow lists the minimum values for the capacitor for several different values of reference resistance. For values not listed use an approximately scaled capacitance relative to the resistances shown.

R_{REF}	C
1K	15 pF
2.5K	37 pF
5K	75 pF
10K	150 pF

In pulsed operation (Fig. 11-7C), it is wise to use low values of reference resistor so that low values of compensation capacitance may be used. High capacitance values deteriorate the bandwidth, thereby decreasing the slew rate; a condition not conducive to the proper operation of a pulse circuit.

At a reference resistance of 1K (C = 15pF) the slew rate of the circuit is approximately 16 mA/μsec. The pulse shown in Fig. 11-7C can have a rise time on the order of 500 nanoseconds.

The DAC-08 may be *strobed* on and off under program or logic control (i.e., by one bit of a microcomputer output port) by using a circuit such as Fig. 11-8 in conjunction with the low impedance output circuit of Fig. 11-5B. In this circuit, the 7404 TTL inverter is used to vary the threshold voltage applied to the DAC-08. When

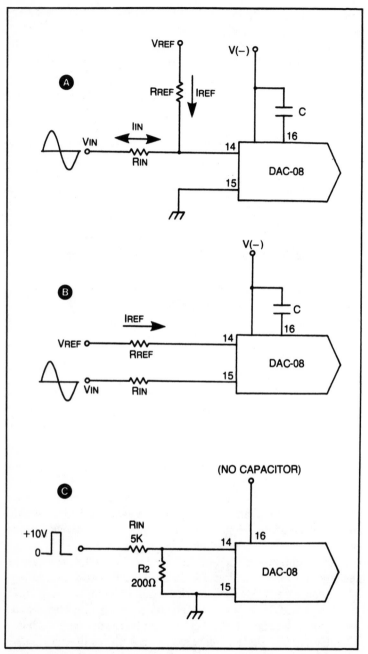

Fig. 11-7. Reference circuits. (A) Simple ac type. (B) High input impedance version. (C) Pulsed reference circuit.

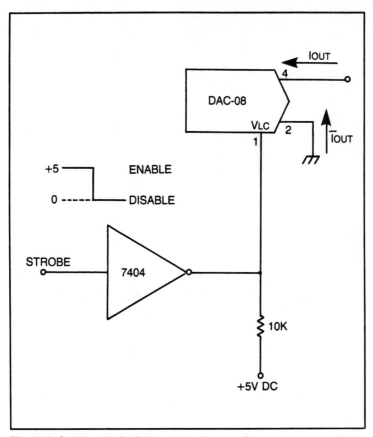

Fig. 11-8. Strobing the DAC-08 under logic control.

the input command applied to the 7404 is low, then the threshold voltage (V_{TH}) of the DAC-08 is raised to $5 + 1.4$, or 6.4 volts. Since the digital input terminals are connected to TTL sources, none will ever exceed the 6.4-volt trip threshold, no matter which state they are in. The DAC-08 thinks it sees all zeros, so all current flow is in I_{OUT} to ground, making $\overline{I_{OUT}}$ zero, hence $E_{OUT} = 0$.

When a high is applied to the input of the 7404, the V_{LC} terminal of the DAC-08 becomes grounded, which is the normal condition for operation with TTL inputs. Keep in mind, however, that the 7404 output terminal might not sink enough current to drop the voltage at pin 1 of the DAC-08 all the way to zero, so the threshold voltage might be a little bit higher than the nominal 1.4 volts normally observed. The minor constraint requires that I be sure of my TTL levels, and that they not be allowed to become sloppy.

THE 1408 D/A CONVERTER

The 1408-type D/A converter is available from Motorola Semi-conductor Products as the MC1408, while PMI and Advanced Micro Devices offer it under the generic 1408-type number. The 1408 is another low-cost current output D/A converter in integrated circuit form. The block diagram for the 1408 is shown in Fig. 11-9A, while the pin layout for the 16-pin DIP is shown in Fig. 11-9B.

The 1408 is provided with both positive and negative reference voltage terminals. The DAC-08, you will recall, used a similar arrangement, except that current inputs are specified.

Both positive and negative reference sources can be accommodated, but the configuration must be changed so that current always flows in to pin 14. A positive reference voltage requires that pin 15 be directed to ground through a resistor, and that the $V_{REF}(+)$ terminal be connected to the reference voltage through a resistor.

The same rules regarding frequency compensation apply to both the 1408 and the DAC-08, so will not be reiterated here. In fact, the similarity between these two devices is so marked that one is lead to the conclusion that the DAC-08 is essentially an improved second generation 1408.

THE DAC-05

The eight-bit D/A converter has become a low-cost favorite because of two factors: many applications are suitable for eight-bit representation, and most common microprocessor chips are in the eight-bit format, so will be easily interfaced with eight-bit D/A converters.

There are situations, however, where eight bits are not sufficient; although most such cases are where the D/A converter is used in the feedback loop of certain types of analog-to-digital converters (see Chapter 12). Consider an eight-bit D/A converter producing output signals in the 0- to 10-volt range. Each step of the digital input word (i.e., the LSB value) produces a 40-millivolt change in the output voltage:

$$
\begin{aligned}
E_{LSB} &= E_{FS}/2^n \\
&= 10/256 \\
&= 40 \text{ mV}
\end{aligned}
\qquad (11.15)
$$

But if a ten-bit D/A converter is used instead:

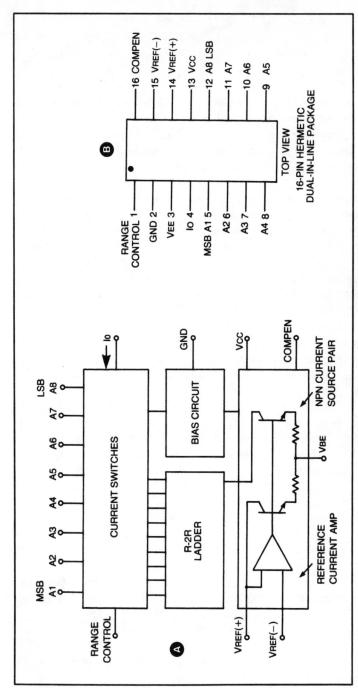

Fig. 11-9. Type 1408 D/A converter. (A) Circuit. (B) Layout. (Courtesy of Precision Monolithics, Inc.)

237

$$E_{\text{LSB}} = 10/2^{10}$$
$$= 10/1024$$
$$= 10 \text{ mV}$$

By the same type òf calculation, I find that a twelve-bit D/A converter will produce LSB values of 2 millivolts, while a sixteen-bit D/A converter will produce a LSB value of only 0.15 millivolts. Clearly, then, the resolution improves markedly as the number of bits increases. Unfortunately, price increases also, with sixteen-bit D/A converters often costing much more than twice as much as the low-cost eight-bit version. Part of this is attributable to the increased demands placed on the accuracy, stability, and other properties of the internal reference components and the electronic switches, as well as the precision of the $R\text{-}2R$ resistor ladder.

While the sixteen-bit D/A converter tends to be very expensive, several companies offer low-cost integrated circuit D/A converters in 10- and 12-bit configurations. An example is the Precision Monolithics DAC-05, shown in Fig. 11-10.

The DAC-05 can be used in either multiplying or nonmultiplying modes because the internal reference source must be tied to the input of the reference amplifier through an external connection. The reference source comes out on pin 17, while the reference amplifier input is on pin 15.

While the DAC-05 has its own internal reference supply, it may require external trimming for full-scale operation. This is provided by a potentiometer, connected as shown in Fig. 11-10C.

Two grounds are provided on the DAC-05, one each for analog and digital sections of the device. In ordinary practice, these two grounds are tied together as close as possible to the DAC-05 package through a heavy ground bus. This is necessary because the heavy dynamic currents in the digital circuits produce voltage drops sufficient to affect the analog circuits. The effects of the digital ground currents cannot be trimmed out because of their dynamic nature.

The DAC-05 is a voltage output device, so contains its own internal operational amplifiers to convert the current from the ladder network into a voltage level. Either 5-volt of 10-volt full-scale operation can be specified (prior to purchase) by adding the appropriate suffix to the type number. The suffix *X1* indicated ± 10-volt operation, while *X2* denotes ± 5-volt operation. Specific

238

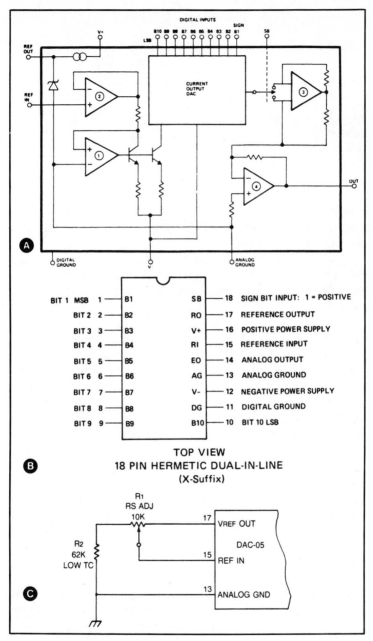

Fig. 11-10. PMI DAC-05. (A) Block diagram. (B) Pin layout. (Courtesy of Precision Monolithics, Inc.) (C) V_{REF} trim.

239

information for various grades of DAC-05 is as follows:

Type Number	Monotonicity (bits)	Temp Range (°C)
DAC-05AX1 (or X2)	10	– 55/ + 125
DAC-05BX1 (or X2)	9	– 55/ + 125
DAC-05CX1 (or X2)	8	– 55/ + 125
DAC-05EX1 (or X2)	10	0/ + 70
DAC-05FX1 (or X2)	9	0/ + 70
DAC-05GX1 (or X2)	8	0/ + 70

Monotonicity is definable as an increase in output for every increase in the input word. Different grades of the DAC-05 come with different guaranteed monotonicity specifications, so be aware of which is being ordered.

If a DAC-05 is on hand, but fewer than ten bits are required, then ground the unwanted input terminals, effectively setting them to zero. An ungrounded input will be noisy and can be taken as a logic 1.

The DAC-05 is a bipolar device and has a *sign* bit (making it effectively an eleven-bit device) that allows the circuit to distinguish positive from negative, which effectively eliminates the resolution span dilemma discussed earlier in this chapter. The information that follows shows the coding for various output levels. Notice that plus and minus full-scale entries have the same code, with only the sign bit being different for positive and negative outputs. Similarly, plus and minus half-scale codes are identical, as are ± zero.

Condition	Sign Bit	Binary Code
+ Full-scale	1	1111111111
+ Half-scale	1	1000000000
+ Zero	1	0000000000
– Zero	0	0000000000
– Half-scale	0	1000000000
– Full-scale	0	1111111111

The sign bit will be *one* for positive and *zero* for negative operation. Unipolar operation results when the sign bit is permanently tied to either + 5 volts dc (positive outputs) or ground (negative outputs).

A question that might fairly be asked is "how do you use a ten-bit or more D/A converter on an eight-bit microcomputer"? The answer lies in using more than one output port. Assume that I have an eight-bit microprocessor, which can address up to 256 different output ports. In the Zilog Z-80 chip, for example, an instruction for output might be *OUT(n),A*. This is a two-byte instruction that will place operand *n* on the lower byte of the address bus (A0 through A7) to select any of 256 possible output devices, then the accumulator contents are placed on the data bus. If the ten-bit D/A converter input is tied to a pair of output ports, then the ten bits could be allocated as eight bits from one port, and the remaining two bits from another port. I could then write a subroutine program to perform the data transfer. Say, for example, that I have a ten-bit output D/A converter latched as in Fig. 11-11. The lower eight bits are connected to output port 4, and the upper two bits are connected to output port 5. My program would have to load the lower eight bits into the accumulator, output them to port 4, load the upper two bits into the accumulator (setting them into the correct slots), then output them to port 5. In some cases an additional latch at the D/A converter input may be desired, so that the

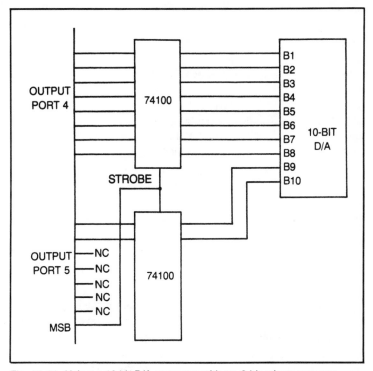

Fig. 11-11. Using a 10-bit D/A converter with an 8-bit microprocessor.

D/A converter will see only the entire ten bits at once, rather than climbing to the final value in two steps.

THE DAC-06

The DAC-06, another PMI product, is shown in Fig. 11-12. It is a straight ten-bit D/A converter (without a sign bit), but responds to two's complement input coding. The complements, or one's complement as it is sometimes called, of a binary number is formed by inverting all of its bits. All *ones* become *zeros*, and all *zeros* become *ones*. The two's complement of a binary number is formed by taking the one's complement, then adding one to the LSB. This was discussed in Chapter 9. The input coding for various output levels is as follows:

Two's Complement

+ FS	0111111111	+ 5
+ LSB	0000000001	+ 0.01
0	0000000000	0
– LSB	1111111111	– 0.01
– FS	1000000001	– 5

One's Complement

+ FS	0111111111	+ 5
+ 0	0000000000	+ 0.01
– 0	1111111111	– 0.01
– FS	1000000000	– 5

Straight Offset Binary

+ FS	1111111111	+ 5
+ 0	1000000000	+ 0.005
– 0	0111111111	– 0.005
– FS	0000000000	– 5

Figure 11-13 shows the adjustment circuit for the DAC-06. The internal reference is used to provide the reference level to both full-scale adjust and bipolar offset adjust input terminals. Of course, an external reference can be provided by disconnecting the top of the potentiometer network from pin 17 of the DAC-06. An exter-

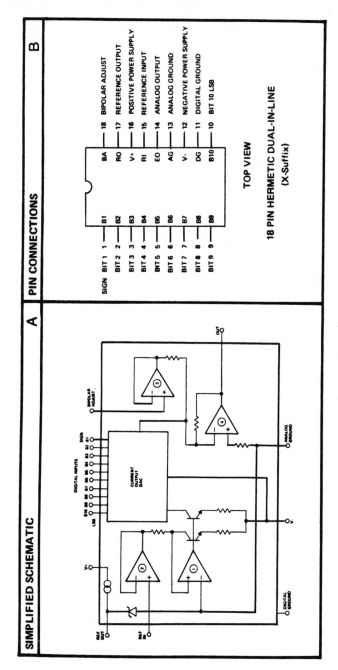

Fig. 11-12. PMI DAC-06. (Courtesy of Precision Monolithics, Inc.)

243

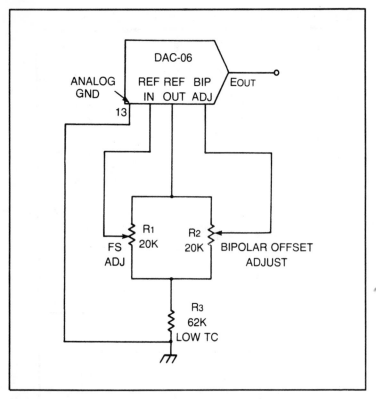

Fig. 11-13. DAC-06 full-scale and bipolar offset adjust circuit.

nal reference is often used to provide superior full-scale temperature coefficient, or to allow several D/A converters in a system to track on the same reference. In the latter case, the internal reference of one D/A converter could be used as a master reference for the system, and all other D/A converters would be slaved to it. Reference output current, however, should not exceed 100 microamperes.

As in other D/A converters containing an analog amplifier, the DAC-06 provides two separate ground pins, one each for analog and digital circuits. Again, the requirement is that these be connected together as close to the package of the D/A converter as possible, and that the ground bus be large.

In addition to grounding, the lowest noise operation requires that the reference input and bipolar adjust input be bypassed to the analog ground through 0.01 μF disc ceramic capacitors. The bypass capacitors will improve settling time.

Adjustment for Two's Complement Operation

1. Turn all bits off ($-V_{FS}$ – LSB) – 1000000000.
2. Adjust bipolar offset adjust $R2$ for an output of $-V_{FS}$ – LSB. For $\pm$ 5-volt operation this is – 5.0098 volts.
3. Turn all bits on ($+V_{FS}$) – 0111111111.
4. Adjust FS; adjust $R1$ for full-scale output (i.e., +5 volts).
5. Repeat the procedure until no further improvement is noted.

Adjustment for One's Complement Code

1. Turn off all bits ($-V_{FS}$) – 1000000000.
2. Adjust bipolar offset adjust $R2$ for – V_{FS} output. For $\pm$ 5-volt operation this is – 5.000 volts.
3. Turn all bits on ($+V_{FS}$ – 0111111111).
4. Adjust FS; adjust $R1$ for the desired full-scale output (i.e., + 5.000 volts). Note that + zero is + 5 millivolts and – zero is – 5 millivolts.

Adjustment for Straight Offset Binary Code

Straight offset binary operation is identical to one's complement, except that the MSB occurs in true binary rather than in complement form (i.e., inverted). The DAC-06 can be operated in straight offset binary if the MSB is inverted. Connect an inverter between the input source and the MSB terminal on the DAC-06, then perform the one's complement calibration procedure.

THE DAC-20

The DAC-20 by PMI is shown in Fig. 11-14. It is billed as a two-digit BCD high-speed multiplying D/A converter. This device uses the standard R-$2R$ resistance ladder, but lacks an output amplifier, so it is a current output D/A converter.

The input coding is a little different on this device in that it accepts two four-bit BCD digits rather than ordinary eight-bit binary coding. In other respects the DAC-20 is very similar to the DAC-08. Nonlinearity specifications as tight as $\pm$ 1/4LSB are possible, and will hold true over the entire $\pm$ 4.5- to $\pm$ 18-volt dc power supply range.

BCD coding is often viewed as being too clumsy for use in digital circuits unless a decimal readout device such as a seven-segment LED display or Nixie tube is desired. But when BCD is the avail-

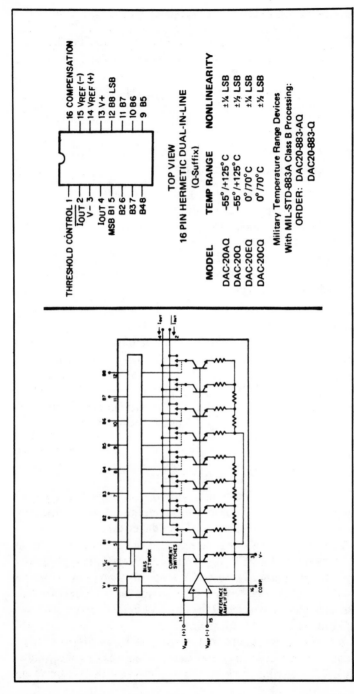

Fig. 11-14. PMI DAC-20 (BCD input coding). (Courtesy of Precision Monolithics, Inc.)

246

able code, or where the use of a prepackaged digital panel meter would prove the best means for digitization, then a BCD D/A converter might be indicated. An example might be to produce a control signal, or possibly a signal to a strip-chart recorder, from a digital panel meter or frequency counter data output connector.

The same hookup and reference current design rules apply also the DAC-20 as applied to the DAC-08. The output levels and input codes for the positive mode of operation are as follows:

| Decimal Input | Binary Input | | I_{OUT}(mA) | E_{OUT}(volts) |
	MSD	LSD		
0	0000	0000	0	0
10	0001	0000	0.20	+ 1.0
20	0010	0000	0.40	+ 2.0
30	0011	0000	0.60	+ 3.0
40	0100	0000	0.80	+ 4.0
80	1000	0000	1.60	+ 8.0
99	1001	1001	1.98	+ 9.9

THE DAC-100

The PMI DAC-100 (see Fig. 11-15) is an eight-bit or ten-bit D/A converter (as ordered) with an internal reference source. It uses the standard R-$2R$ resistance ladder technique, but lacks the internal output amplifier. The DAC-100 is a current output device.

The input coding for the DAC-100 is complementary binary for unipolar operation and offset complementary binary for bipolar operation.

In previous cases any unused digital inputs were tied to ground (i.e., logic 0), but since complement coding is used in the DAC-100, unused inputs should be tied to a voltage source that is greater than + 2.2 volts dc.

Full-scale adjustment is provided by placing a 200-ohm potentiometer that is rheostat connected (i.e., one end strapped to the wiper terminal) between $V(-)$ and the full-scale adjust terminal (pin 15).

For operation in the bipolar mode it is necessary for a half-scale current to be summed with the D/A converter output current. This is done by connecting pin 1 through a 5000-ohm zero adjust potentiometer (also rheostat connected) to a precision + 6.4-volt dc

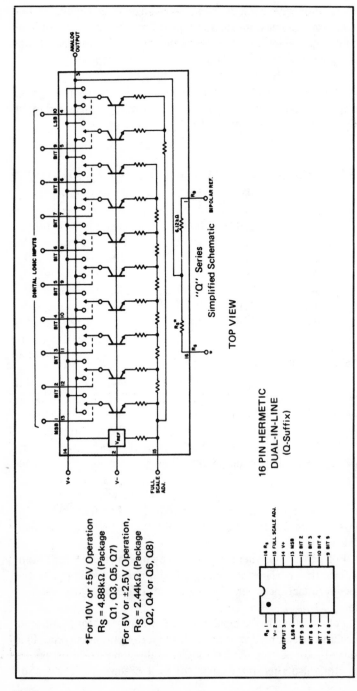

Fig. 11-15. PMI DAC-100. (Courtesy of Precision Monolithics, Inc.)

source. Pin 16 of the DAC-100 goes to an internal resistance (R_s) that can be used as the feedback resistor of an external operational amplifier inverting follower. Such a circuit would provide voltage output capability.

SOME PRACTICAL D/A CIRCUITS

In this section I will discuss some practical circuits involving digital-to-analog converters. These can be constructed as is, or modified by the user for a specific application which I cannot predict.

Figure 11-16 shows an example of a D/A converter circuit using the Ferranti ZN425E integrated circuit. An operational amplifier is used to convert the current output of the D/A converter to a voltage. A 741 or other low-cost operational amplifier can be substituted for the type shown if suitable pin changes are made to the schematic.

The particular circuit is shown in Fig. 11-16 has the logic select terminal (pin 2) grounded, so the internal binary counter is disabled. Pins 3 (counter reset) and 4 (clock) are not connected.

The internal voltage reference output (pin 16) is connected to the reference amplifier input (pin 15) in order to eliminate the need for an external reference source. The full-scale output is +3.84 volts.

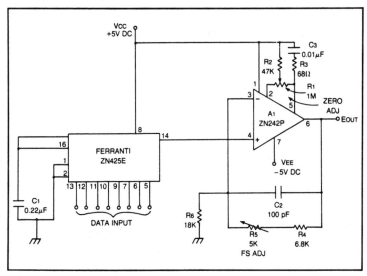

Fig. 11-16. Voltage output D/A converter circuit using the ZN-425E.

249

Adjustment Procedure

1. Ground all input bits (set them to 00000000).
2. Adjust zero adjust pot $R1$ for 0.000 volts output.
3. Set all input bits high (11111111).
4. Adjust potentiometer $R5$ for V_{FS} – LSB, which in this case, where LSB is approximately 15 millivolts is +3.825 volts.
5. Repeat the procedure several times until no further improvement is possible.

The converter will now read full scale voltage when all bits are high, and zero when all bits are low. Applying a binary word between 00000000 and 11111111 will result in an output voltage that is between the high and low limits.

A D/A converter circuit based on the DAC-08 is shown in Fig. 11-17. This circuit is designed along the lines given earlier in this chapter for TTL compatibility, so pin 1 (V_{LC}) is grounded.

Three power supplies must be provided. Two are VCC and VEE, while the third is a +5-volt dc supply for the 74100 TTL integrated circuit quad latch. In addition, a PMI REF-01 is used to provide the +10.00-volt dc reference potential, which is derived from the +15-volt VCC line.

V_{REF} adjustment is provided by potentiometer $R3$. This control is used to trim V_{REF} to exactly 10.00 volts. Potentiometer $R2$ provides full-scale adjust, per Fig. 11-3B given earlier in this chapter. Zero adjust is provided by summing an offset current with the DAC-08 output current I_{OUT} in the inverting input of an operational amplifier that is used as a current-to-voltage converter ($IC3$).

The DAC-08 input lines are connected to a 74100 TTL dual quad latch to provide a one-byte memory that will hold the input word until a new data word is supplied. As long as the strobe line on the 74100 is low (i.e., zero volts), the word stored previously is held on the output lines, so the DAC-08 sees a constant input word. When the strobe line is brought high, however, data appearing on the 74100 input lines are transferred to the output lines. This new condition will be held on the output lines if the strobe line is now made low.

The 74100 data latch can be disabled if desired by connecting the strobe line to a +5-volt dc point, keeping the strobe permanently high, or it can be deleted altogether. This might be the desirable way to go if the D/A converter circuit is used with a minicom-

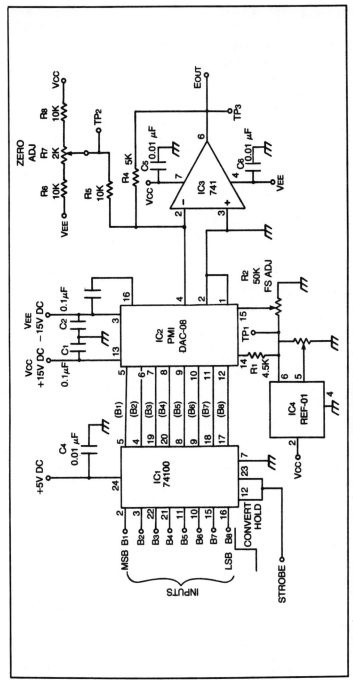

Fig. 11-17. Latched D/A converter circuit based on the DAC-08.

251

puter that provides its own output latching. If used directly with a microprocessor chip, on the other hand, the data bus will contain the eight-bit D/A converter input word for only a few hundred nanoseconds, so that latch will be required.

Some microprocessor chips have an OUT or OUT line that can be used for strobing the D/A converter. The popular 8080A, for example, has a write (WR) line that goes low when the data on the output data bus is stable and ready for use. This line can be inverted, then applied to the strobe terminal of the 74100. As long as the WR line is high, data on the data bus is not appropriate for output, but when the WR line drops low the 8080A is telling me that the data is to be outputted.

An alternative strobe technique might be to connect the strobe line to the output of an address decoder. A TTL eight-input NAND gate such as the 7430, for example, can be used to output a high condition when the output port address selected for the D/A converter is seen on the lower byte of the address bus.

Adjustment

1. Turn on the circuit and allow to warm up for 5 minutes.
2. Adjust V_{REF} adjustment $R3$ for + 10.00 volts at test point $TP1$.
3. Set all inputs low (00000000).
4. Adjust zero adjust $R7$ for 0 volts at test point $TP2$.
5. Set all bits high (11111111).
6. Adjust FS adjustment $R2$ for + 9.96 volts at test point $TP3$.

The maximum output is nominally + 10.00 volts, but an eight-bit D/A converter can only produce 2^8 or 256 different states, one of which is 0 volts for a code of 00000000. The maximum actual output voltage, then, is (by Eq. 11.1):

$$E_{OUT} = 10.00 \, (255/256)$$
$$= 9.96 \text{ volts}$$

If the full-scale output voltage is required (only occasionally in this case), I can redefine zero as + 0.040 volts, then adjust $R7$ so that an input word of 00000000 produces an output of + 0.040 volts and a word of 11111111 produces an output of + 10.00 volts.

Chapter 12

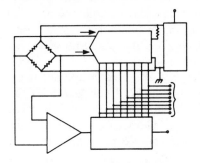

A/D Converters:
Some Real Products

THE ANALOG-TO-DIGITAL CONVERTER PRODUCES A BINARY OR BCD output that represents an analog current or voltage applied to its input. The availability of low-cost, digital-to-analog converter integrated circuits allows the implementation of low-cost A/D converter circuits using the D/A converter chip in the negative-feedback loop. A/D converter designs that allow this include the binary counter, or ramp, type and the successive-approximation register (SAR). The latter types are made even easier than was previously the case, when individual digital IC shift registers and latches had to be provided, because at least two companies (Motorola and Advanced Micro Devices) market successive-approximation registers in integrated circuit form.

Examples of D/A-based binary counter and successive-approximation A/D converter circuits will be given in this chapter. I will also give an example of a simple BCD-encoded 3 1/2-digit integration A/D converter intended for use in digital voltmeter circuits.

FERRANTI ZN425E-BASED DESIGN *BINARY COUNTER*

Figure 12-1 shows an analog-to-digital converter that takes advantage of some of the interesting features of the Ferranti Semiconductors ZN425E integrated circuit D/A converter. Recall from Chapter 11 that this chip contains the ordinary voltage output *R-2R* resistor ladder, *plus* an eight-bit binary counter that is connected to the digital inputs of the resistor ladder when the logic select terminal (pin 2) is high. Under this condition clock pulses from the external oscillator will increment the counter once for each negative-going-from-positive transition applied to pin 4. The counter state is not only applied to the inputs of the R-$2R$ resistance ladder, but also appears on the ZN425E input terminals (pins 5,6,7,9,10,11,12, and 13).

The circuit in Fig. 12-1 uses, in addition to the D/A converter, an operational amplifier ($IC2$) and a 7400 TTL quad two-input NAND gate ($IC3$). Pin 2 of the ZN425E is wired to $+5$ volts dc through a 1000-ohm pull-up resistor, thereby setting pin 2 to a permanent logic 1 condition. Keeping pin 2 high turns on the internal binary counter.

Operational amplifier $IC2$ is a Ferranti type ZN424P, although it could just as easily be almost any good quality operational amplifier, or since it is used as a voltage comparator, an integrated circuit comparator such as the 311, 710, AMD AM686, or Precision Monolithics CMP-01 or CMP-02.

A voltage comparator examines two input voltages and issues an output that tells the outside world whether they are equal or unequal. In the case of $IC2$, a sample of the analog input voltage is applied to the noninverting input of the operational amplifier through a resistor voltage divider ($R2/R6$). This voltage is nominally:

$$E2 = E1 \times \frac{R6}{R2 + R6} \qquad (12.1)$$

and in the special case covering Fig. 12-1, Eq. 12-1 breaks into:

$$E2 = E1 \times \frac{15K}{15K + 15K} \qquad (12.2)$$

$$E2 = 1/2E1 \qquad (12.3)$$

where $E1$ is the analog input voltage and $E2$ is the voltage applied to the noninverting input of operational amplifier $IC2$.

The voltage comparator is made using an integrated circuit operational amplifier with no negative feedback. The gain of the comparator, then, is the *open-loop* gain of the operational amplifier. Operational amplifier manufacturers list typical values for open-loop gain between 20,000 and over 1,000,000 depending upon qual-

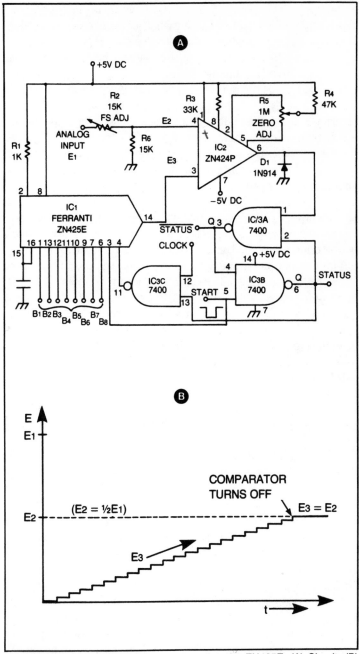

Fig. 12-1. Binary counter A/D converter using the ZN425E. (A) Circuit. (B) Waveform.

255

ity and price level. Even the sloppiest operational amplifiers provide sufficient open-loop gain that, when it is used as a comparator, it will saturate and produce an output close to either $V(+)$ or $V(-)$ if more than a few millivolts of potential exists between the inverting and noninverting inputs.

Voltage $E3$ is the output of the ZN425E, which is a ramp rising from zero by a voltage increment equal to LSB every time the binary counter increments by one clock pulse. When voltage $E3$ is less than voltage $E2$, the comparator ($IC2$) sees it as a positive differential potential ($E2 - E3$) applied to the noninverting input, so the operational amplifier output will be a high positive voltage approximating $V(+)$.

The $V(+)$ voltage in this circuit is $+5$ volts dc, so the voltage at the output of the comparator will be just a little less than $+5$ volts when $E2$ and $E3$ are not equal. When $E2$ and $E3$ are equal, on the other hand, the comparator sees a differential input potential of zero, so its output will also be zero. The condition where $E3$ is greater than $E2$ would cause a high negative output [i.e., $V(-)$, or close to -5 volts dc] from $IC2$. But since that condition is not required for the conversion process, and might easily destroy NAND gate $IC3$ (TTL) if it did inadvertently occur, diode $D1$ is used to clamp the output of $IC2$ for negative voltages. A negative output voltage would be limited to one diode drop, or 0.6 to 0.7 volts negative. A positive output will reverse bias diode $D1$, so is not affected.

Three of the four gates in the 7400 ($IC3$) are used in this circuit. $IC3A$ and $IC3B$ form an RS flip-flop in which section B forms the set terminal, and section A forms the reset terminal. The Q output of the RS flip-flop (pin 6 of the 7400 are here configured) drives one input of the third gate.

The 7400 is a NAND gate, so will set (i.e., force Q high and $\overline{Q}$ low) when the set input is brought low. Similarly, it will reset when the reset terminal is brought low. These rules follow from the normal rules for TTL RS flip-flops using NAND gates.

Section C of the 7400 will pass clock pulses to the input of the ZN425E internal binary counter when pin 12 (the RS flip-flop Q output) is high.

Conversion begins when a negative-going trigger pulse is applied to the start terminal, which is connected to the set input of the RS flip-flop and the reset pin on the ZN425E binary counter (chip pin 3). This pulse sets the flip-flop and resets the binary counter to 00000000_2. Since Q is now high, gate $IC3C$ will begin to pass clock pulses to the counter in the ZN425E, and this causes the output voltage to begin ramping upwards.

If the analog input voltage sample ($E2$) is greater than the ramp output voltage $E3$, then the voltage comparator output remains high. When the voltage from the ramp reaches a level equal to the analog input voltage sample (i.e., $E2$), as in Fig. 12-1B, then the comparator output goes low, resetting the RS flip-flop. The Q output goes low and the $\overline{Q}$ goes high, shutting off the stream of clock pulses into the binary counter. The binary counter output appears on terminals $B1$ through $B8$, and is a bit pattern representing the analog input voltage. The output word will be 00000000_2 at 0 volts, and 11111111_2 at full-scale input.

The maximum clock frequency that can be accommodated by this circuit is 100 kHz, but that can be extended to approximately 400 kHz if a faster comparator is used at $IC2$. The conversion time is a function of the clock frequency selected and the value of the analog input voltage. Since the D/A converter "ramps up" toward the analog input voltage as the binary counter increments, it stands to reason that the conversion time is shorter for low-value analog input voltages than it is for full-scale voltages. The general expression for conversion time is

$$T_C = \frac{2^n}{f_{CLK}} \qquad (12.4)$$

where T_C is the conversion time in seconds, f_{CLK} is the clock frequency in hertz, and n is the binary counter output word representing the analog input voltage.

Example 12-1
Find the full-scale conversion time (i.e., $n = 8$) if the clock frequency in the circuit of Fig. 12-1 is 100 kHz.

Solution:
$$
\begin{aligned}
T_C &= 2^n/f_{CLK} \\
&= 2^8/10^5 \\
&= 256/10^5 \\
&= 0.00256 \text{ seconds} \\
&= 2.56 \text{ msec}
\end{aligned}
$$

Calibration Procedure
1. Apply a continuous train of pulses to the clock terminal.

2. Apply a voltage equal to the full-scale voltage minus 1.5LSB. (in this case 4 volts in the full-scale voltage, so LSB is 4/256, or 15.63 millivolts; 1.5LSB, then, is 23.45 millivolts. The voltage to apply if you duplicate the circuit shown is 4.00 − 0.02345, or 3.9765 volts).
3. Adjust potentiometer $R2$ (i.e., FS adjust) until all bits except the LSB are high (i.e., 11111110), and the LSB bobbles back and forth between high and low.
4. Apply a voltage equal to 1/2LSB (i.e., 1/2 × 15.63, or 7.82 millivolts) to the analog input.
5. Adjust potentiometer $R5$ (i.e., zero adjust) for all bits low except the LSB (i.e., 00000001), and the LSB bobbling back and forth between high and low.
6. Repeat steps 2 through 5 until no further improvement can be obtained. These adjustments on all A/D converter circuits are usually somewhat interactive.

The full-scale voltage (i.e., 4 volts) is applied to a voltage divider that delivers only a sample ($E2$) to the comparator input. I may conclude, then, that the maximum output voltage of the D/A converter ramp circuit is around 2 volts (in fact, it is closer to 2.5 volts). The full-scale range of the A/D converter can be extended by changing the values of the resistors forming the input voltage divider to allow 2 volts to appear at the comparator input when the desired full-scale analog voltage is applied to the analog input.

THE PMI AD-02 A/D CONVERTER (SAR)

The block diagram to the Precision Monolithics, Inc. AD-02 monolithic A/D converter is shown in Fig. 12-2. This chip is a twelve-bit A/D converter in a forty-pin package (DIP). It can be connected in six- or eight-bit configurations, and the eight-bit conversion time can be as low as 1 microsecond per bit, or 8 microseconds.

Almost unique with the AD-02 is the availability of multiple range analog input terminals. Pin 33, for example accepts input voltages up to ± 10 volts, pin 32 accepts ± 5 volts or $+ 10$ volts, and pin 34 accepts signals between ± 2.5 volts or $+ 5$ volts. The AD-02 can be adjusted to output in straight binary, offset binary, or two's complement. Both parallel and serial outputs are provided.

The AD-02 will operate at $V(+)$ and $V(-)$ supplies of ± 18 volts dc, and the V_{CC} supply can be up to $+ 7$ volts dc, although it is usually set to +5 volts for TTL compatibility. Each digital output can sink up to 4.8 milliamperes of current.

The conversion technique used is the successive-approximation method (see Chapter 10). The AD-02 contains twelve switch-controlled current sources, a ground-referenced comparator, and the latched-output shift registers required to implement a successive approximation operation.

Figure 12-3 shows the basic connections and bypassing required to place the AD-02 chip in service. All three power supply terminals are bypassed to ground. All three ground terminals (i.e., reference, digital, and signal) should be joined together on a wide ground bus as close as possible to the AD-02 package. The power supply return should also be connected to this point, as should be the analog input.

This A/D converter chip contains its own built-in reference power supply and reference amplifier, both of which are accessible to the designer through package pins. Pin 37 is the dc reference supply output, while pin 36 is the reference amplifier input. Full-scale adjustment is provided by potentiometer $R1$. One end of $R1$ is grounded and the other goes to the reference supply output (pin 37). The potentiometer wiper is used as a voltage divider and feeds the reference amplifier input. This point is bypassed to ground through a 0.1 μF capacitor.

The start and end-of-conversion (called end-of-encode or EOE in PMI literature) pins drop low when it is active, so are labeled START and EOE. For synchronous operation a negative-going pulse applied to pin 27 (i.e., start) will begin the conversion process at the next clock pulse. The EOE terminal will go high at that time, and will remain high until the conversion is completed. When the encoding process is finished, and the data is ready for use, the EOE terminal goes low, and this event can be used to strobe any external device that is to receive the data.

Continuous, or asynchronous conversions can be performed by strapping together the START and EOE terminals, so that the EOE pulse at the end of one cycle will become the start pulse for the next cycle.

One pitfall in this situation is that the output data is valid only during one clock period, which is very short at the maximum clock speed (i.e., 1 microsecond or less). An output latch such as the 8212 or the 74100 should be used to catch the data while it is valid. The START/EOE pulse will have to be inverted, then applied to the strobe pins of the latch chip.

PIN CONNECTIONS

N/C	1		40	N/C
N/C	2		39	N/C
V+	3		38	BIPOLAR OFFSET ADJ
BIT 12	4		37	REFERENCE OUT
BIT 11	5		36	REFERENCE IN
BIT 10	6		35	SIGNAL / REF GND.
BIT 9	7		34	SIGNAL IN (±2.5 V OR +5 V)
BIT 8	8		33	SIGNAL IN (±10 V)
BIT 7	9		32	SIGNAL IN (±5 V OR +10V)
BIT 6	10		31	N/C
BIT 5	11		30	V–
BIT 4	12		29	POWER GND.
BIT 3	13		28	POWER GND.
BIT 2	14		27	START IN
BIT 1	15		26	CLOCK IN
BIT 1	16		25	N/C
Vcc	17		24	COMPARATOR OUT
Vcc	18		23	N/C
SERIAL	19		22	N/C
EOE	20		21	N/C

NOTES:

Power ground (pins 28 and 29) is not connected internally to Signal/Reference Ground (pin 35). Best results will be obtained if these grounds are connected together at the AD-02 package, so that digital currents do not flow through the analog ground path.

N/C – No connection to internal circuit.

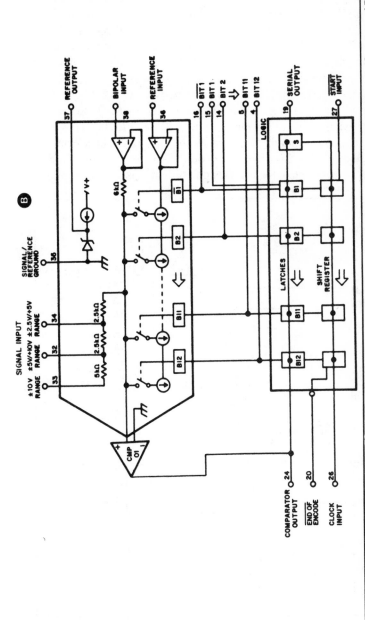

Fig. 12-2. Precision Monolithics, Inc. AD-02 successive-approximation A/D converter. (A) Block diagram. (B) Pin layout. (Courtesy of Precision Monolithics, Inc.)

With only two exceptions the data output format is standard for twelve-bit operation. Bits 12 (LSB) through 1 (MSB) occupy pins 4 through 15, respectively. These pins form a twelve-bit parallel output. Pin 16 produces an inverted bit 1. It will always be the complement of the MSB, going low whenever the MSB goes high. The other exception noted above is the serial data output which can be used to drive a data communications channel, or systems allowing serial data transfer in nonreturn to zero format.

The adjustment circuit shown as part of Fig. 12-3 is for full-scale adjustments only. A similar circuit (see Fig. 12-4) can be used for zero adjustment. The circuit in Fig. 12-4A is for bipolar operation, and is essentially the same as the full-scale adjustment circuit, except that the wiper of the potentiometer goes to pin 38 of the AD-02, which is for bipolar adjust input biases. This terminal

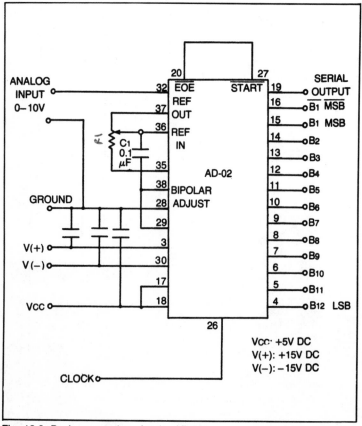

Fig. 12-3. Basic connections for the AD-02.

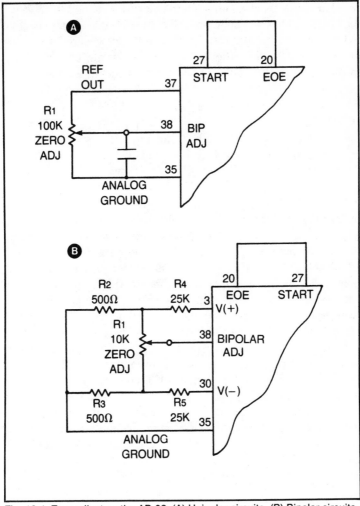

Fig. 12-4. Zero adjust on the AD-02. (A) Unipolar circuits. (B) Bipolar circuits.

had been grounded in Fig. 12-3. The adjustment procedure is as follows:

1. Apply a voltage equal to $V_{FS(-)} \times 1/2\text{LSB}$ to the analog input.
2. Adjust potentiometer $R1$ for all bits ($B1$ through $B11$) low and bit 12 to bobble back and forth between high and low.

The circuit in Fig. 12-4B is used for zero adjustment in the

263

unipolar mode. In this case, potentiometer $R1$ is connected between $V(+)$ and $V(-)$ power supplies, while the wiper remains connected to the bipolar adjust terminal. The adjustment procedure is the same as that for the bipolar mode in Fig. 12-4A. Adjustment for full-scale, in both bipolar and unipolar, continuous modes is as follows:

1. Apply a voltage equal to V_{FS} – 1.5LSB to the analog input.
2. Adjust potentiometer $R1$ (in Fig. 12-3) so that bits $B1$ through $B11$ are high (i.e., 111111111110) and bit $B12$ (LSB) bobbles between high and low.

Short-cycle operation of an analog-to-digital converter is the operation at less than the full complement of bits. For a twelve-bit analog-to-digital converter, then, short-cycle operation is used whenever the output is 6, 7, 8, 10 or any bit length less than 12. Figure 12-5A shows the simplified circuit used when only six-bit operation is desired. Bits $B1$ through $B6$ are the data bits. If the successive-approximation register were allowed to attempt a full twelve-bit conversion it would waste a lot of time. At any given clock speed it takes twice as much time to go through twelve bits as it does six bits. To overcome this problem a NAND gate and an inverter (i.e., NAND gate with both inputs tied together) are used. The 7400 NAND gate output will go high whenever either input goes low. If the $\overline{EOE}$ terminal of the AD-02 drops low, indicating that encoding is completed, then the output of the NAND gate will go high. The NAND gate output can be used as an EOC strobe, and is inverted so that it will drive the $\overline{START}$ terminal on the AD-02 chip. The output of the inverter stage will also serve as a $\overline{EOE}$ terminal, if desired. If the conversion process gets to a point where the seventh bit is activated (i.e., goes low), then the same EOC sequence takes place. The identical situation is used in Fig. 12-5B, except that eight bits are active, and bit $B9$ is used to drive the NAND gate.

Figure 12-6 shows a complete eight-bit, microcomputer-compatible A/D converter based on the AD-02 chip. Central to the circuit is the eight-bit, short-cycle configuration of Fig. 12-5B. An RC TTL astable multivibrator serves as the clock, although you might want to substitute a crystal oscillator or the computer's system clock. In the case shown, however, the clock frequency is given by:

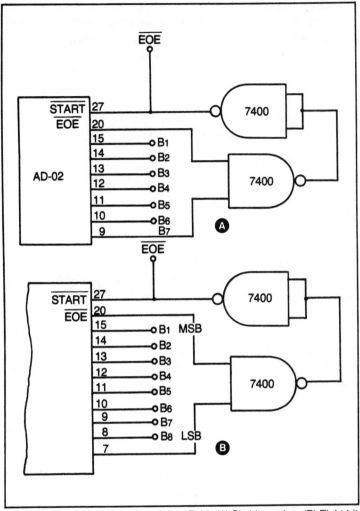

Fig. 12-5. Short-cycle operation of the AD-02. (A) Six-bit version. (B) Eight-bit version.

$$f_{\text{CLK}} = \frac{3.3 \times 10^5}{RC} \qquad (12.5)$$

where f_{CLK} is the frequency in hertz, R is the resistance in ohms, which will be between 150 ohms and 1500 ohms, and C is the capacitance in microfarads (μF). (*Note:* for C in picofarads use 3.3×10^{11} in the numerator of the equation instead of 3.3×10^5).

Example 12-2

Find the frequency of oscillation in an astable multivibrator such as shown in Fig. 12-6 if the resistor has a value of 220 ohms, and the capacitance is 0.0015 μF (i.e., 1500 pF).

Solution:

$$
\begin{aligned}
f_{CLK} &= 3.3 \times 10^5/RC \\
&= 3.3 \times 10^5/(220)\,(0.0015) \\
&= 10^6 \text{ hertz} = 1 \text{ MHz}
\end{aligned}
$$

$FF1$ and $FF2$ are part of a 7474 TTL dual D-type flip-flop. Recall the basic operating rules for the D-type flip-flop:

1. Data transfers from D to Q only when the clock terminal is high.
2. A low pulse placed on the preset input immediately causes Q to go high, and $\overline{Q}$ to go low, regardless of the conditions applied to D and clock.
3. A low pulse on the clear input causes Q to go low, and $\overline{Q}$ goes high, regardless of the conditions applied to D and clock.

Returning to Fig. 12-6, we are now ready to discuss the operation of the circuit. The data (i.e., D) and preset terminals of $FF1$ are tied high, so that rule 2 above will never apply, and Q will go high when the clock is high, unless the clear is low.

1. The clock terminal of $FF1$ is used as the start line. When the clock terminal goes high, transferring a high to the Q output, $\overline{Q}$ is forced low.
2. The low $\overline{Q}$ is seen by the AD-02 as a negative-going $\overline{\text{start}}$ pulse, so conversion begins. This low forces the preset terminal of $FF2$ low, so the Q terminal of $FF2$ goes high (rule 2 applies in $FF2$). This is used as a status or busy signal by outside-world devices. This terminal is high when the A/D converter is busy, and drops low when the data is valid.
3. When the conversion is completed, indicated by either an overflow (bit $B9$ active) or $\overline{\text{EOE}}$ going low, a low is placed on the clear input of $FF2$ by the inverter. By rule 3 this forces the Q low and $\overline{Q}$ high, which in turn applies a high on the clear input of $FF1$, disabling it.
4. Parallel data is available on the output lines when the status terminal drops low.

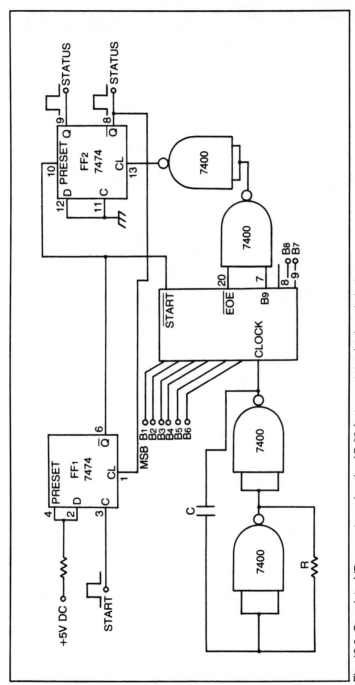

Fig. 12-6. Complete A/D converter using the AD-02 features status latch and clock.

The AD-02 monolithic analog-to-digital converter chip is competitive with most twelve-bit function module A/D converters in the same quality range. The AD-02 is available in several different grades, as follows:

Model	Temp Range (°C)	Linearity (%)	Temp Coef (ppm)
AD-02AW	– 55/ + 125	0.2	± 60
AD-02W	– 55/ + 125	0.2	± 120
AD-02-883AW	– 55/ + 125	0.2	± 60
AD-02-883W	– 55/ + 125	0.2	± 120
AD-02-EW	0/ + 70	0.2	± 60
AD-02-CW	0/ + 70	0.2	± 120

MC14559/MC1408-BASED DESIGNS (SA)

Motorola Semiconductor's MC1408 integrated circuit D/A converter (see Chapter 11) and MC14559 successive-approximation register can be paired to make a successive-approximation (SA) type analog-to-digital converter. The MC1408 was discussed at length in the previous chapter, so the details of that chip will not be repeated here. The MC14559 is a complete CMOS integrated circuit successive-approximation logic block that includes switching, shift registers, output latches, and the internal control logic that makes the whole thing work together. This chip allows me to implement an SA type A/D converter using a minimum of external components.

Figure 12-7 shows the circuit for a well-behaved A/D converter using the MC1408 and MC14559 combination. This circuit requires power supplies to provide ± 15 volts dc, and + 5 volts dc. It is also necessary to provide a well-regulated reference power supply, two of which were given in Chapter 11.

The clock can be a Motorola MC4024 as shown, or an astable multivibrator as used with the AD-02 in Fig. 12-6. The clock frequency must be less than 500 kHz (2-microsecond period), so if a 220-ohm resistor is used for R in Fig. 12-6, then the capacitance will be 0.003 μF (use the same equation).

Two operational amplifiers are used in this circuit. Operational amplifier $A1$ serves as an analog input amplifier, while $A2$ serves as a comparator. Both are semipremium 301A devices; be aware that a 741-class device should not be used, especially in the $A2$ slot, unless clock speeds are very low.

The reference current I_{REF} is set by the reference voltage and potentiometer $R1$. The reference current must be between 0.5 and

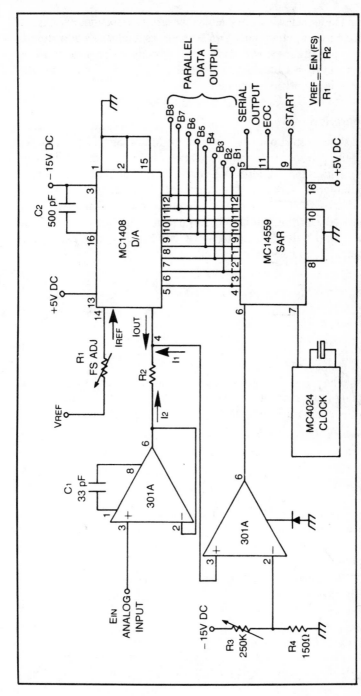

Fig. 12-7. Successive-approximation A/D converter using the Motorola MC1408 D/A converter and MC14559 SA register.

269

4 milliamperes, with 1 to 2 milliamperes recommended. The reference current potentiometer ($R1$) serves as a full-scale adjustment, while potentiometer $R3$ serves as a zero adjust. This latter potentiometer works by shifting the operating point of amplifier $A2$, which is nominally grounded.

Operation

When a start pulse is received, the MC14559 will begin to step through the successive-approximation routine (see Chapter 10), incrementing the MC1408 D/A converter output current as it goes. Amplifier $A2$ serves as a comparator that is ground referenced. If $I1$ is zero, then the comparator output is low, and the SAR action is halted. Current $I1$ is the algebraic sum of the D/A converter output current and the analog input current (I_{OUT} and $I2$, respectively). In other words:

$$I1 = I_{OUT} + I2 \qquad (12.6A)$$

and at $I1 = 0$,

$$I2 = -I_{OUT} \qquad (12.6B)$$

When the D/A converter output current becomes equal to the analog input current, the current to the comparator will be zero, so the comparator output drops low, shutting off the SAR.

Current I_{OUT} is given by:

$$I_{OUT} = I_{REF} \times \frac{2^n - 1}{256} = \frac{V_{REF}}{R1} \times \frac{2^n}{256} \qquad (12.7)$$

where I_{OUT} is the D/A converter output current, I_{REF} is the reference current between 0.5 and 4.0 mA, and n is the binary word at the D/A converter input terminals (which also serve as the A/D converter output terminals.

Current $I2$ is given by:

$$I2 = E_{IN}/R2 \qquad (12.8)$$

So, by substituting Eqs. 12.7 and 12.8 into Eq. 12.6B:

$$\frac{E_{IN}}{R2} = \frac{V_{REF}}{R1} \times \frac{2^n - 1}{256} \qquad (12.9)$$

From this equation (Eq. 12.9) I obtain the basic design equation for this circuit (assuming an offset of 1/2LSB):

$$\frac{E_{IN(FS)}}{R2} = \frac{V_{REF}}{R1} \qquad (12.10)$$

Provided that I_{REF} is between the limits given earlier, 0.5 to 4.0 mA.

Example 12-3
Let the reference current I_{REF} be 2 milliamperes, and V_{REF} be 10.00 volts dc. Assume that the full-scale analog voltage is +2.56 volts, so that the LSB potential is 10 millivolts. Calculate the resistance values for $R1$ and $R2$.
Solution:

$$
\begin{aligned}
R1 &= V_{REF}/I_{REF} \\
&= 10.00/0.002 \\
&= 5000 \text{ ohms}
\end{aligned}
$$

and

$$R2 = E_{IN(FS)}/I2$$

But at full-scale current $I2$ equals I_{REF}, so

$$
\begin{aligned}
R2 &= +2.56/0.002 \\
&= 1280 \text{ ohms}
\end{aligned}
$$

Adjustment

1. Apply a voltage equal to $E_{in(FS)}$ – LSB to the analog input. In the example above this would be $2.56 - 0.01$, or 2.55 volts.
2. Adjust $R1$ so that bits $B1$ through $B7$ are high, and bit $B8$ (LSB) bobbles between high and low.
3. Apply a voltage equal to 1/2LSB to the analog input. In the example given above this would be 1/2 (10), or 5 millivolts.
4. Adjust potentiometer $R3$ so that bits $B1$ through $B7$ are low, and bit $B8$ bobbles back and forth between high and low.

5. Repeat steps 1 through 4 until no further improvement is obtained.

DAC-100/AM2502 DESIGNS (SA)

Figure 12-8 shows the PMI DAC-100 paired with the Advanced Micro Devices AM2502PC successive-approximation register. The AM2502 device is similar to the Motorola MC14559 in function, if not form.

The AM2502PC is an eight-bit register, so the last two bits of the DAC-100, a ten-bit device, must be tied permanently high by connecting them to the +5 volt dc source.

The DAC-100 has an internal reference supply, so a 200-ohm potentiometer serves as the full-scale adjustment control. No zero control is provided, and it probably won't be needed. A circuit similar to that in Fig. 12-7 could be used on the noninverting input of the comparator should a zero control be desired.

Several special features are provided by this design. Among them are fast operation (i.e., eight bits in 6 microseconds), a special complemented MSB, and a serial output. The high speed is achieved by using a fast clock speed, but be cautioned that layout and construction practices become more critical at fast speeds, and sloppiness can ruin the performance of this, or any other high-speed digital circuit.

Two digital output formats are available from this A/D converter circuit. The parallel data consists of nine parallel lines, of which one is the complemented most significant bit. The serial output is synchronized to the clock, and is in standard nonreturn-to-zero (NRZ) format.

The comparator is a premium PMI type, but an ordinary type 311 can be substituted if slower operating speeds are anticipated. The 311 is not exactly pin-for-pin compatible, and a minor circuit change is required; a 1.5 to 2.7K pull-up resistor must be connected between the output terminal and the +5 volt dc power supply.

Comparators are normally considered voltage input devices, but in most A/D converter applications their current comparison capability is used. The analog input voltage becomes a current by applying it to the comparator input through a precision resistor. In the circuit of Fig. 12-8 one of the internal 4.88K resistors of the DAC-100 (pin 16) is used for this purpose. Input current $I2$ is generated by analog input voltage E_{IN} and resistor R_S. Current $I2$ will have a magnitude of $E_{IN}/4880$, and is summed in the output mode

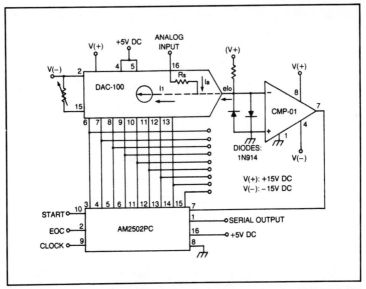

Fig. 12-8. Successive-approximation A/D converter using the PMI DAC-100 and Advanced Micro Devices AM-2502PC SA register.

with the regular DAC-100 output current $I1$. When the two currents are equal, the DAC-100 output current, as seen by the external comparator, is totally canceled, so has a magnitude of zero.

The reason for using the current mode of the comparator is to increase the operating speed of the A/D converter. If I applied the analog voltage directly to the comparator input, then an operational amplifier current-to-voltage converter would be required at the output of the DAC-100. Most operational amplifiers are very slow compared with the other A/D converter components, so will cause a decrease in the maximum attainable operating speed. Even high-slew-rate, high-frequency, operational amplifiers require compensation to avoid oscillation, and that in itself will slow the circuit down markedly. The best solution is to eliminate the operational amplifier altogether, and that requires summing currents in the comparator input circuit.

The DAC-100 can be operated in the bipolar mode (see Chapter 11), so the A/D converter using the DAC-100 will also provide bipolar operation. Use the remaining 4.88K resistor and an external 500-ohm potentiometer to apply a half-scale current from an external 6.4-volt dc precision reference source. The procedure was discussed in the previous chapter.

Adjustment

1. Apply an input voltage of V_{FS} – 1.5LSB (in this example + 9.941 volts) to the analog input terminal.
2. Adjust FS adjust control $R2$ for bits $B1$ through $B7$ low and bit $B8$ bobbling back and forth between low and high.
3. No zero adjust is necessary in most cases. The required 1/2LSB bias for the comparator is provided by resistor $R1$.

Short-Cycle and Long-Cycle Operation

Six-bit short-cycle operation of this A/D converter circuit is essentially the same as that for the AD-02. A NAND gate is connected with one input to the conversion-completed terminal of the successive-approximation register, and the other to bit $B7$. The output of the NAND gate is then inverted and used to drive the start terminal on the SAR.

Ten-bit long-cycle operation is possible by substituting the Advanced Micro Devices AM2504PC twelve-bit SAR integrated circuit for the eight-bit AM2502PC used in the example. Use the ten most significant bits on the SAR to drive the ten-bit DAC-100 input. Disregard bits $B11$ and $B12$ on the AM2504PC.

DAC-08 A/D CONVERTER

The Precision Monolithics DAC-08 can be used in essentially the same circuit to form one of the lowest-possible-cost A/D converters. The necessary modifications are shown in Fig. 12-9, but otherwise the circuit is the same as Fig. 12-8.

In this circuit current I_{OUT} (pin 2) from the DAC-08 is summed at the comparator input with a sample of the analog input signal. Current I_{OUT} (pin 4) is returned to the analog input terminal in order to maintain a constant input impedance for the analog source. Remember that the two DAC-08 output currents produce complementary currents. Refer to Chapter 11 for the proper power supply and reference current connections for the DAC-08. The AM2502PC connections are the same as in Fig. 12-8.

ANALOG DEVICES 13-BIT AD7550

Some of the A/D converter circuits which I have considered have been constructed of several discrete integrated circuits, and at least one was a monolithic integrated circuit that was completely self-contained. The Analog Devices (Route 1, Industrial Park, P.O.

274

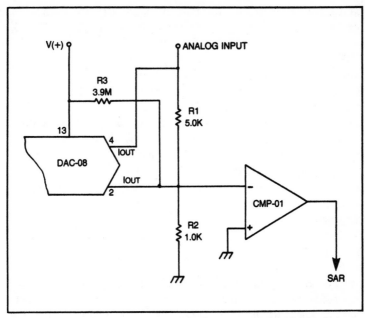

Fig. 12-9. Circuit of Fig. 12-8 using the DAC-08 instead of DAC-100. Only circuitry shown need be changed; all other circuitry is the same as in Fig. 12-8.

Box 280, Norwood, MA 02062) type AD7550 shown in Fig. 12-10
is an example of another monolithic A/D converter chip. This chip
is actually part of a small family of different, but similar devices,
so you are advised to contact the manufacturer for a more complete short-form catalogue, especially if the description below
sounds a lot, but not quite, like what you need in a particular application.

The AD7550 is a monolithic device made using the CMOS
process. As such, it will require somewhat less operating current
than many devices of similar bit size, but may tend to be static sensitive. The critical inputs are protected by zener diodes, but the
manufacturer still recommends that standard CMOS handling
procedures be followed. This, incidentally, is good advice for any
CMOS device, even though diode protected. The fact is that some
static charges that are quite ordinary can still blow the chip before
the zener diodes have a chance to do their protective function.

The AD7550 uses a special Quad-Slope conversion system that
is patented by Analog Devices. This technique is similar to the more
traditional dual-slope technique discussed in Chapter 10.

This chip is perhaps unique in that the outputs terminals (digi-

275

tal data) are tri-state, and use two enable pins to turn them on or off. Pin 18 enables the high five bits (i.e., *high byte enable*, HBEN), while pin 19 enables the lower eight bits (i.e., *low byte enable*, LBEN). This feature makes the AD7550 compatible with a large number of devices and circuits, including eight-bit microprocessors. The HBEN and LBEN are active when high. If either enable terminal is low, then the output bits that are controlled by that line (i.e., high or low byte) are tri-stated to float at a high impedance.

Four other control pins used on the AD7550 are status enable (STEN), overrange (OVRG), BUSY, and $\overline{\text{BUSY}}$. The status enable terminal controls the remaining control pins. When STEN is low, the OVRG and BUSY lines float tri-state style at a high impedance. These output signals become active only when the STEN terminal is high.

The overrange (OVRG) is used to signal when the analog input voltage (A_{IN}) exceeds the positive or negative full-scale limits by more than 1/2LSB.

The BUSY and $\overline{\text{BUSY}}$ terminals are complements of each other. The busy terminal is high when a conversion is taking place, and drops low when the conversion is completed (indicating that the data is valid). The $\overline{\text{BUSY}}$ terminal is the exact complement. It will be low when the conversion is taking place and snaps high when the conversion is completed. Either of these terminals can be used to strobe an external device or microcomputer input port that the data on the output lines is now valid and ready for use.

There are three power supply terminals (*VSS, VDD,* and *VCC*), and two inputs for reference supplies. The *VSS* terminal goes to the negative power supply, and should be kept at a potential of -5 volts dc to -12 volts dc. The *VDD* supply terminal goes to a positive power source, and must be kept between $+10$ volts and $+12$ volts dc. The *VCC* terminal is called the logic supply, and is set to $+5$ volts dc if the circuit is to be TTL compatible, and some potential between $+10$ volts and *VDD* if the A/D converter is to be CMOS-logic compatible.

The two reference voltage terminals are labeled V_{REF1} and V_{REF2}. The V_{REF1} terminal should be connected to a well-regulated reference voltage source, examples of which are given in Chapter 11. The magnitude of the reference potential must be between *VSS* and *VDD*.

V_{REF2} is set to a potential of $1/2 V_{REF1}$, and is usually derived from a resistor voltage divider connected across the first reference supply, V_{REF1}. The second reference supply, in any event, must be

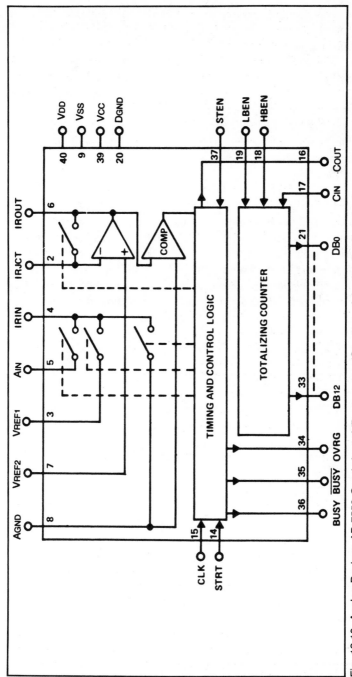

Fig. 12-10. Analog Devices AD-7550 Quad-slope A/D converter. (Courtesy of Analog Devices, Inc.)

kept between a zero reference equal to the potential of the analog ground (nominally zero if proper layout grounding procedures are followed) and V_{DD}.

The maximum value for the analog input voltage (A_{IN}) is set by the first reference supply, and is given by $V_{REF1}/2.125$.

Separate access is provided for several components of the A/D converter through package pins. The integrator input, output, and summing junction, for example, are accessible through pins 4, 6, and 2 respectively. The resistor and capacitor required to make the operational amplifier operate as an integrator are connected to these pins.

The START terminal requires an 800 nanosecond (i.e., 0.8 ms) positive pulse to initiate the conversion cycle. On the leading edge of the start pulse the internal logic is set (i.e., initialized) and the status flags, BUSY and $\overline{\text{BUSY}}$, are latched in their respective conditions indicating that a conversion is taking place. When the start pulse returns low following its trailing edge, the actual conversion cycle will begin. Asynchronous continuous operation is possible by returning the START terminal to ground through a capacitor.

The operation of the Quad-Slope integrators is shown in Figs. 12-11 and 12-12. During period $T1$ switch $S1$ is closed, applying V_{REF1} to the input of the integrator. The other input of the opera-

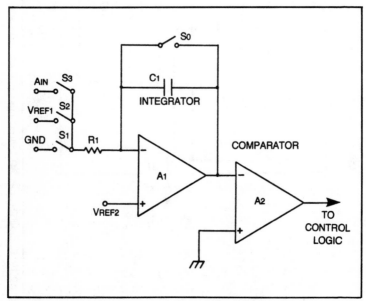

Fig. 12-11. Simplified diagram showing the Quad-Slope technique.

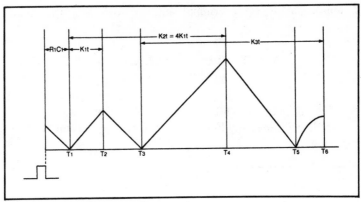

Fig. 12-12. Quad-Slope waveforms.

tional amplifier serving as the integrator is connected permanently to V_{REF2}, which is $1/2 V_{REF1}$.

During period $T1$, which lasts exactly one RC time constant ($R1C1$), the integrator ramps down to zero. When it reaches zero the comparator will toggle, telling the control logic section that the period is finished, so it can start the binary counters ($K1$ through $K3$) and reset the switches. At this time switch $S1$ closes and $S2$ is opened.

During period $T2$, with switch $S1$ closed, the integrator input is grounded, so the integrator ramps from zero to V_{REF2} for a specific length of time defined as the accumulated count in counter $K1$ times the clock period, t, or $K1 \times t$. When period $T2$ expires, switch $S1$ opens and $S2$ closes.

At period $T3$ the integrator input is again connected to V_{REF1} (its input voltage is $V_{REF1} - V_{REF2}$). This period requires a period of $K1 + n$ times t, where the n term is the error count, if any, caused by integrator offsets, comparator hysteresis, etc. At the conclusion of $T3$ switch $S2$ is opened and $S3$ is closed. This turns on counter $K3$ and connects the analog input (A_{IN}) to the input of the integrator.

The analog input voltage ramps the integrator during period $T4$ until the $K2$ counter reaches a count that is four times greater than the $K1$ counter. When this period expires switch $S2$ is again closed and $S3$ is opened.

During period $T5$ the integrator ramps down to zero under the influence of the $V_{REF1} - 1 V_{REF2}$ potential. The period $T6$ between the zero crossing of the integrator at the comparator input is equal to twice the count ($2N$) that represents the analog input voltage.

N appears at the C_{OUT} terminal, and is defined as

$$N = 2^{12} \left[\left(\frac{A_{IN}}{V_{REF1}} \times 2.125 \right) + 1 \right] \qquad (12.11)$$

The information to follow summarizes the output code under assorted conditions, and Fig. 12-13 shows a typical connection arrangement.

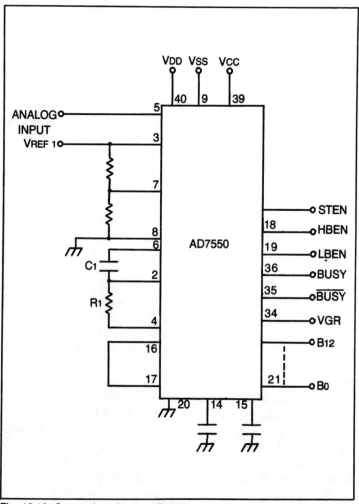

Fig. 12-13. Connections for the AD7550.

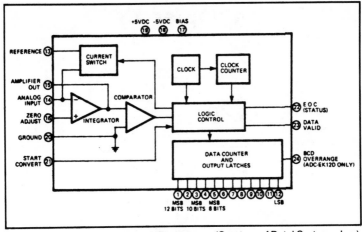

Fig. 12-14. Datel ADC-EK series A/D converter. (Courtesy of Datel Systems, Inc.)

Condition	N	Parallel Data Output	OVRG	DB12
$+\mathrm{OVRG}$	—	111111111111	1	0
$+V_{FS}(1-2^{-12})$	8191	111111111111	0	0
$+V_{FS}(2^{-12})$	4097	000000000001	0	0
Zero	4096	000000000000	0	0
$-V_{FS}(2^{-12})$	4095	111111111111	0	1
$-V_{FS}$	0	000000000000	0	1
$-\overline{\mathrm{OVRG}}$	—	000000000000	1	1

THE DATEL ADC-EK SERIES

The Datel ADC-EK series are monolithic CMOS integrated circuit analog-to-digital converters using the dual-slope integration method for performing the conversion. Various models are available that offer 8-, 10-, or 12-bit binary output coding, or 3 1/2-digit BCD coding, which is popular in digital voltmeter circuits. The block diagram to the ADC-EK A/D converter is shown in Fig. 12-14. This chip contains the integrator, comparator, clock, counter, and all logic for control of a dual-slope converter circuit.

Conversion times will be between 1.8 and 24 milliseconds. The linearity is $\pm 1/2$LSB minimum, with better than $\pm 1/4$LSB being more typical. Power consumption at ± 5 volts dc is on the order of 20 milliwatts (current drain 2 milliamperes). If -5-volt supplies are used, then a 1.22-volt reference supply is required, and if a -15-volt supply is used, the reference supply should be -6.4 volts.

Chapter 13

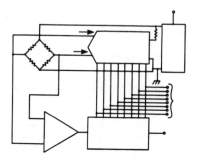

Some Data
Converter Applications

THE TRADITIONAL APLLICATION FOR A/D CONVERTER CIR-
cuits and D/A converter circuits is to convert data for use in
a computer, or to control some external device on command of a
computer. The D/A converter, for example, might be used to plot
graphic displays of data on an oscilloscope or strip-chart recorder.

In the A/D converter case, an analog signal is converted to a
binary representation that can be used by the computer, while the
D/A converter is used to convert back to an analog representation.

Ordinarily, an A/D converter should be preceded by a low-pass
filter that limits the bandwidth of the signal being converted to the
minimum required, while a D/A converter should be succeeded by
the same type of filter to smooth out the ripple caused by the conversion process. Recall that analog signals are continuous in both
range and domain, but a digital signal is allowed only discrete states
in either range or domain. For a 0-to 10-volt output D/A converter,
for example, the minimum step (assuming eight-bit operation) is
40 millivolts. The apparent ruggedness in the trace caused by these
steps is removed by a low-pass filter.

Data conversion, however, covers a lot of territory, and is not
strictly limited to the traditional applications given above. There
are quite a few other applications, which are limited mostly by your
own design imagination and your ability to manipulate the rules
and constraints placed on any given device. In this chapter, I will
discuss some of the more interesting of these applications.

TWO-QUADRANT MULTIPLICATION

D/A converters such as the Precision Monolithics DAC-08 are known as multiplying D/A converters because they require an external reference voltage or current, and have a transfer function of the form

$$I_{\text{OUT}} \; = \; I_{\text{REF}} \left(\frac{A1}{2} + \frac{A2}{4} + \; . \; . \; . \; + \frac{A8}{256} \right) \quad (13.1)$$

where $A1$ through $A8$ are either *one* or *zero* depending upon the binary word applied to the DAC-08 digital inputs. Equation 13.1 becomes a voltage mode expression if I_{OUT} and I_{REF} are known in terms of a reference voltage and resistance, and an output resistor.

By applying a digital word to the inputs and a 0- to 10-volt signal to the reference input (i.e., 0- to 2-milliampere current), I obtain a digital-by-analog, two-quadrant multiplier circuit. The digital values can be supplied by a computer, while the analog voltage would be from some other circuit.

An application of the two-quadrant multiplier is a digitally programmable attenuator with response down to dc. The signal applied to the reference input is the signal that is attenuated an amount dictated by the binary word at the digital inputs.

FOUR-QUADRANT MULTIPLICATION

Figure 13-1 shows an example of two DAC-08 devices used to perform four-quadrant multiplication. A 1.5-milliampere reference current is applied to the $+I_{\text{REF}}$ inputs of the two DAC-08s, while the analog input voltage is applied differentially across the $-I_{\text{REF}}$ inputs. The digital word is applied to both DAC-08s in parallel.

The I_{OUT} of DAC-08 #1 is connected to the $\overline{I}_{\text{OUT}}$ output of DAC-08 #2, and vice versa. The combined current lines form a differential output current to any balanced load. Table 13-1 shows the expected output for a variety of input conditions.

AC-COUPLED AUDIO ATTENUATOR

The compensation capacitor terminal (i.e., pin 16) of the DAC-08 is normally bypassed to V_{EE} through 0.1 μF, but if I use the regular DAC-08 voltage-output circuits (see Chapter 11), and instead connect pin 16 through 0.1 μF to an ac signal source, then

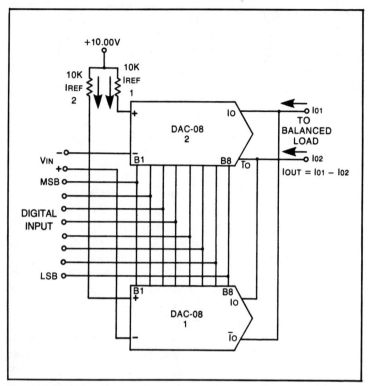

Fig. 13-1. Four-quadrant analog-by-digital multiplier. (Courtesy of Precision Monolithics, Inc.)

I obtain an ac-coupled attenuator for 0- to 1-volt signals. The DAC-08 will be flat to 200 kHz and will operate out to 1 MHz.

RATIOMETRIC A/D CONVERSION

A *ratiometric* instrument is one that has an output voltage that is proportional to the ratio of two input signals, i.e.,

$$E_{OUT} = k(V_x/V_y) \qquad (13.2)$$

It is often the case that ratiometric operation will either improve, make easier, or even make possible some instrumentation process.

For example, it might be true that in some cases the data of interest is the ratio of two measured parameters. Transducers or other devices can be used to acquire the appropriate signals, but

Table 13-1. Output States for Circuit of Fig. 13-2.

DIGITAL INPUT	$V_{IN}(+)$	$V_{IN}(-)$	V_{IN} DIFF.	I_{REF} #1 (mA)	I_{REF} #2 (mA)	I_O #1 (mA)	$\overline{I_O}$ #2 (mA)	I_{O1} (mA)	I_O #2 (mA)	$\overline{I_O}$ #1 (mA)	I_{O2} (mA)	I_{OUT} DIFF
1111 1111	+5V	-5V	+10V	2.000	1.000	1.992	0	1.992	0.996	0	0.996	0.996 mA
1000 0000	+5V	-5V	+10V	2.000	1.000	1.000	0.496	1.496	0.500	0.992	1.492	0.004 mA
0111 1111	+5V	-5V	+10V	2.000	1.000	0.992	0.500	1.492	0.496	1.000	1.496	-0.004 mA
0000 0000	+5V	-5V	+10V	2.000	1.000	0	0.996	0.996	0	1.992	1.992	-0.996 mA
1111 1111	0V	0V	0V	1.500	1.500	1.494	0	1.494	1.494	0	1.494	0.000 mA
1000 0000	-10V	-10V	0V	2.500	2.500	1.250	1.240	2.490	1.250	1.240	2.490	0.000 mA
0111 1111	+10V	+10V	0V	0.500	0.500	0.248	0.250	0.498	0.248	0.250	0.498	0.000 mA
0000 0000	0V	0V	0V	1.500	1.500	0	1.494	1.494	0	1.494	1.494	0.000 mA
1111 1111	-5V	+5V	-10V	1.000	2.000	0.996	0	0.996	1.992	0	1.992	-0.996 mA
1000 0000	-5V	+5V	-10V	1.000	2.000	0.500	0.992	1.492	1.000	0.496	1.496	-0.004 mA
0111 1111	-5V	+5V	-10V	1.000	2.000	0.496	1.000	1.496	0.992	0.500	1.492	0.004 mA
0000 0000	-5V	+5V	-10V	1.000	2.000	0	1.992	1.992	0	0.996	0.996	0.996 mA

there are two approaches to handling the data once it is acquired. In one method, two A/D converter circuits are used, and the computer inputs data from each, then performs a software division of one by the other. The alternate approach is to build a ratiometric A/D converter, as shown in Fig. 13-2. Note that this is a simplified drawing. The actual conversion logic selected can be either successive-approximation (SA) or binary counter types, although SA is recommended.

The two D/A converters are connected in a circuit similar to Fig. 13-1, with the exception that the respective $-I_{REF}$ terminals are grounded and the $+I_{REF}$ inputs see a summation of two currents: $\pm(V_y/5000) + 2$ mA. The output currents ($I1$ and $I2$) applied to the comparator inputs are determined by the level of $+I_{REF}$ and the binary word applied to the digital inputs of the DAC-08s.

Also applied to the comparator inputs are currents $I3$ and $I4$, generated by V_x, and equal to $V_x/5000$. The comparator, then, sees

$$I(+) = I1 + I4 \qquad (13.3)$$

at the noninverting input, and

$$I(-) = I2 + I3 \qquad (13.4)$$

at the inverting input.

If the following equality holds true, then the comparator output is zero, otherwise it is one, and a conversion can take place.

$$I(-) = I(+) \qquad (13.5)$$

$$(I1 + I4) = (I2 + I3) \qquad (13.6)$$

TRANSDUCER NULLING

Wheatstone-bridge strain-gauge transducers should have a zero output voltage when the stimulus parameter is also zero, but manufacturing tolerances often cause a slight offset potential to exist at the output.

Similarly, there can be a static value of the stimulus parameter always applied to the transducer, creating an output offset. It

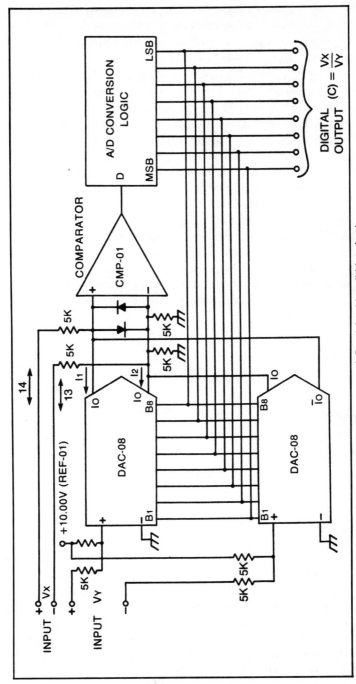

Fig. 13-2. Ratiometric (i.e., X/Y) A/D conversion. (Courtesy of Precision Monolithics, Inc.)

is changes above and below the static value that are of interest in those cases.

If there is plenty of dynamic range in the instruments used to process the transducer output signal, then a nulling circuit may not be needed. Even in computerized instruments, where reading interpretation of an offset is not possible, the offset need not be a problem. A program segment can be written to null the offset in software. The output of the A/D converter is read under conditions of known zero stimulus, and the result is then stored in a register or memory location. Subsequent data will represent the sum of the offset and the real data, so the program merely subtracts the offset from each new data that is entered.

In cases where the offset must be actually nulled out, there must be provided a means for adjusting the transducer output to zero when the stimulus parameter is also zero. The traditional method (shown in Fig. 13-3) uses a potentiometer.

The transducer is a Wheatstone bridge with four equal valued arms of resistance R. The transducer is excited by a dc potential E. By the ordinary method of analysis we know that in the null condition:

$$I1R1 \ = \ I2R3 \qquad (13.6A)$$

$$I1R2 \ = \ I2R4 \qquad (13.7)$$

and if $R1 \ = \ R2 \ = \ R3 \ = \ R4 \ = \ R$, then,

$$I1 \ = \ I2 \qquad (13.8)$$

but if any R ($R1$ through $R4$) is not equal, then

$$I1 \ \neq \ I2 \qquad (13.9)$$

and an output offset voltage will exist across the output terminals.

A balance circuit consisting of a potentiometer ($R5$) and a fixed resistor ($R6$) injects a current into one node that reestablishes equality Eq. 13.8, reducing E_{OUT} to zero under zero stimulus conditions.

A digital method using a D/A converter is shown in Fig. 13-4A. In this example, the D/A converter is used as a digitally controlled potentiometer and accomplishes essentially the same job as the manual method of Fig. 13-3. If an offset voltage greater than a very small amount exists when the stimulus is known to be zero, then

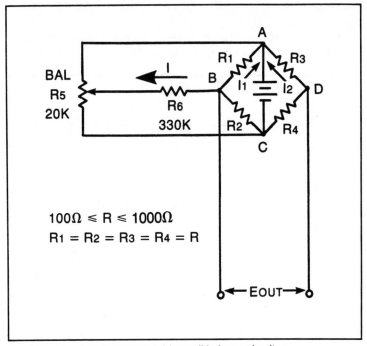

Fig. 13-3. Manual Wheatstone bridge null-balance circuit.

the D/A converter can be incremented until the comparator senses that the offset has been nulled out.

A variation on this idea that makes use of the complementary output currents of the DAC-08 is shown in Fig. 13-4B. In this circuit the ground node is broken, and the I_{OUT} and $\overline{I}_{OUT}$ of the DAC-08 are used to drive the two halves. Although this technique looks good, it often proves very difficult to actually implement with real transducers, because most commercial transducers are constructed in a manner that precludes breaking any node. Some industrial strain gauges are sold in a nonpackaged form, often mounted on a sheet of plastic or mica. In these, the technique of Fig. 13-4B will work.

Figure 13-5 shows a type of A/D converter that will directly convert the output of a transducer. The circuit works as a continuously seeking null-balance circuit. In the configuration shown the digital output will track any changes in the transducer output voltage.

The technique of Fig. 13-5 can be ex.ended to other A/D converter methods, such as the successive-approximation circuit, by rearranging the circuit of Fig. 13-5 according to the rules and designs given in Chapter 10.

Offset null can be by subtraction in software, or by connecting a second DAC-08 in parallel with the DAC-08 of Fig. 13-5. The bi-

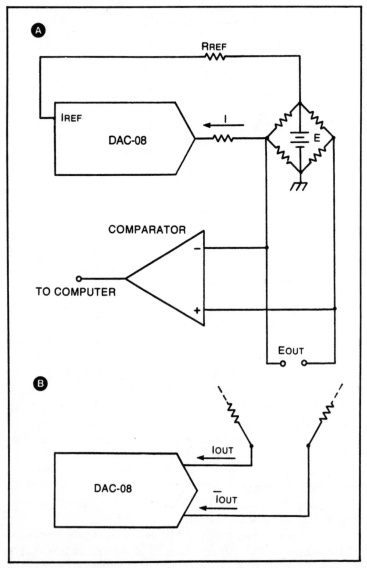

Fig. 13-4. (A) D/A converter null circuit in which the D/A converter replaced the potentiometer of Fig. 13-3. (B) Method that takes advantage of complementary current output of the DAC-08.

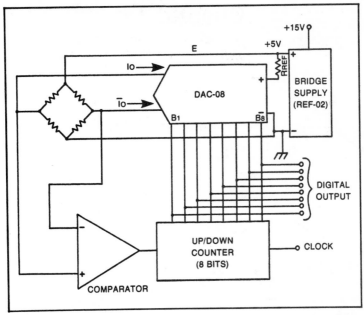

Fig. 13-5. Null-tracking A/D converter for Wheatstone bridge. (Courtesy of Precision Monolithics, Inc.)

nary word applied to the A/D converter DAC-08 would be the code that sets $I_{OUT} = \bar{I}_{OUT}$, which is 10000000 in unipolar operation.

The circuit of Fig. 13-5 also offers the advantage that it tracks drift in the power supply voltage to the transducer because E is also used as the DAC-08 reference voltage, E_{REF}.

Be aware that the REF-02 shown in the figure may not be able to drive some Wheatstone bridge transducers, especially those in which R has a low value. The REF-02 is rated by the manufacturer at a load current of 20 milliamperes, of which up to 2 milliamperes be needed for I_{REF}. The 18 milliamperes left over for the transducer may limit the selection of transducers to comply with the expression

$$5 \text{ volts}/R \leq 18 \text{ mA}$$

where R is the resistance looking back into the transducer, which in a Wheatstone bridge (only) is the resistance of an arm. This expression tells me that the arm resistance, if the REF-02 is used, must be 278 ohms or greater. Many transducers have lower value resistances.

291

Chapter 14

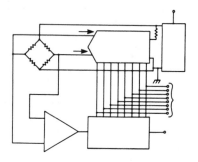

Analog & Digital Multiplexing

MULTIPLEXING IS THE MIXING TOGETHER OF TWO OR more signals in a single communications channel or path in such a manner that they can be easily separated at the receiver end. Two types of multiplexing are commonly employed: *frequency domain* and *time domain.*

Frequency domain multiplexers use different frequencies or frequency bands for different signals. In carrier telephony, for example, each voice band (i.e., 300 to 3000 hertz) occupies approximately 3 kHz of spectrum space. If a single-sideband system using audio-range carriers is employed, then two or more channels can pass over the same pair of wires or radio link.

Figure 14-1 shows how this can be accomplished. Channels 1 and 2 are both 300 to 3000 hertz voice band communications channels, and if transmitted over the same pair of wires *as is,* would interfere with each other. But if I send channel 1 over the wires in baseband (i.e., unchanged), then I must frequency-translate channel 2 before it can be sent. This is done by using channel 2 signals to modulate an upper-sideband generator with a carrier of, say, 6000 hertz (see Fig. 14-1B).

Another, but simpler, type of frequency division system, used in computer and mechanical (i.e., teletypewriter) systems using binary character representation, is to select different tone pairs for the *one* and *zero* logic levels of different machines.

An example is shown in Fig. 14-2. In this hypothetical system

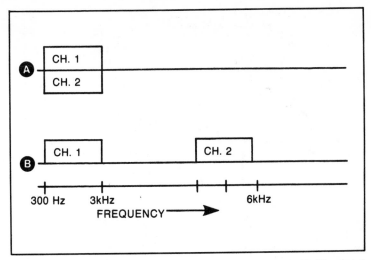

Fig. 14-1. Multichannel operation. (A) Channels overlapping. (B) Channel 1 in voice band with channel 2 frequency-translated to a higher frequency.

I have two channels, 1 and 2. Each channel passes binary data in which the logic 1 and logic 0 states are represented by different tones. Four tones are required because there are four distinct states: channel 1 logic 1, channel 1 logic 0, channel 2 logic 1, and channel 2 logic 2.

In channel 1 I have defined logic 1 as a 1070 Hz tone, and logic 0 as a 1280 Hz tone. Channel 2 is similarly defined with logic 1 being represented by 2025 Hz, and logic 0 being 2225 Hz. These tones can be summed in a linear audio mixer, then applied to the communications system (i.e., radio or telephone channels).

At the receiver end of the system the composite audio tone created by mixing is applied simultaneously to a set of narrow bandpass filters that give an output only when the proper tone passes through the system. The decoders at the output detect the existence or nonexistence of each tone. The decoder is usually some sort of phase-locked loop and comparator circuit, or a rectifier and comparator system. The comparator senses the existence information and produces a logic level output indicating the state.

Time domain multiplexing is a little easier to implement in some cases because of the ready availability of integrated circuits that do the majority of the work. In the time domain technique, a channel is sampled at separate times, and the system sees a stream of interleaved samples in the transmission path. Time division requires that both receiver and transmitter ends be synchronized to the same

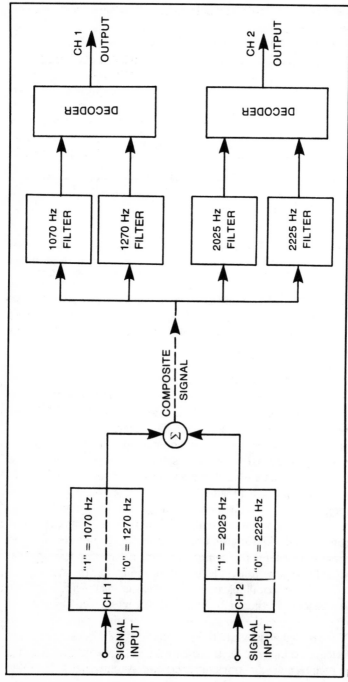

Fig. 14-2. Frequency division digital transmission system.

294

clock, otherwise the data would remain interleaved and be a confusing mess in that case.

A simplified representation of a six channel time domain multiplexed system is shown in Fig. 14-3. Switches S1 and S2 are two-pole-six-position rotaries that are ganged together. Switch S1 is connected so that one channel at a time feeds the transmission path, while S2 is connected so that the transmission path drives one channel at a time. It is necessary to insure that the transmit and receive switches are connected to the same channel at the same time.

When S1-S2 are in position 1, then channel 1 in the transmitter is connected to channel 1 on the receiver. Moving S1-S2 to position 2 similarly connects channel 2 in the transmitter to channel 2 in the receiver, after disconnecting channel 1, of course. This action sweeps through all six positions in sequence. Switches S1-S2 are 360-degree types, so that rotating them clockwise produces a continuous 1, 2, 3, 4, 5, 6, 1, . . . sequence.

An example of a simple two-channel analog signal multiplexer is shown in Fig. 14-4. This circuit is based on the CMOS 4016 or 4066 (updated version of 4016) quad bilateral switches. The 4066 is a newer, updated, version of the 4016 and is preferred if easily available. The 4016, however, is widely available through mail-order and hobbyist retail electronics stores.

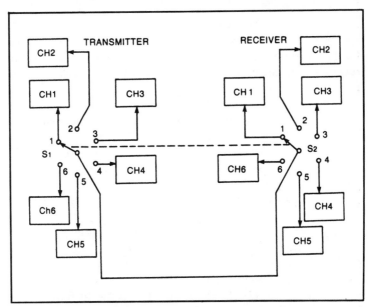

Fig. 14-3. Simplified schematic of time-division multiplexing.

Each switch in either 4016 or 4066 is independent from the others, except for a common power supply. When the voltage applied to the control pin for any given switch (i.e., pins 5, 6, 12, or 13) is equal to the voltage on pin 7, the switch will be off. That is to say it will present a very high series impedance. In this case pin 7 is grounded, so a 0-volt condition on a control will turn off that switch. The pin 7 voltage can be 0 to -5 volts.

Applying a voltage equal to the pin 14 voltage to a control pin will turn on the switch, that is, cause its series impedance to the signal path to drop very low, often less than 100 ohms, and in all cases less than 2000 ohms.

In the circuit of Fig. 14-4, I use just two of the four switches ($S1$ and $S2$). The circuit is configured so that one terminal of each switch (i.e., pins 2 and 3) are tied together to become the output. Pin 1 is the channel 1 input, while pin 4 is the channel 2 input.

The circuit is clocked by a square wave signal driving a JK flip-flop ($FF1$). In a JK flip-flop, the Q and $\overline{Q}$ are complementary, so one is high while the other is low (see the timing diagram in Fig. 14-4B).

During period $T1$, the Q is high and $\overline{Q}$ is low. This turns on switch $S1$ and turns off $S2$ because pin 5 of the 4066 is high and pin 13 is low. The channel 1 signal will appear on the output line during this period.

Following the next clock pulse, during period $T2$, the situation reverses; $S1$ is now open and $S2$ is now closed. The channel 2 signal will appear on the output during this period. The situation reverses following each clock pulse, so channels 1 and 2 are alternatively connected to the output line.

This action causes the two signals to be interleaved with each other, and would result in a hopeless mess at the other end of the system unless a demultiplexer is provided. A suitable circuit would be another 4066 connected the same as Fig. 14-4A, except that the output becomes the input and the two inputs are redesignated as outputs to the respective channels. The 4016 and 4066 are *bilateral* switches, meaning that they don't care which direction the signal travels.

The circuit in Fig. 14-4 has been billed as a multiplexer (which is an accurate designation), but it is also occasionally seen under the name "electronic oscilloscope switch," or something similar. In that case, it is used to make a single-channel oscilloscope into a dual-channel model. The chopping action of the 4066 will not be noted on the screen of the CRT if the switching frequency is high

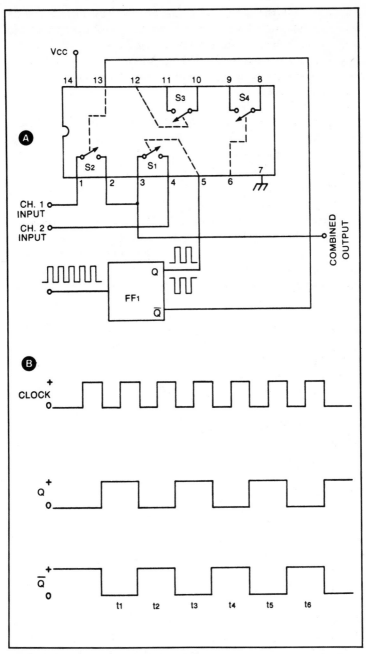

Fig. 14-4. Simple two-channel time-division multiplexer using the 4016 or 4066 CMOS switch. (A) Circuit. (B) FF1 waveforms.

compared with the frequency of the applied waveforms. The dc bias on each channel serves as a position control to separate the two beams to different places on the CRT screen.

Three other common CMOS 4000-series chips are also used extensively as analog or digital multiplexers/demultiplexers. These are the 4051, 4052, and 4053.

The 4051 chip is called a 1-of-8 switch, and is analogous to a single-pole-eight-position rotary switch. The 4052 is a dual 1-of-4 switch, and is analogous to a pair of ganged single-pole-four-position switches. The 4053 is billed as a triple 1-of-2 switch, so is analogous to three ganged SPDT switches.

All three chips have an inhibit pin that turns off all switches in that package. When the inhibit is low, then the select pins control which switches are on or off. The 4051 and 4053 use three select pins, while the 4052 uses two select pins. These are driven from binary sources as was done in the example of Fig. 14-4.

PMI MUX-88

Figure 14-5 shows the Precision Monolithics MUX-88 multiplexer-demultiplexer chip. It is a monolithic eight-channel JFET analog switch circuit. Since it uses JFETs rather than CMOS transistors in the switch circuits, it is not susceptible to static charge blowout when it is handled. It will operate from either TTL or CMOS logic level signals, and provides the make-before-break switching action.

One terminal of each switch is connected to the output (i.e., drain), while the other terminal on each switch goes to separate pins on the IC package. This configuration provides SP8T action in response to the three-bit select code (see truth table). An octal (i.e., three-bit) counter can be used to generate the select code. As in the 4051-4053 devices, there is a chip select pin that enables the MUX-88 when high and inhibits it when low.

DATEL MX SERIES

The Datel MX-series 4-, 8-, and 16-channel CMOS multiplexers are shown in Fig. 14-6. These devices are compatible with TTL, DTL, and CMOS logic levels, and are driven through 2-, 3-, or 4-bit binary address select codes. The MX-series chips are monolithic IC technology.

These multiplexers can boast of 0.01% transfer accuracy at sample rates of 200 kHz and signal swings over ±10 volts. The

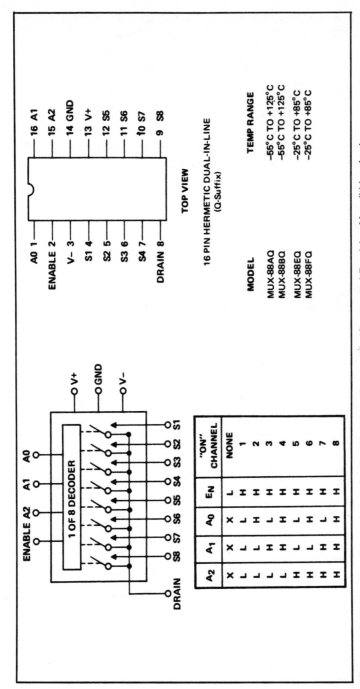

Fig. 14-5. Precision Monolithics, Inc. MUX-88 eight-bit multiplexer IC. (Courtesy of Precision Monolithics, Inc.)

299

channel-off resistance is typically 1.5K at room temperature and less than 2K over the rated temperature range.

The power supply is ±5 volts dc to ±20 volts dc, and power consumption is only 7.5 milliwatts in standby condition. At a 100 kHz sampling rate the power consumption is still down as low as 15 milliwatts. Total package dissipation is 725 milliwatts for the MX-808 and MSC-409 and 1200 milliwatts (i.e., 1.2 watts) for the MX-1606 and MXD-807.

The MX-1606 is a 1-of-16 channel single-ended device, analogous to an SP8T switch. It uses a four-bit (i.e., $2^4 = 16$) address select code. The MXD-409 and MXD-807 are 2-of-4 and 2-of-8 devices, respectively. The four-channel model uses a two-bit address select code, while the eight-channel model uses a three-bit code.

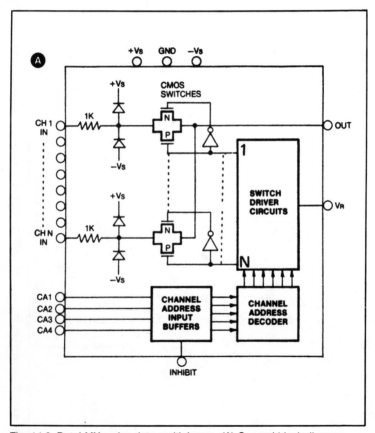

Fig. 14-6. Datel MX-series data multiplexers. (A) General block diagram.

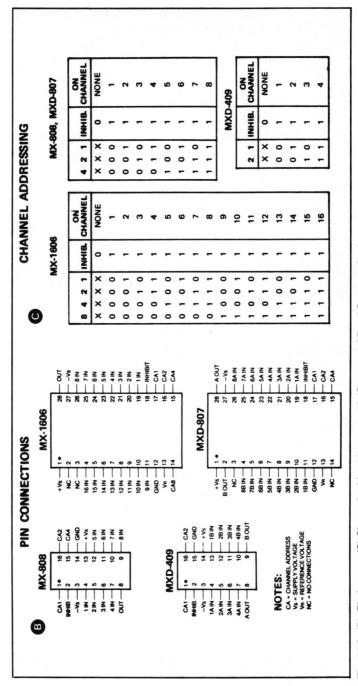

Fig. 14-6. (B) Pin layouts. (C) Channel addressing protocol. (Courtesy of Datel Systems, Inc.)

The transfer accuracy of any electronic switching multiplexer depends upon the source and load resistances. The output voltage will be given by

$$E_{\text{OUT}} = \frac{E_{\text{IN}} R_L}{R_S + R_{\text{ON}} + R_L} \qquad (14.1)$$

where E_{OUT} is the switch output voltage (across load R_L), E_{IN} is the switch input voltage (i.e., the open-circuited output voltage of the source), R_L is the load resistance seen by the switch, R_S is the output impedance of the signal source, and R_{ON} is the series-on resistance of the switch.

Equation 14.1 suggests that it is wise to keep R_S and R_{ON} as low as possible, while keeping R_L as high as possible. The former criteria is met by using a low output impedance source such as an operational amplifier at the input of the switch. The output load should be another operational amplifier with an input impedance greater than 10^7 ohms, with greater than 10^8 ohms preferred. Note the BiMOS and BiFET (RCA devices) operational amplifiers can boast input impedances of over 10^{12} ohms, so are often specified for use in this application.

Example 14-1
Find the output voltage in terms of input voltage if $R_S = 100$ ohms, $R_{\text{ON}} = 2000$ ohms, and $R_L = 20M$.
Solution:

$$E_{\text{OUT}} = \frac{E_{\text{IN}} R_L}{R_S + R_{\text{ON}} + R_L}$$

$$= \frac{E_{\text{IN}} (2 \times 10^7)}{(10^2) + (2 \times 10^3) + (2 \times 10^7)}$$

$$\frac{E_{\text{OUT}}}{E_{\text{IN}}} = \frac{2}{2.0007}$$

$$= 0.9999$$

The error, then, is

$$S = \frac{1 - 0.9999}{1} \times 100\%$$

$$S = (0.0001)(100\%)$$
$$S = 0.01\%$$

This example tells me that 0.01% error is possible, given normal operational amplifier output impedance levels and a worse case 2K switch resistance, if a 20M or larger load impedance is main-

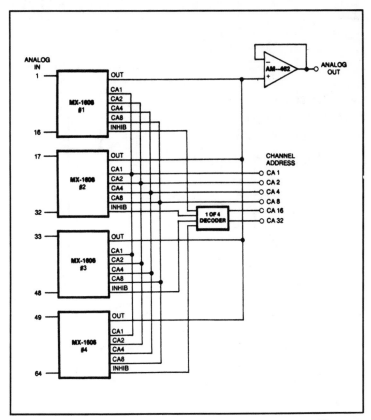

Fig. 14-7. Using four MX-1606 multiplexers in a 64-channel system. (Courtesy of Datel Systems, Inc.)

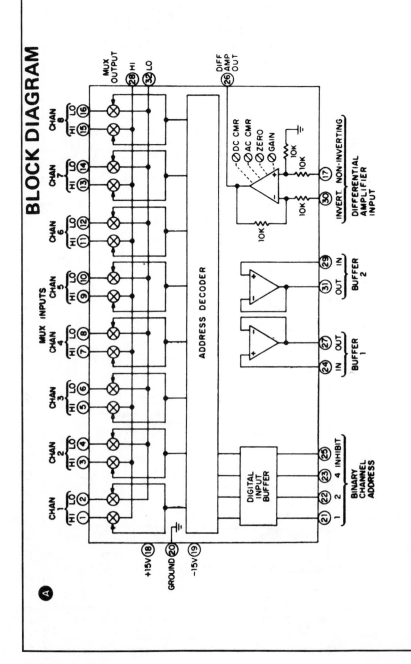

BLOCK DIAGRAM

304

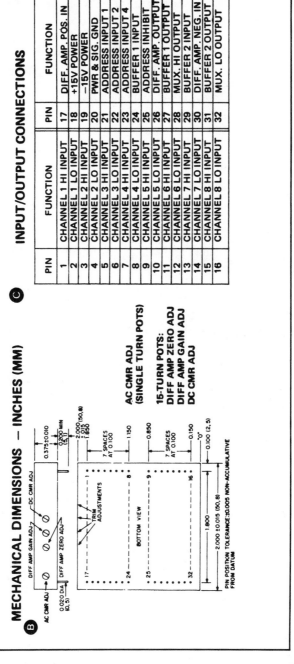

MECHANICAL DIMENSIONS — INCHES (MM)

B

AC CMR ADJ

DIFF AMP GAIN ADJ — DC CMR ADJ

DIFF AMP ZERO ADJ

0.020 DIA.
(0,5)

0.375±0.010

0.200 MIN
(5,1)

2.000 (50,8)
1.800

7 SPACES
AT 0.100

1.150

TRIM
ADJUSTMENTS

BOTTOM VIEW

0.850

7 SPACES
AT 0.100

0.150

"0"

0.100 (2,5)

1.800

2.000 ±0.015 (50, 8)

PIN POSITION TOLERANCE:±0.005 NON-ACCUMULATIVE
FROM DATUM

AC CMR ADJ
(SINGLE TURN POTS)

15-TURN POTS:
DIFF AMP ZERO ADJ
DIFF AMP GAIN ADJ
DC CMR ADJ

C

INPUT/OUTPUT CONNECTIONS

PIN	FUNCTION	PIN	FUNCTION
1	CHANNEL 1 HI INPUT	17	DIFF. AMP. POS. IN
2	CHANNEL 1 LO INPUT	18	+15V POWER
3	CHANNEL 2 HI INPUT	19	−15V POWER
4	CHANNEL 2 LO INPUT	20	PWR & SIG. GND
5	CHANNEL 3 HI INPUT	21	ADDRESS INPUT 1
6	CHANNEL 3 LO INPUT	22	ADDRESS INPUT 2
7	CHANNEL 4 HI INPUT	23	ADDRESS INPUT 4
8	CHANNEL 4 LO INPUT	24	BUFFER 1 INPUT
9	CHANNEL 5 HI INPUT	25	ADDRESS INHIBIT
10	CHANNEL 5 LO INPUT	26	DIFF. AMP. OUTPUT
11	CHANNEL 6 HI INPUT	27	BUFFER 1 OUTPUT
12	CHANNEL 6 LO INPUT	28	MUX. HI OUTPUT
13	CHANNEL 7 HI INPUT	29	BUFFER 2 INPUT
14	CHANNEL 7 LO INPUT	30	DIFF. AMP. NEG. IN
15	CHANNEL 8 HI INPUT	31	BUFFER 2 OUTPUT
16	CHANNEL 8 LO INPUT	32	MUX. LO OUTPUT

Fig. 14-8. Datel MMD-8 hybrid function module multiplexer. (Courtesy of Datel Systems, Inc.)

305

tained. This is obtained by specifying a high input impedance operational amplifier such as the LF156 or RCA CA3160 for the follower stage after the multiplexer.

Figure 14-7 shows how four sixteen-channel multiplexers such as the Datel MX-1606 can be connected to form a 64-channel multiplexer. To address 64 different channels requires a six-bit address code (2^6 = 64). In the circuit of Fig. 14-7, this is created by tying together in a four-bit parallel bus format the $CA1$, $CA2$, $CA4$, and $CA8$ lines from all four MX-1606 devices. The remaining two address lines are fed to a 1-of-4 decoder that sequentially activates the inhibit lines of each MX-1606 device. These form $CA16$ and $CA32$ address lines. The common output lines from the four devices are connected together at the input of a high impedance unity gain follower.

DATEL MMD-8 ANALOG MULTIPLEXER

Figure 14-8 shows a hybrid analog multiplexer function module. This device is an eight-channel differential multiplexer. It has a three-bit address bus input and an inhibit terminal.

The MMD-8 contains three different isolated operational amplifiers. Two are unity-gain noninverting followers used primarily as input buffers. The third is a unity-gain dc differential amplifier with zero offset, gain, and both ac and dc common-mode rejection (CMR) adjustment capability. It can handle signals of ± 10 volts, and offers switching times on the order of 500 nanoseconds (without amplifiers, which tend to slow down any circuit). Settling time for 0.01% of full-scale output, with amplifiers, is on the order of 4 microseconds. This device would be selected in most critical applications that could not be handled by the lower MX-series devices.

SOME DIGITAL MULTIPLEXER DEVICES

Digital multiplexers are usually called *data selectors* in manufacturer's catalogues. In CMOS such devices are the 4019, 4539, and 4512; while in TTL the 74150 through 74160 series of devices are available. Any of these will translate an input logic level or its reverse to the output on receipt of a command from the address bus. There are too many devices to enumerate here, so I refer to the following list and any TTL data book:

74147	1-of-8
74148	8-to-3

74150	16-to-1
74151	8-to-1
74152	8-to-1
74153	dual 4-to-1
74154	4-to-16
74155	dual 2-to-4
74157	quad 2-line
74158	quad 2-line

Chapter 15

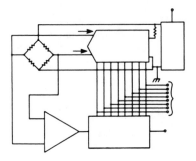

Data Acquisition Systems

DATA ACQUISITION SYSTEMS ARE AVAILABLE FROM SEVERAL
different manufacturers and, on first blush, appear to be very
costly. But designing and building any electronic assembly can be
fraught with problems: errors in concept, unexpected properties
of the electronic components used, glitches due to differences in
the propagation or closing times of allegedly identical gates, power
supply problems, and so forth. In general, it is a long way from
the initial motivation to design and build a product and actually placing the finished product into service. In a small data acquisition
system these problems are not too terrible, but as the number of
channels increases, say to more than six or seven, then the severity
of even simple problems also increases dramatically. It is simply
not enough to pick an A/D converter and place it on a PC board;
many of these systems are simply not of the plug-and-chug variety.

Fortunately, several manufacturers offer printed circuits or hybrid function modules that are complete data acquisition systems;
including all necessary data multiplexers, A/D converters, and control logic. Models are available that provide 8, 16, 32, or 64
channels.

There are three basic approaches to making these systems: hybrid function modules, universal printed-circuit card models, and
PC boards designed for use in a specific brand or model computer.

Which is most appropriate is determined by your own application and resources.

An example of the hybrid system is the Burr-Brown (International Airport, Industrial Park, Tucson, AZ 85734) MP-10/MP-11 and MP-20/MP-21 modules.

The MP-10 and MP-11 analog output modules are shown in Fig. 15-1. Both of these subassemblies are completely self-sufficient and require no external logic other than the computer. The

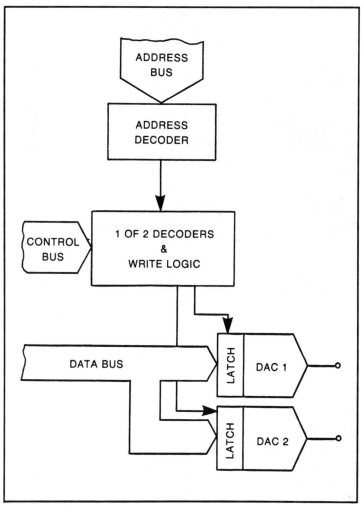

Fig. 15-1. MP-10/MP-11 data output system by Burr-Brown.

microprocessor sees these modules as a memory location rather than as an I/O port. This means that only a single instruction is needed to latch a new word into the D/A converter of each of the two channels.

The MP-10 and MP-11 require approximately 90 milliamperes of +5 volts dc and ±30 milliamperes of ±15 volts dc. Placed on a circuit card with a standard plastic package three-terminal regulator (750 to 1000 milliamperes), these devices leave plenty of current capacity for other circuits.

The MP-10 is directly compatible with the following microprocessor chips: 9080, 8080A, Z-80, 8048, and SC/MP. The MP-11 works with the 6800, F8, and 650X series.

The MP-20 and MP-21 are data acquisition input modules. They are also treated as memory. Programming is by hard-wiring a certain set of programming terminals (i.e., $\overline{A}4$ through $\overline{A}15$) that determine the code that the address bus inputs will recognize.

The address bus is divided into two sections. $A0$ through $A3$ are used as the channel select bits. Channel 0 is selected if this low-order half-byte (i.e., nybble) is 0000_2, while channel 15 is selected if the code is 1111_2 (i.e., $F16$).

The $\overline{A}4$ through $A14$ lines determine the memory address recognized as being the correct address for the module. For example if we want to select an address in which $\overline{A}14$ is logic 1, then the $\overline{A}14$ terminal is set to logic 0 (i.e., it is grounded). Alternatively, if we want $A14$ to respond to a logic 0, then the $\overline{A}14$ terminal is connected to +5 volts through a 1K pull-up resistor.

Let me consider an example in which the sixteen analog channels are located at a base address (i.e., the address of the lowest-order channel, channel 0) of $F970_{16}$, which in binary is

To encode this address into the MP-20 or MP-21, I would set the $\overline{A}4$ through $\overline{A}14$ inputs as follows.

$$\ldots 0 \quad 0 \quad 0 \quad 0 \quad 1 \quad 1 \quad 0 \quad 1 \quad 0 \quad 0 \quad 0 \quad \ldots$$

$$A14 \qquad\qquad\qquad A4$$

Note: $A15$ is wired internally to respond to only a high.

310

Whenever the programmed code appears on the address bus of the microcomputer (i.e., on $A4$ through $A15$), along with a $\overline{MEMR}$ pulse, then the MP-20/21 will convert the analog signal applied to the channel indicated by the lower-order four bits to a digital form.

Pin $A3$ is grounded for eight-channel operation. The MP-20 and MP-21 are capable of sixteen-channel single-ended operation or eight-channel differential operation. In the sixteen-channel mode a full conversion sequence in 8080A/Z-80 systems would be of the form:

```
LDA     F970
(40 μsec)
LDA     F971

STA

(ADDR)

(40 μsec)

LDA     F972

STA

(ADDR + 1)

(40 μsec)

        .
        .
        .

LDA     F97F

STA

(ADDR + 15)
```

This program will initiate a channel 0 conversion by the MP-20 when a LDA F970 instruction is encountered in the program. The processor must then handle other chores, or loop, or go through some no-op (nonoperational) steps, for at least 40 microseconds to allow the MP-20 time to perform the analog-to-digital conversion. When the LDA F971 command is encountered, the MP-20 transfers the result of the first conversion into the computer's accumulator and begins the second conversion (i.e., of the analog data on channel 1, at location F971).

The next instruction, which can be executed while the data conversion is taking place, as part of the 40 microseconds, saves the data from the first conversion, which is now in the accumulator, by transferring it to a memory location given by the next byte (an address location), or two. This procedure will continue until all sixteen channels are converted. Note that it may not be necessary to simply store the data, and you might want to do some processing during the 40-microsecond conversion interval.

Figure 15-2 shows how the MP-10 and MP-20 can be interfaced with a Z-80 microprocessor chip and memory system. Note that for systems less than 32K in size, there is no possibility of erroneously using a real memory location for the data acquisition system,

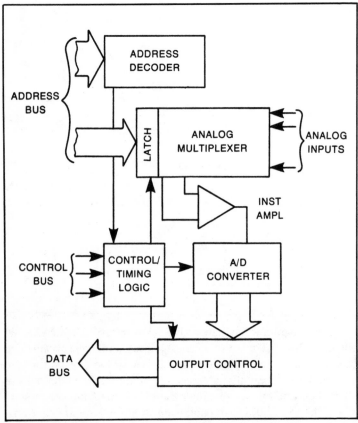

Fig. 15-2. MP-20/MP-21 data acquisition system by Burr-Brown.

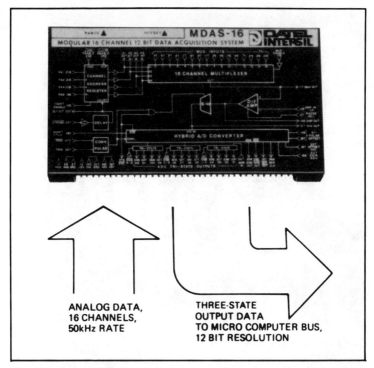

Fig. 15-3. Datel MDAS-16 hybrid data acquisition system. (Courtesy of Datel Systems, Inc.)

because in the Burr-Brown devices $A15$ only responds to a logic 1. This condition places all possible memory locations above 32K (i.e., 2^{15} = 32K). The lowest address that the modules ordinarily recognize is 10000000 (i.e., 32K). They can be forced to respond to addressed locations in the lower 32K of memory by connecting an inverter between the address line and the $A15$ inputs on the data acquisition modules.

Another hybrid data acquisition module is the Datel MDAS-8 and MDAS-16 system shown in Fig. 15-3. These devices also allow the user to select eight differential or sixteen single-ended channels, although in different modules. The MDAS-8 is the eight-channel differential model, while the MDAS-16 is the sixteen-channel model.

Output coding is straight binary in unipolar operation, and offset binary or two's complement in bipolar operation.

The full-scale input range is pin programmable for 0 to +5 volts, 0 to +10 volts, ±2.5 volts, ±5 volts, and ±10 volts.

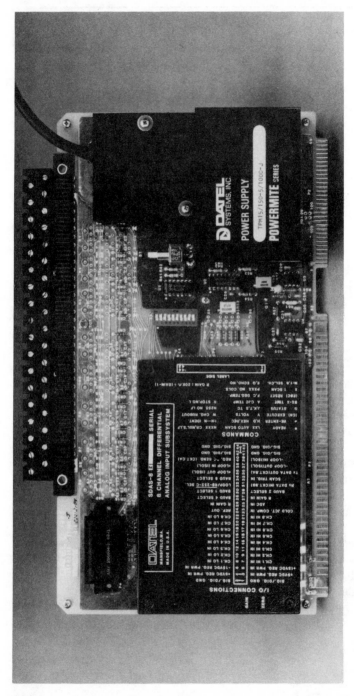

Fig. 15-4. A specialized data acquisition system by Datel (courtesy of Data Systems, Inc.).

The Datel model offers 12-bit resolution at a 50 kHz rate, but they can be short-cycled if less resolution is tolerable. The short-cycle throughput rates are

Bits	Rate (kHz)
12	50
10	53
8	57
4	67

The MDAS-8 and MDAS-16 modules are housed in a painted, black steel, shielded case, and have a 72-pin connector along one edge so that they can be installed on a PC board.

Both Burr-Brown and Datel models can be mounted on one of the universal prototyping boards that are compatible with your microcomputer. Digital Group makes their own board available, while Vector Electronics and others make the S-100 prototyping cards available. This construction technique allows building a data acquisition system that plugs into your mainframe mother board.

Where the computer mother board is filled, or there are other compelling reasons not to use the above method, then build the data acquisition system in a prefabricated cabinet or rack-mounted chassis and panel combination.

Figure 15-4 shows the type of specialized PC data acquisition system available from Datel. Although intended to be plugged into the mainframe of a specific computer, they can often be adapted for other models by appropriate consideration of the circuitry and clever design.

Chapter 16

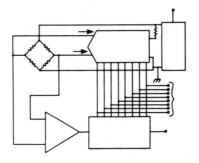

Readout & Display Devices

THE HOBBY COMPUTERIST ORDINARILY USES A VIDEO DISPLAY to read data out to the operator. In other cases, though, some other means is more appropriate, or may be required. Where the computer is an integral part of another instrument, for example, I may require either a simple digital display or an analog recorder as the output device.

Hard copy can be produced by any of several devices, but perhaps the lowest in cost is the government or commercial surplus Teletype machine. These devices date back to World War II and before, and are often available at low cost. One must be cautioned, though, that the price of even old clunkers is rising as amateur computerists find them and place them back into service. Once, only a minority in amateur radio were interested in these machines, and certain older models sold for as little as $25 in relatively good condition. Today, however, prices are several times that amount for what is actually little more than junk.

The older Models 15 and 28 are usually encoded in one version or another of Baudot code, while newer Model 33-series machines may well be encoded in ASCII.

All Teletypes, however, use serial transmission of the code bits in either a 20-milliampere (recent) or 60-milliampere current loop.

The signal terminal board for the Model 33 contains eight screw terminals, labeled 1 through 8, left to right. Their terminal functions are:

Pin	Function
1	110 volts ac
2	110 volts ac
3	Send minus
4	Send plus
5	Receive minus
6	Receive plus
7	—
8	—

Although it has become standard, if somewhat sloppy, practice to call all teletypewriters by the generic "teletype," that word is properly applied only to products of The Teletype Corporation, 5556 Touhey Avenue, Skokie, IL 60076, of which the word "teletype" is a registered trademark.

The IBM Selectric electric typewriter is available in a solenoid-controlled model that is used in computer printout service. These machines use their own Selectric correspondence code, rather than either ASCII or Baudot.

There are also a lot of other teletypewriter computer terminals manufactured by Olivetti, Digital Equipment Corporation, and others, but their costs are usually too high for amateur use and they have not been around sufficiently long for any significant number to have shown up on the surplus market. Within a few years, however, you can expect to see some of the better printers and typewriter terminals available used at good prices.

Figure 16-1 shows a low cost alphanumeric printer manufactured by Datel especially for the microcomputer applications market.

Various input electronics options are available and allow use of the printer with a variety of instrumentation and computer-based devices. Direct microcomputer computer compatibility is provided by an eight-bit parallel interface. Also available are serial interfaces in either RS-232C or 2-milliampere teletypewritten current-loop formats. The serial data transmission options include the standard speeds from 110 baud to 9600 baud, and there is on-board buffer memory for either 120 or 200 characters.

NUMERICAL DISPLAYS

The digital computer produces binary outputs, usually in an

Fig. 16-1. Datel Model APP-48 printer (courtesy of Datel Systems, Inc.).

eight-bit format. The same type of seven-segment display used in frequency counters and other noncomputer digital instruments can also be used as an output for the computer if a suitable means for converting the eight-bit data at the output into binary-coded decimal (BCD) format is provided.

An alternative approach is to design a hardware circuit that will accept eight-bit binary and convert it into BCD.

Analog Outputs

A digital-to-analog converter can be used to convert the eight-bit binary output of a microcomputer to a proportional voltage or current level. The availability of low-cost eight-bit D/A converters in integrated circuit form makes this approach even more appealing, but more about that is found in Chapters 10 and 12.

Although not strictly intended for computer interface service, the digital panel meter (DPM) coupled with a voltage output D/A converter can make a very effective readout device. An example of the digital panel meter is shown in Fig. 16-2. Most digital panel meters are offered in models that cover either 0 to 1.999 volts (most common) or 0 to 19.99 volts. Many DPMs have a feature that allows the user to reposition the decimal point, or eliminate it altogether. This ability allows you to conform to the numerics of the units being measured. A pressure monitor, for example, might use 0 to 1 volt to represent pressures from 0 to 100 torr. A 0- to 1999-millivolt digital panel meter would read *.500* at, for example, a pressure of 50 torr. By repositioning the decimal point we could make the display read *50.0* when a 500-millivolt input (representing 50 torr) is applied.

Chart recorders are probably the most common and popular of all analog readout instruments. The reason is that they can be used to provide a hard-copy readout in graphical form. An obvious

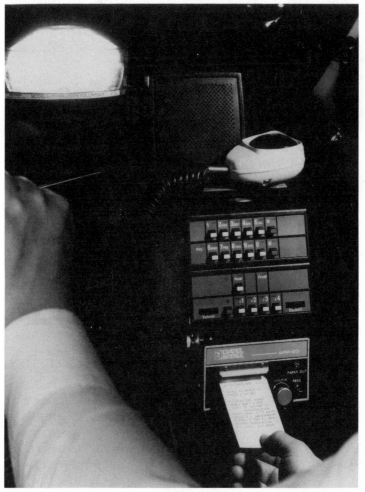

Fig. 16-2. Datel APP-20 printer (courtesy of Datel Systems, Inc.).

advantage is that humans are more used to interpreting graphics than columns of numbers.

Drum recorders are built with a cylindrical paper carrier, most of which are designed to accept the standard engineering and scientific graph paper sizes. To some users this is an advantage in its own right.

Most drum recorders are of the incremental type. The horizontal axis is controlled by the stepwise rotation of the drum paper carrier. A stepper motor is pulsed by an electronic time base circuit, so that the chart rotates in a precisely controlled manner.

A pen is mounted on an assembly that sweeps the vertical axis, again in a stepwise manner. The size of the step will depend upon the digital word at the vertical input. Drum recorders are available which produce small enough increments (i.e., 0.01 inch or 0.25 mm) to create graphs that appear continuous.

A related, but different, type of recorder is the strip-chart recorder. This type of machine will produce a graphical display on a continuous strip of paper stored in either roll or Z-fold form.

Strip-chart recorders make use of several techniques for writing the analog waveform onto the paper. These include both galvanometer and servomechanism systems, and such writing media as thermal, ink jet, ink pen, and several varieties of optical recording.

The three basic systems for positioning the analog waveform along the vertical axis include the stepping motor just described, permanent magnet galvanometers, and servo-control methods.

The permanent magnet moving coil (PMMC) galvanometer is a mechanism that very nearly resembles the classical D'Arsonval analog meter movement, except that a pen is used instead of a pointer. A moving coil bobbin is positioned in the field of a large permanent magnet, and is electrically connected to the input signal. When no signal currents flow in the coil, the pointer (i.e., writing stylus) is at a rest, or equilibrium, position. A signal current flowing in the coil will create a magnetic field that opposes or aids the magnetic field of the permanent magnet, depending on the current's polarity. The deflection of the coil bobbin is proportional to the strength of the signal current, so the deflection of the stylus is also proportional to the signal amplitude.

In the servo system type of recorder, there is a transducer, either a potentiometer or ac motor, that produces a pen position signal. This signal is compared with the input signal being recorded, and an error signal is established; this signal drives a pen motor in such a manner that the error signal is cancelled to zero. A pen is ganged to the pen motor, so accurately follows the input signal waveform.

The actual writing process can take several forms. For narrow strip charts, such as the common medical electrocardiograph machine thermal recording is the predominant method. The paper, usually 50 mm wide single or 100 mm wide dual-channel paper, is treated with paraffin. The writing stylus has a heated tip, so a black mark is made on the paper wherever the tip touches the paper.

Some recorders, in both wide and narrow chart versions, use actual ink pens. Some wide chart recorders use a disposable felt tip, or a refillable reservoir type of cartridge. At least one model requires an ordinary cartridge type fountain pen (see Fig. 16-3). Most narrow paper strip-chart recorders (i.e., those similar to the medical EKG machine) using ink pen galvanometers have a hollow stylus pen connected to an ink reservoir through a thin, hollow, capillary tube.

The frequency response of any pen type of writer is limited by the mechanical inertia of the pen assembly. High-frequency limits for such devices are usually in the 100- to 200-hertz region, and certainly under 500 hertz. The recording of frequency components out to several kilohertz requires other means, all but one being optical techniques.

The one mechanical method is the high-velocity ink jet PMMC galvanometer. A nozzle is mounted to the PMMC bobbin, and this nozzle is connected to a high-pressure ink reservoir. Ink is sprayed onto the paper at high velocity. The low-viscosity ink is designed to dry rapidly, so little splatter takes place.

The primary advantage of the high-velocity ink jet system is that the nozzle can be made very light weight, so there is less inertia to slow down the frequency response.

The main optical methods are the Polaroid oscilloscope camera, a slower oscilloscope camera for strip-chart recording, and the mirror galvanometer.

The majority of oscilloscopes above the hobbyist or service grade can be equipped with a special Polaroid camera to photograph waveforms. A high-speed film (i.e., Polaroid Type 107 with an ASA rating of 3000) is used to make the images. Where an expensive camera for a particular model is not available, or where the cost is prohibitive, it might be wise to settle for the hand-held Polaroid Model CR-9 oscilloscope camera. A variety of light-tight hoods are available that adapt the CR-9 to almost any standard oscilloscope.

Another type of oscilloscope camera is the CRO-based system used by Electronics-for-Medicine (E-for-M) in their series of physiological monitoring equipment. In their system, a sheet of photosensitive paper is passed over a cathode-ray tube screen that displays the waveform of interest. The paper will fade somewhat if not developed properly or stored in a light-tight box. This system is capable of providing very good frequency response, limited only by the writing speed of the paper (the CRT, or course, is capable of a lot higher speed than the paper).

The last class of optical recorders which I will consider is the mirror galvanometer. This system uses a PMMC galvanometer with a light-weight mirror mounted on the bobbin instead of a heavy pen assembly. A thin light beam is directed at the mirror, which reflects the beam onto the surface of the photosensitive paper. If an input signal perturbs the bobbin, the mirror deflects the light beam across the paper an amount proportional to the amplitude of the signal. Like the CRO method the paper can be dry developed in an ultraviolet light, or wet developed for more permanence.

SOME ACTUAL PRODUCTS

A low-cost servo-type strip-chart recorder is shown in Fig. 16-3. This machine is the Heath Model IR-18M, which is unique in that it is available in both assembled and kit (you assemble) versions. Having built one of these, I can attest to the ease of construction. All but the most mechanically inept should be able to successfully follow the unusually well-written instructions in the assembly manual.

The pen in this model is an ordinary cartridge type fountain pen, although I have found that many of the felt type of pens will work at least as well.

Chart speeds can be selected over the range of 5 inches per minute to over 200 inches per minute, through judicious selection of chart speed push-buttons. The paper is sprocket driven from a stepper motor that is controlled by a digital speed circuit.

A related, but somewhat more sophisticated, model also by Heath (but not in kit form) is the two-channel strip-chart recorder shown in Fig. 16-4. This instrument is the Heath Model SR-206. It is similar to the IR-18M in basic function, but boasts considerably greater versatility due to added features.

Up until now all of my examples and discussions have concentrated on recorders that assume that the variable being recorded

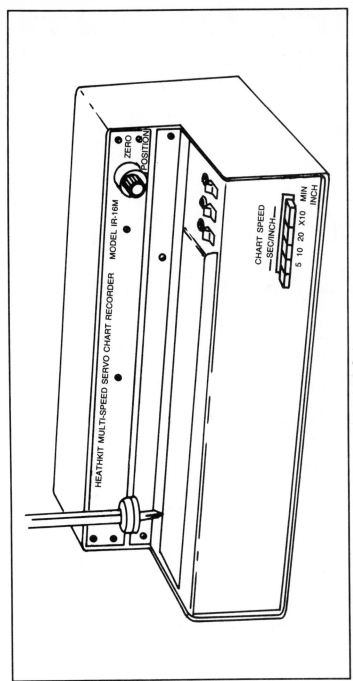

Fig. 16-3. Low-cost strip-chart recorder. (Courtesy of the Heath Co.)

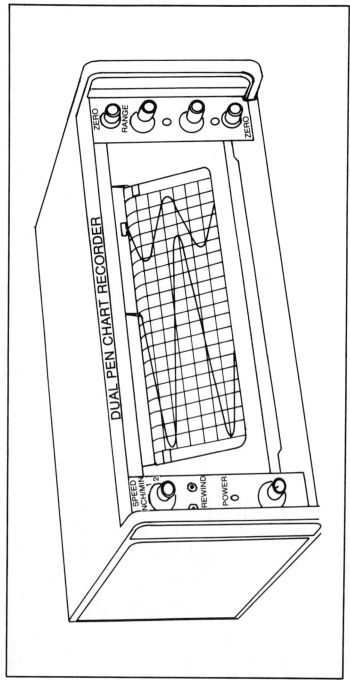

Fig. 16-4. Professional two-channel strip-chart recorder. (Courtesy of Heath Co.)

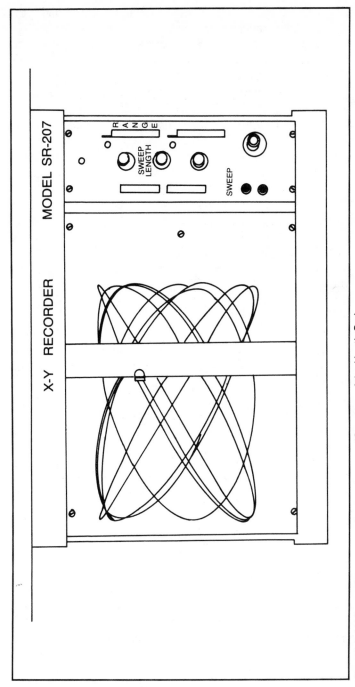

Fig. 16-5. X-Y recorder uses regular graph paper. (Courtesy of the Heath Co.)

is time dependent. A motor is used to drag the paper under the writing pen, automatically creating a time base on the X-axis which has an accuracy equal to that of the paper speed. Although the CRO optical methods are sometimes used for recording of nontime-dependent variables, it is usual to use an X-Y recorder for such work.

The Heath Model SR-207 X-Y recorder is shown in Fig. 16-5. This type of instrument will record on standard sizes of graph paper, so the display format is especially useful for human interpretation.

The write pen is fastened to a vertical bar. The Y-amplifier drives the pen up and down the bar, while the X-amplifier drives the bar left and right along the paper's horizontal edge. If a Y-time (instead of Y-X) display is required, then it is merely a matter of connecting a ramp function voltage to the X-input. If the slope of the ramp is precisely known, then a time base is created. The ramp amplitude must be adjusted so that the bar is driven all the way to the right after precisely the time selected for the full scale X-value.

REAR END CIRCUITRY

Let me ask you a personal question: "Have you got rear end troubles?" Now, I'm not so bold as to take the place of your doctor. What I am talking about is your electronic instrumentation projects (or at least some of them). One of the major failings of hobbyist-designed projects, plus not a few supposedly professionally designed instruments, is inadequate attention to the rear end circuitry. In most instruments, the really neat circuits are in the front end, so the rear end is merely the output section. But proper design of the rear (or "output") end of an electronic instrument/control circuit can make the difference between "ho-hum" working and a really useful device.

First, though, let me get some terminology straight. What do "rear end" and "front end" mean, if not human anatomy? In the electronics field, the front end of a device is the section where the input signals are received. The RF amplifier in a communications receiver, the transducer amplifier in a temperature monitor, or the differential amplifier in an ECG amplifier or Wheatstone bridge transducer circuit. Almost all of the signals processing or shaping is done in the front end of the circuit. The rear end, on the other hand, is less glamorous but none the less important. The rear end of the circuit is the output stages.

Figure 16-6 shows a semi-universal rear end circuit that I have

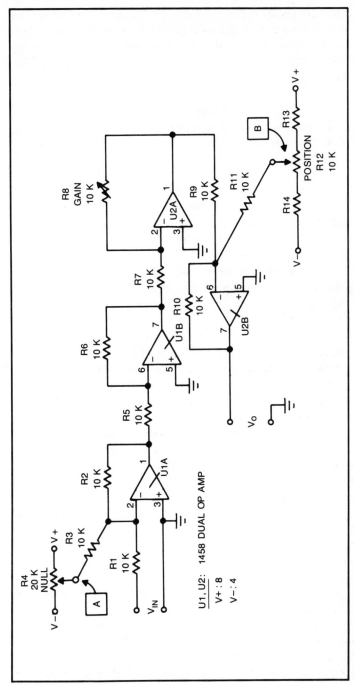

Fig. 16-6. Universal "rear-end."

used in any number of cases. During the years when I worked in a medical school/hospital environment, I built this circuit into several physiological (biopotentials) amplifiers, quite a few transducer amplifiers, and numerous other projects. The circuit provides several features:

1. Gain (if $R2 > 10$ kohms), otherwise unity gain
2. Gain control (0 to 1, or 0 to A_v)
3. Dc balance control/offset null
4. Position control

The circuit is constructed from two type 1458 dual operational amplifiers. The 1458 contains two 741-family operational amplifiers inside of an 8-pin miniDIP package. Any other standard operational amplifier will also work; the 1458 is used to reduce the parts count and wiring time.

The dc power supply for this circuit must be bipolar. In other words, it requires both positive-to-ground $(V+)$ and negative-to-ground $(V-)$ dc power supplies, as with any other operational amplifier circuit that does not use special techniques. Potentials between $-/+4.5$ and $-/+15$ volts dc will serve nicely.

The circuit of Fig. 16-6 is made up of four inverting followers in cascade, so it serves overall, as a noninverting follower. Removing any one stage (U1B recommended) will make the circuit into an inverting follower, since there would be an odd number of inverting stages.

The gain of an inverting follower operational amplifier stage is set by the ratio of the feedback and input resistors. For example, the gain of stage U1A is $-R2/R1$, where the negative sign indicates that inversion (180-degree phase reversal) is taking place. Similarly, the gain of U1B is $-R6/R5$; of U2A, $-R8/R7$; and of U2B, $-R10/R9$. The overall gain of the entire circuit is simply the product of all the individual gains:

$$A_{vt} = A_{V1} \times A_{V2} \times A_{V3} \times A_{V4}$$

Since all gains in the circuit as shown in Fig. 16-6 are unity (i.e., one), the overall gain is one. But I can increase or decrease the overall gain by varying the stage gains. The recommended procedure is to vary the gain of the fixed stage, U1B. In this case, I can assume that the gain of the overall circuit will be $-R6/R5$. Leave $R5$ equal to 10 kohms in most cases, unless it is impossible to find a value of $R6$ that will result in the correct gain without modifying the value of $R5$. In any event, do not let $R5$ become less than 1000 ohms. The rules for changing the gain are the following:

1. Unity gain: leave as is ($R5$ = $R6$ = 10 kohms).
2. Less than Unity Gain: $R5 > R6$ (e.g., $R5$ = 10 kohm, $R6$ = 2000 ohms, yields a gain of 2,000/10,000 or 0.20).
3. Greater than Unity Gain: $R5 < R6$ (e.g., $R5$ = 10 kohms, $R6$ = 100 kohms, yields 100/10 = 100).

The *Gain Control* ($R8$) is used to vary the overall gain of the circuit from zero to full gain. If the values are as shown in Fig. 16-6, then $R8$ varies the gain from 0 to 1. This potentiometer is usually a front panel control, making it easily accessible.

The *Null Control* is used to cancel the effects of dc offsets created both in this circuit and in previous stages. It also provides the dc balance effect noted earlier. The "Dc Balance Control" on some instruments is used to cancel the change of output baseline as the sensitivity control is varied—a most disturbing effect when trying to make a measurement! Potentiometer $R4$ is adjusted using a dc output meter at V_o, and is adjusted until there is no shift in dc output when the Gain Control is varied through its full range. If there are dc offsets present in the input signal (V_{in}), then there will be such a shift noted in the output. The function of $R4$ is to provide an equal, but opposite, polarity offset signal to cancel the offset from all other sources. In some cases, there might be 10-kohm resistors (similar to $R14$ and $R15$ near $R12$) between the ends of the potentiometer and the power supply potentials. These resistors reduce the offset range, while increasing the resolution of the adjustment. Use those resistors only if there is a problem homing in on the correct value.

The *Position Control, R12,* is optional, and is normally used when the output signal is to be displayed on an analog paper chart recorder, or dc CRT oscilloscope. Potentiometer $R12$ provides an intentional offset to the final stage (U2B) that is quite independent of the input signal. The effect is to position the output waveform anywhere on the oscilloscope or chart recorder vertical axis that you might desire.

In some cases, the range of the potentiometer is too great. Only the slightest adjustment of the potentiometer will send the trace offscreen. I can counter this problem with the simple expedient of selecting values for $R13$ and $R14$ (note: $R13$ = $R14$) that allow the trace to just disappear off the top when the potentiometer reaches the limit of its upward travel, and off the bottom when the potentiometer reaches its lower limit.

Adjustment

The adjustment of this circuit requires either a dc voltmeter or a dc-coupled oscilloscope that has a grid on the screen so that potentials can be read. If the oscilloscope is used, then short the input with the oscilloscope switch marked for that purpose ("GND" in "AC-GND-DC" on some models), and set the trace to exactly the center of the vertical lines on the grid. Select a sweep speed that yields a nonflickering line. Next place the switch into the "dc" position. The vertical deflection factor should be around 0.5 volts/division. The procedure is as follows:

1. Disconnect V_{in} from the front-end circuit and short this input to ground.
2. Using a dc voltmeter, set the potential at point A to 0.00-volts.
3. Similarly, set the potential at point B to 0.00-volts.
4. Set $R8$ to maximum resistance (highest gain).
5. Make all adjustments to the front end circuits as needed, and then return to the rear end circuit.
6. Adjust $R8$ through its range from 0 to 10 kohms several times while watching the output indicator (oscilloscope or meter). If the output potential shifts, then adjust $R4$ until the shift is cancelled. You will have to continually run $R8$ through its range while adjusting $R4$. This adjustment is somewhat interactive, so try it several times, or until no further improvement is attainable.
7. Check the range of the position control.

The universal rear end circuit is a simple project that can give your instrument and control projects that final, "professional" touch that makes them more useful to all concerned. The display analog output will be more easily controlled if the circuit of Fig. 16-6 is used to "dress up" your project.

Chapter 17

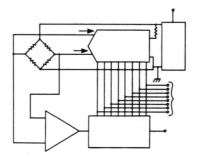

Serial Data Transmission

IT IS OFTEN THE CASE THAT DATA ORIGINATED AT ONE LOCA-
tion must be transmitted to another location before it can be used.
Some instrument, transducer, or other sensor at a remote location,
for example, might send data to a computer or other digital processor at another location.

Technical difficulties preclude sending analog voltages or currents over the lines. Unless the analog source and the computer
are located relatively close to each other problems will interfere
with the data. An analog-to-digital converter should be located as
close to the analog source as possible.

If an eight-bit data format were determined, so that it would
be compatible with microprocessors, then a parallel data communications system will require not less than eight parallel lines, and
some popular schemes would require even more. Across the room,
or even down the hall, this is not too terrible, but when you connect to public common carrier lines such as the telephone, Telex,
or Western Union, the cost of eight parallel channels would be pro-
hibitively high. It would also represent a serious waste of resources
that might be better spent elsewhere.

The same situation pertains when connecting a remote CRT
terminal or teletypewriter to the main or central computer, or when
setting up a communications network that would allow two or more
computers to talk to each other. The cost and lack of practicality

of using a *parallel* data transmission format makes the *serial* mode of transmission a very attractive alternative.

THE TELEPHONE SYSTEM

Most serial data transmission takes place over the telephone lines. In this chapter, I will assume that your application will require such a system, although the same techniques would also suffice if you used radio telemetry or twisted-pair local wire connections. The most important thing is the interfacing techniques.

The United States telephone system consists of wire cables, coaxial cables, microwave relays, and satellite links. In most ordinary channels that you can access from an ordinary residential or business telephone receiver, the bandwidth of the system is limited to the range 300 to 3000 hertz. This limitation sets the maximum data transmission rate, forcing the use of audio representations for data states that fall inside of these limits.

Sound level on the telephone system is measured in a decibel scale called *volume units* (VU), which is defined using the power decibel equation $VU = 10 \log 10 (P1/P2)$. The term dBm refers to a standardized system in which the zero-VU value is defined as 1 milliwatt of audio power dissipated in a 600-ohm load at 1000 hertz. The originating source is expected to create a sound level on the line of approximately -5 VU, while the receiver must be prepared to accept signals between -45 VU and -10 VU.

In data communications the high and low logic conditions (i.e., *one* and *zero*) cannot be connected directly to the telephone system, but first must be converted to audio tones that fall inside of the 300 to 3000 hertz band pass of the telephone system. This same trick must also be applied when transmitting by radio or tape recording the data signal.

There are two ways to apply the audio tones representing data states to the telephone system, direct and acoustical coupling. Each have their own respective advantages and disadvantages, so both will be considered here.

The telephone companies are understandably concerned about people connecting devices to their system over which they have no control. It is important to keep the sound level on the line within specifications, and to prevent dc from getting on the line; both situations can cause service problems and equipment damage.

Until recently, only local telephone company policy, which was often very conservative, determined whether or not an individual could interconnect a foreign attachment to telephone company-

332

owned lines. Amateur (ham) radio operators often connected their phone patches so that they could perform public service chores for distant amateurs, servicemen overseas, etc. These phone patches sometimes had the blessing of the phone company because of the public service, not-for-money angle. In other cases, the amateurs would elect to ignore the phone company. They could get away with this so long as they kept a low profile and their equipment was of such a design that it did not draw attention to itself.

More recently, the Federal Communications Commission (FCC) and the U.S. Courts have allowed certain foreign attachments without the blessing of the phone company. Perhaps the first really important case in this area was the now famous *Carterfone* decision. Interconnection is still considered a gray area however, in that the telephone company (as of this writing) can still require you to rent a device called a *voice coupler*. These are little gray boxes, connected between your equipment and the telephone line, containing two capacitors (one for each wire) and a voltage dependent resistor (VDR) across the lines on the subscriber side. The VDR limits the audio level so that you don't overdrive the lines.

Even more recent decisions allow commercially made equipment, that is certified to comply with phone company technical specifications, to interconnect without a voice coupler. But amateur and hobbyist equipment, as well as laboratory instruments or commercial equipment not so-certified, can be required to interconnect through a voice coupler or other telephone company-specified device.

Perhaps the easiest way to perform a direct interconnect is to use an amateur radio phone patch device. The Heath Company (Benton Harbor, MI 49022) offers a quality device in kit form that is popular among ham operators. Or alternatively, you may wish to construct your own hybrid phone patch from the ground up (no pun intended).

The circuit in Fig. 17-1 is representative of a large class of published circuits. It is essentially a Wheatstone bridge. If only a single receiver or transmitter device is required, then only a single transformer would be required. This circuit allows the parallel connection of two devices.

An alternate form is shown in Fig. 17-2. A single transformer is used to match two devices to the line. The secondary of the transformer should be 1000 ohms, while the primary should match (as best it can) the two devices connected to lines 1 and 2.

Switch S1 can be a mechanical switch if the application allows

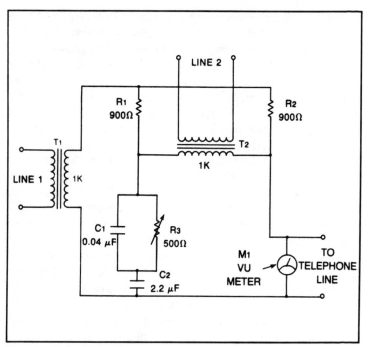

Fig. 17-1. Phone patch circuit.

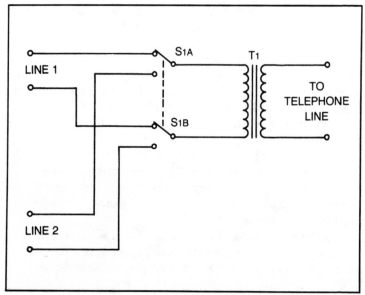

Fig. 17-2. Simple half-duplex telephone connection.

334

slow hand selection of the channel. But a CMOS or other electronic switch should be selected if the switching speed is somewhat faster, or if it is desired to switch lines under more automatic control. The latter is considered more elegant, but the former is simple and has one engaging aspect: it is cheap.

An *acoustical coupler* is the method for connecting to the telephone lines more often employed in slow-to-moderate speed data communications systems. Figure 17-3 diagrams the basic type of acoustical coupler. A microphone is placed against the telephone earpiece to pick up incoming signals, while a small loudspeaker is placed against the telephone mouthpiece to send tones to the line.

Commercial acoustical couplers use large rubber cups to house the microphone and loudspeaker. These cups are constructed such that a standard telephone handset will fit snugly, sealing out extraneous room noises.

Although usually quite costly (the cost has dropped markedly in the past few years) when purchased new on the open market, the data communications industry is now old enough that acoustical couplers and other hardware is appearing on the surplus scene. Additionally, it is noted that various suppliers used by electronic

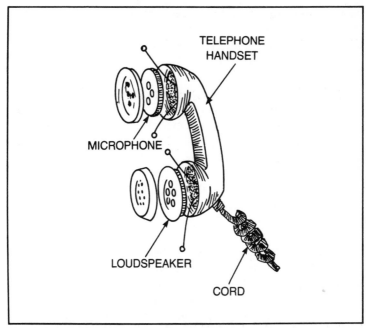

Fig. 17-3. Acoustical coupling to phone system.

hobbyists (i.e., Radio Shack and Lafayette) have offered telephone room amplifiers that are fitted with the same type of rubber cups. Some have only the microphone cup, while others have both. In one model there was a loudspeaker cup and a dummy cup where the microphone would go. It is a simple matter to modify this type of device to permit the construction of a telephone pickup unit.

There are three types of communications circuits, as classified by direction of transmission: simplex, half-duplex, and full duplex. These are different from each other in the manner in which data is passed over the circuit.

There are actually two meanings for the word simplex. One definition maintains that a simplex network is one that can pass only a single message at a time. But more commonly accepted in data communications is the definition that maintains that a simplex circuit is a one-way circuit. A transmitter will be located on one end of the circuit and a compatible receiver is at the other.

A duplex channel allows messages to pass in both directions at once; a feat ordinarily requiring two pairs of wires, two separate radio frequencies, or a multiplexing scheme.

The half-duplex channel also allows two-way message traffic, but only in one direction at a time. The circuit of Fig. 17-2 could be a half-duplex system if one line were connected to a receiver and the other to a transmitter. The position of switch S1 would then determine the direction of transmission.

Two common problems associated with communications networks such as the telephone system are crosstalk and echo. These can be troublesome in a data communications system.

Crosstalk exists when signals from one channel show up on another channel. Wire communications systems often have bundled pairs of wires handling many channels. There can be, for example, as many as 900 pairs of wires, each handling a channel, bundled together in a single cable that is less than 3 inches in diameter.

Crosstalk occurs as a result of induction between two or more of the wire pairs. The amount of crosstalk varies as the signal strength of the offending line and the length of the run over which the two lines are parallel to each other.

Echo is the reflection of a signal back to the source due to circuit anomalies, much after the manner of the reflected waves (i.e., SWR) on a ham or CB antenna system. Telecommunications networks use echo suppression equipment to eliminate this problem. The suppression works similarly to a VOX circuit on a ham transmitter. When a person begins to speak, the return path is short circuited. When the party on the other end of the line begins to speak, then the echo suppressor is disengaged (within 10 milliseconds). This results in several interesting phenomena that have implications in data communications networks.

You may, for example, notice on long-distance calls that the line noise (i.e., hiss) stops almost immediately after you begin speaking. If the other party begins speaking too soon after you stop, or if the echo suppressor hangs on the line too long, then the first syllable spoken by the other party may be clipped off, causing you to ask for a repeat. In a data communications system this could cause the loss of several data bits, especially if the data transmission rate is high.

It is, however, possible to disable the echo suppressors by applying a 2025-hertz tone burst of at least 300 milliseconds duration to the line during periods when no data is being transmitted (or by using 2025 hertz as one of your data tones). Any interval of 100 milliseconds or longer in which the disabling tone is interrupted will reactivate the echo suppression equipment.

The rate at which data is transmitted over a channel is called the *baud rate,* although the use of the word rate is redundant because baud is a rate. The baud rate is usually defined in terms of bits per second or the reciprocal of the time in seconds required for the shortest bit used in creating the data word.

THE UART

Until recently the design of serial data network equipment was rather difficult and was considered a chore best suited to clever digital design engineers. But today we have a specialized class of LSI chips that will do the job for us: universal asynchronous receiver-transmitters, or UARTs.

Synchronous transmission requires that the clocks at both ends of the circuit track together because data is sent at precise times. One authority claims that the respective baud rates must be within ± 0.01% of each other. This requirement adds complexity and cost. Asynchronous transmission sends the data in a steady stream, so is a lot more flexible than synchronous systems.

Figure 17-4 shows the block diagram to the standard Western Digital (3128 Red Hill Avenue, P.O. Box 2180, Newport Beach, CA 92663) TR1602A/B UART. This device is essentially the same as more common, but now considered obsolete, devices such as the AY-1013. It is capable of either full-duplex or half-duplex opera-

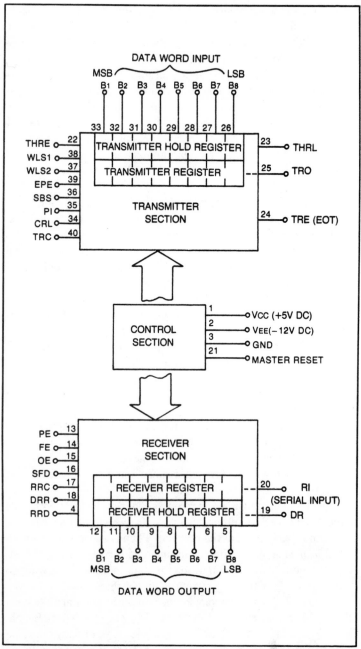

Fig. 17-4. Block diagram to a universal asynchronous receiver/transmitter (UART).

338

Table 17-1. Compatible UARTs.

S1883	2536	TMS6012
AY-5-1013	2502	—
AY-1014	TR1602	—

tion because the receiver and transmitter circuits are totally independent, except for the power supply connections.

The UART chip is particularly useful, earning its nickname universal, because it can be externally programmed for word length, baud rate, parity (odd-even, receiver verification, transmitter generation), parity inhibit, and stop bit length (i.e., 1, 1.5, or 2 stop bits). It also provides six different status flags: transmission completed, buffer register transfer completed, received data available, parity error, framing error, and overrun error. Table 17-1 lists compatible UARTs.

The clock speed is up to 320 kHz for A and B versions, 480 kHz for A03/B03 versions, 640 kHz for A04/B04 versions, and 800 kHz for A05/B05 versions. The receiver output lines are tri-state logic (high impedance when inactive), so can be directly connected to a data bus in a computer or other digital instrument.

The transmitter section has an eight-bit input register that accepts data from a keyboard, computer output port, A/D converter, or any other similar data source. It will convert these to a serial output word that contains the eight-bit input word plus start, parity, and stop bits.

The receiver is conceptually the mirror image of the transmitter. It receives a serial input word containing start bits, data, parity, and stop bits. This data is converted to a parallel eight-bit output word; the transmission is checked for validity by comparison with parity for the existence of stop bits.

The UART data format is shown in Fig. 17-5. The transmitter serial output (pin 25) or receiver serial input (pin 20) will be

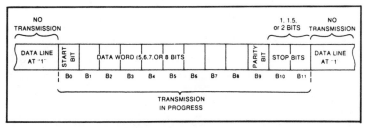

Fig. 17-5. Output word produced by UART.

at a high level (i.e., logic 1) unless data is being transmitted or received, respectively. Start bit $B0$ is always low, and tells the system that a data transmission is about to take place. Bits $B1$ through $B8$ are the data bits loaded into the transmitter on the sending end. All eight bits of the maximum word length format are shown in the figure. Unused bits in shorter formats are ignored, with $B1$ lost in the 7-bit format, bits $B1$ and $B2$ for the 6-bit format, and bits $B1$ through $B3$ are lost in 5-bit operation.

Example 17-1

Design a serial data transmission system around a TR1602A UART that has the following specifications:

> Data rate.......300 baud
> Word length.....8 bits
> Stop bits.......2
> Parity..........even

Solution:

The circuit is shown in Fig. 17-6. The transmitter is shown in Fig. 17-6A, while the receiver is shown in Fig. 17-6B. Some connections are shown in both drawings to enhance clearness.

In both the receive and transmit modes the data rate is 300 baud, so the clock frequency is

$$300 \text{ baud} \times 16 = 4800 \text{ hertz}$$

In most cases it is not wise to use a 555 or any other RC-controlled oscillator as the clock. Most of these circuits are simply not stable enough. The ability of the UART to function depends in part upon the accuracy of the clock at both ends of the system. The only time where the clock loses its importance is in the case where the transmitter and receiver sections of the same chip form a closed-loop communications system.

It is suggested that a crystal oscillator be used as the clock, with a binary counter following the oscillator to reduce the frequency to the required 4800 hertz. One solution would be a string of counters such as the 7490, but a better solution is the single-chip circuit of Fig. 17-6C. This circuit uses a CMOS baud-rate-generator chip and a crystal to achieve the needed frequency.

The required eight bit word length is set by applying the correct code to WLS1 and WLS2 (i.e., pins 37 and 38). By the chart given earlier I know to set both pins to a high logical condition. This programs the UART receiver and transmitter to the eight-bit word length.

Similarly, a two-bit stop code is provided by connecting SBS (i.e., pin 36) high, and even parity is selected by setting EPE (pin 39) high.

In the transmitter part of the circuit (Fig. 17-6A) only the eight-bit data input, clock, and serial output are necessary. The TRE, THRE, and THRL signals are status flags, and are optional. They can be used if the information they convey is needed elsewhere in the circuit. Review the meanings of these flags to determine whether or not they would be useful in your case.

A similar situation exists in the receiver section. Only the clock, serial input, and eight-bit parallel output are needed in this simplified example, but optional DR, OE, FE, and PE flags are available should they be needed.

Notice that the inverter from the *data received* terminal (pin 19) resets the DRR terminal (pin 18), telling it to be ready for the next character.

One of the appealing things about the LSI UART chip is that its two sections can be used either separately or together as required by the application. In a simplex circuit, a UART at the originating end would be wired as the transmitter, while another at the receiver end would be wired, naturally, as a receiver.

If half-duplex operation is desired, then both sections are used at both ends, and the status flags can be used to tell the channel which direction to transmit. Full-duplex operation is also possible, but a second channel may be required.

If both send and receive functions are to be programmed to the same specifications, then the control programming pins are hard-wired in place, as shown in Fig. 17-6. But if different specifications are imposed under different situations, then some sort of external controls are needed. One alternative is to connect the programming pins to a computer output port and the status flags to an input port. This would allow specification changes under program control. Table 17-2 lists the UART pin functions.

RS-232 SYSTEMS

One term that you will hear bandied about in the data commu-

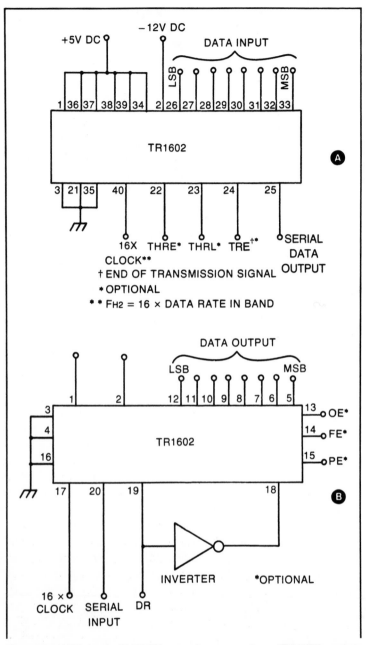

Fig. 17-6. UART circuit. (A) UART transmitter connections. (B) UART receiver connections. (C) CMOS baud-rate generator for use with UART. (Continued on next page.)

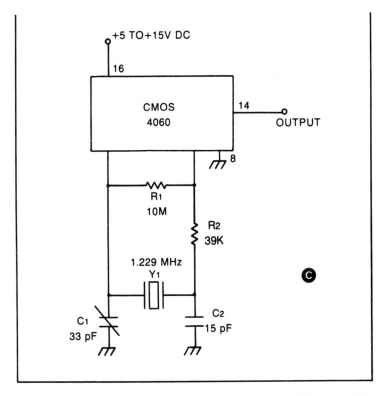

nications field is RS-232 interface. This refers to the Electronic Industries Association (EIA) RS-232 standard for serial data transmission systems. Modems, CRT terminals, and other peripherals are often designs with an RS-232 I/O compatibility.

The RS-232 standard is an older set of specifications, and it predates even TTL integrated circuit technology. Because of this situation, the voltage levels that represent logic 1 and logic 0 in the RS-232 standard appear a little difficult to handle. The voltage levels and the load impedances are fixed by the standard, so that many different devices from an awfully large number of manufacturers can be connected together with at least some small hope that they will operate together. Because of the old voltage levels, however, some translation is required.

Figure 17-7 shows the standard RS-232 voltage levels. Note that there are actually two versions shown, RS-232B and RS-232C. The B-version is older, and uses a wider spread between the voltages representing the two different logic states. These limits were tightened up in the C-version, presumedly to speed things up. In

Table 17-2. UART Pin Functions.

Pin No.	Mnemonic	Function
1	Vcc	+5 volts DC power supply.
2	Vee	– 12 volts DC power supply.
3	GND	Ground.
4	RRD	Receiver Register Disconnect. A high on this pin disconnects (i.e., places at high impedance) the receiver data output pins (5 through 12). A low on this pin connects the receiver data output lines to output pins 5 through 12.
5	RB8	LSB
6	RB7	
7	RB6	
8	RB5	Receiver data output lines
9	RB4	
10	RB3	
11	RB2	
12	RB1	MSB
13	PE	Parity error. A high on this pin indicates that the parity of the received data does not match the parity programmed at pin 39.
14	FE	Framing Error. A high on this line indicates that no valid stop bits were received
15	OE	Overrun Error. A high on this pin indicates that an overrun condition has occurred, which is defined as not having the DR flag (pin 19) reset before the next character is received by the internal receiver holding register.
16	SFD	Status Flag Disconnect. A high on this pin will disconnect (i.e., set to high impedance) the PE, FE, OE, DR, and THRE status flags. This feature allows the status flags from several UARTs to be bus-connected together.
17	RRC	16 × Receiver Clock. A clock signal is applied to this pin, and should have a frequency that is 16 times the desired baud rate (i.e., for 110 baud standard it is 16 × 110 baud, or 1760 hertz).
18	DRR	Data Receive Reset. Bringing this line low resets the data received (DR, pin 19) flag.
19	DR	Data Received. A high on this pin indicates that the entire character is received, and is in the receiver holding register.
20	RI	Receiver Serial Input. All serial input data bits are applied to this pin. Pin 20 must be forced high when no data is being received.
21	MR	Master Reset. A short pulse (i.e., a strobe pulse) applied to this pin will reset (i.e., force low) both receiver and transmitter registers, as well as the FE, OE, PE, and DRR flags. It also sets the TRO, THRE, and TRE flags (i.e, makes them high).
22	THRE	Transmitter Holding Register Empty. A high on this pin means that the data in the transmitter input buffer has been transferred to the transmitter register, and allows a new character to be loaded.
23	THRL	Transmitter Holding Register Load. A low applied to this pin enters the word applied to TB1 through TB8 (pins 26 through 33, respectively) into the transmitter holding register (THR). A positive-going level applied to this pin transfers the contents of the THR into the transmit register (TR), unless the TR is currently sending the previous word. When the transmission is finished the THR→TR transfer will take place automatically even if the pin 25 level transition is completed.
24	TRE	Transmit Register Empty. Remains high unless a transmission is taking place, in which case the TRE pin drops low.
25	TRO	Transmitter (Serial) Output. All data and control bits in the transmit register are output on this line. The TRO terminal stays high when no transmission is taking place, so the beginning of a transmission is always indicated by the first negative-going transition of the TRO terminal.
26	TB8	LSB
27	TB7	
28	TB6	
29	TB5	Transmitter input word.
30	TB4	
31	TB3	
32	TB2	
33	TB1	MSB

Table 17-2. (Continued from previous page.)

34	CRL	Control Register Load. Can be either wired permanently high, or be strobed with a positive-going pulse. It loads the programmed instructions (i.e., WLS1, WLS2, EPE, PI, and SBS) into the internal control register. Hard wiring of this terminal is preferred if these parameter never change, while switch or program control is preferred if the parameters do occassionally change.
35	PI	Parity inhibit. A high on this pin disables parity generation/verification functions, and forces PE (pin 13) to a low logic condition.
36	SBS	Stop Bit(s) Select. Programs the number of stop bits that are added to the data word output. A high on SBS causes the UART to send two stop bits if the word length format is 6, 7, or 8 bits, and 1.5 stop bits if the 5-bit teletypewriter format is selected (on pins 37-38). A low on SBS causes the UART to generate only one stop bit.
37 38	WLS₁ } WLS₂ }	Word Length Select. Selects character length, exclusive of parity bits, according to the rules given in the chart below: **Word Length WLS1 WLS2** 5 bits low low 6 bits high low 7 bits low high 8 bits high high
39	EPE	Even Parity Enable. A high applied to this line selects even parity, while a low applied to this line selects odd parity.
40	TRC	16 × Transmit Clock. Apply a clock signal with a frequency that is equal to 16 times the desired baud rate. If the transmitter and receiver sections operate at the same speed (usually the case), then strap together TRC and RRC terminals so that the same clock serves both sections.

any given circuit, it takes longer to make a -15- to $+15$-volt transition than it does to make a -5- to $+5$-volt transition.

The B-version will recognize any voltage between $+5$ and $+25$ volts as a logic 0, and any voltage between -5 and -25 volts as

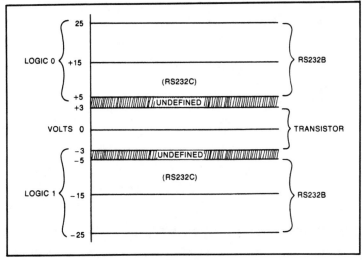

Fig. 17-7. RS-232B and C voltage levels.

a logic 1. This is exactly the opposite of what one might expect. These levels assume an impedance between 3000 and 7000 ohms.

Designing RS-232C circuits can be a real chore because there are a number of other considerations, including driver output impedance and a 30 volt/microsecond slew-rate specification. Fortunately, the chore is made easier by special integrated circuit RS-232C line drivers and receivers. The Motorola MC1488 driver and MC1489 receiver are prime examples.

The standard 25-pin D-type connector is specified for RS-232C, with the following pinout functions assigned:

Pin No.	RS232 Name	Function
1	AA	Chassis ground
2	BA	Data from terminal
3	BB	Data received from modem
4	CA	Request to send
5	CB	Clear to send
6	CC	Data set ready
7	AB	Signal ground
8	CF	Carrier detection
9	undef	
10	undef	
11	undef	
12	undef	
13	undef	
14	undef	
15	DB	Transmitted bit clock, internal
16	undef	
17	DD	Received bit clock
18	undef	
19	undef	
20	CD	Data terminal ready
21	undef	
22	CE	Ring indicator
23	undef	
24	DA	Transmitted bit clock, external
25	undef	

TELETYPEWRITERS AND PRINTERS

One of the earliest forms of data communications with peripherals was the teletypewriter machine. These were typewriter-

346

like machines that used a mechanism of electrical solenoids to pull in the type bars, or to position the type cylinder. The original devices used Baudot code and a 60-milliampere current. Later versions of the teletypewriter machine used a 20-milliampere loop and were generally more sophisticated than previous designs. Some modern teletypewriters use dot matrix printing and contain an eight-inch floppy disk to store a magnetic copy of the data transmitted and received.

Figure 17-8 shows the basic elements of a teletypewriter or other printer based on the 20-milliampere current loop. The keyboard and printer are actually separate, and they usually have to be wired together if a local loop is desired (i.e., where the keystroke on the keyboard produces a printed character on the same machine). This circuit is actually grossly simplified. In a real teletypewriter, there will be an encoder wheel or circuit that produces

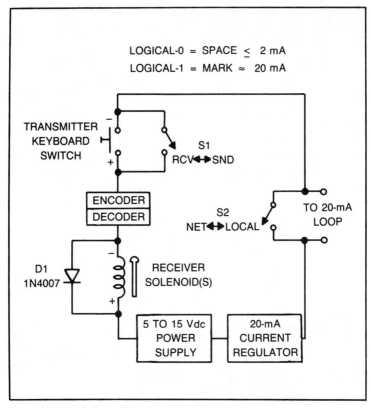

Fig. 17-8. 20-mA loop teletypewriter/printer circuit.

the Baudot code output. The keyboard consists of a series of switches (that actuate the encoder). Since these switches and their associated encoder are in series with the line, a LOCAL switch must be provided to bypass the transmitter section on receive.

The receiver consists of a decoder and the receive solenoids which actually operate the typebar mechanism. Note in Fig. 17-8 that a 1N4007 diode is in parallel with the receive solenoid. This is to suppress the inductive spike that will be generated when the reactive solenoids are de-energized. The diode is placed in the circuit such that it will be reverse biased under normal operation. But the counter electromotive force produced as a result of "inductive kick" forward biases the diode. Under this condition, the diode damps the spike to a harmless level. In some older machines, the inductive spike was safely ignored because the mass of the mechanism effectively integrated the spike to nothingness. But modern solid-state equipment does not move the mechanism directly with the 20-mA loop. The solid-state components can be damaged by the high voltage spike, so it is recommended that a 1N4007 be used even if the original design ignored it.

When the loop is closed, the circuit of Fig. 17-8 will produce a readable signal. Another similarly designed teletypewriter will be able to read the current variations produced by the machine.

Some modern printers are designed to operate with a 20-mA loop. Although most engineers would agree that the 20-milliampere loop is obsolete for modern designs, there are still large amounts of older equipment on the market, and in place at user sites, that are based on the current loop concept. When replacing older equipment, it is prudent, in most cases, to simply buy a new printer that operates from a 20-milliampere loop rather than to redesign the whole system. Also, hobbyists and smaller users may well want to take advantage of older 20-mA loop equipment that comes on the surplus market at low cost when larger users upgrade their systems and no longer need the old machine.

Figures 17-9 and 17-10 shows how to interface 20-milliampere equipment to TTL-compatible serial outputs from computers. The circuit in Fig. 17-9 shows the transmitter arrangement. The assumption is that there is a single TTL-compatible bit from either a serialized-parallel output, or a UART IC. The TTL level is applied to an open collector TTL inverter, which has as its collector load an LED inside of an opto-isolator. When the LED is turned on, the phototransistor is turned on hard, since this transistor operates as an electronic switch in series with the 20-milliampere current loop.

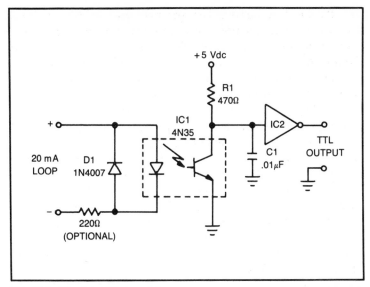

Fig. 17-9. 20-mA-to-TTL translator.

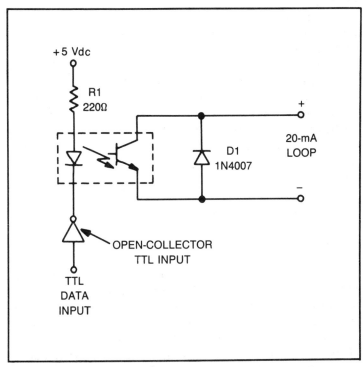

Fig. 17-10. TTL-to-20-mA translator.

Thus, when the TTL bit is high, the LED is on and the transistor is saturated. In that condition, the current loop transmits a "mark" sign (equivalent to a logical 1 in binary).

The receive end of the current loop-to-TTL interface is shown in Fig. 17-10. In this case, the opt-isolator is still used, but in reverse. Here the LED is connected in series with the current loop. Thus, when a mark is transmitted, the LED will be turned on. When a space is transmitted the LED is turned off. During the mark periods, the opto-isolator phototransistor is saturated, and the input to the TTL inverter is low. This condition results in a high on the output to the computer. Again, a mark is a logical-1 (high) and a space is a logical-0 (low). The 0.01 μF capacitor is used for noise suppression.

Chapter 18

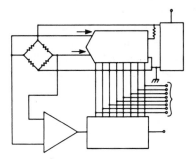

Tape Recorders & Data Loggers

A NALOG AND DIGITAL DATA CAN BE STORED ON ORDINARY AU-
dio recording tape. Of course, the data density in the digital
case is greater if digital tape systems are used, but audio recorders
are very low in cost, so are attractive to many users. Most of these
low-cost systems cannot easily locate specific bundles of data, but
where the data is reviewed or used in the same sequence as it was
recorded, the audio recorder is often a viable alternative.

Several different types of recording are used: direct, amplitude-
modulated carrier, frequency-modulated carrier, and frequency-shift
keying (FSK). Any of these systems can be implemented on
machines employing reel-to-reel, cassette, eight-track cartridges,
or four-track cartridges.

Direct recording applies the analog waveform directly to the
record inputs of an unmodified tape recorder. The frequency response of a recorder is on the order of 50 to 8000 hertz for low-
cost machines, and 40 to 18,000 hertz for high-grade stereo models.
The fundamental frequency of the data signal must fall within the
bandpass limits of the recorder or data will be lost. The direct
method is of limited application because most signals that are
amendable to recording on low-cost machines are low frequency,
i.e., 0.01 to 100 hertz.

The two carrier methods (AM and FM) offer the best means
for recording analog data on an audiotape recorder. Both AM and
FM methods use an audio-frequency carrier that is inside of the

bandpass of the tape recorder. Carrier frequencies between 400 and 8000 hertz are commonly used.

Most tape recorders suffer a severe loss of high-frequency response if the head cluster alignment is not exact, or if a thin layer of iron oxide stripped off the tapes is covering the head gap. Additionally, once a tape is recorded, it tends to loose some of the high frequencies each time the tape is played. For these reasons, most systems employing audiotapes for analog data storage use carrier frequencies in the 1000- to 2000-hertz range.

A simple amplitude modulator is shown in Fig. 18-1A. Any circuit can be used, provided that it mixes the carrier and the analog signal together in a device that is nonlinear, in this case diode $D1$. The point is to select a device that will show a changing impedance over cyclic excursions of the input signals.

The output of the simple amplitude modulator is connected to the record signal input on the tape recorder. If a tape recorder microphone input is used, then a voltage divider is needed that will reduce the output voltage from the modulator to -40 dBm or less.

Another, and far more elegant, circuit is shown in Fig. 18-1B. This circuit is based on the standard three-transistor dc differential amplifier. RCA Semiconductor Division makes a number of linear integrated circuit versions of this circuit in their CA3000-series devices (e.g., CA3028).

The dc bias networks hold the bases of $Q2$ and $Q3$ constant. In the case of $Q3$, that also establishes collector current $I3$ as a constant. Current $I3$ is the source for the collector currents in the differential pair $Q1$ and $Q2$ and since $I3$ is held constant, the following relationship holds true.

$$I3 = I1 + I2 = k \qquad (18.1)$$

where $I1$ is the collector current in $Q1$, $I2$ is the collector current in $Q2$, and $I3$ is the collector current in $Q3$.

The oscillator supplies a 1 kHz carrier signal to the base of $Q3$, so current $I3$ is modulated at a 1 kHz rate. The analog data signal is applied to the base of $Q1$. For my purposes, the initial analysis will assume that $I3$ is dc, which is not terribly unreasonable, since $I3$ will be a sine wave of constant amplitude.

When the analog data signal is positive, transistor $Q1$ will turn on harder. This causes current $I1$ to increase, but since $I3$ is a constant, the increase in $I1$ causes current $I2$ to decrease. If $I2$ decreases, then the collector voltage on $Q2$ will increase.

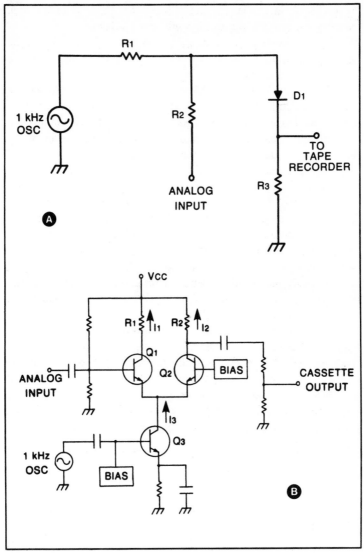

Fig. 18-1. Amplitude modulators. (A) Simple modulator for cassette recorder. (B) Modulator using an IC differential amplifier.

Similarly, when the analog input voltage is negative, the $I1$ current decreases. This change causes the $I2$ current to increase, so that Eq. 18.1 remains satisfied. When $I2$ increases, the $Q2$ collector voltage decreases. The collector voltage of $Q2$ will be an amplified rendition of the analog input waveform.

If the 1 kHz oscillator signal is present at the base of $Q3$, then the output on the collector of $Q2$ will be a 1 kHz carrier that is amplitude modulated by the analog input signal. This signal must be reduced in amplitude by a voltage divider to a level compatible with a tape recorder input.

The circuits of Figs. 18-1A and 18-1B are used to encode the signal on an audio-frequency carrier so that it can be recorded on audiotape. The circuit in Fig. 18-2, on the other hand, can be used to demodulate the recorded signal. It is essentially an envelope detector such as those used in AM broadcast receivers.

Alternate schemes for AM encoder circuits involve the use of one of the linear IC balanced modulators now on the market. Devices such as the MC1495 and MC1496, or the XR-205 will work nicely. These chips are also billed as analog multipliers, since amplitude modulation is essentially a multiplication process.

The same IC devices also work as a product detector on the decode side of the system if the roles of the oscillator and signal are reversed. The oscillator would be applied differentially to the bases of $Q1$ and $Q2$ (Fig. 18-1B), while the modulated signal would be applied to the base of $Q3$.

Frequency-modulated carrier systems are more popular than amplitude-modulated systems. In fact, most instrumentation recorders use FM encoding. The use of FM, however, places greater constraints on the wow and flutter (i.e., speed stability) specifications of the tape recorder. Low-cost voice-grade machines are not usually adequate for this type of recording, but most of those in the over-$100 range will work properly.

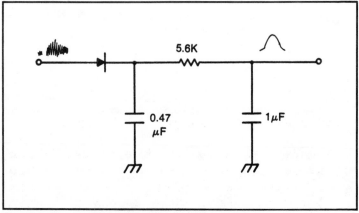

Fig. 18-2. Simple envelope detector for AM decoding.

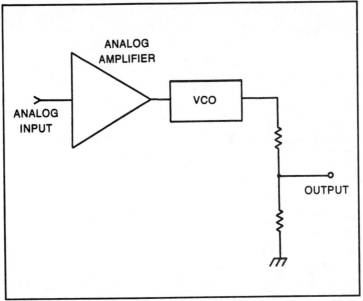

Fig. 18-3. Basic FM tape system.

A voltage-controlled oscillator (VCO) is used as the FM oscillator (Fig. 18-3). The resting frequency of the VCO will be the unmodulated carrier frequency. This frequency should be around 1 or 2 kHz for most moderate-grade tape recorder systems.

An example of such a circuit is shown in Fig. 18-4. This circuit is based on the Signetics 566 VCO, or function generator integrated circuit. Amplifier $A1$ is an operational amplifier used as a level shifter. The control voltage input of the 566 must be given a quiescent level, which is provided by the offset network $R5$ through $R7$. The analog input signal sees a gain of unity, and causes the $A1$ output to vary about the quiescent point:

$$E_{\text{OUT}} = \frac{V_{CC}R7}{R6 + R7} + E_{\text{IN}} \qquad (18.2)$$

Varying the control voltage about the quiescent point (i.e., E_{OUT} when $E_{\text{IN}} = 0$) causes the running frequency of the VCO to shift an amount proportional to the amplitude of the analog input voltage E_{IN}. The result is a frequency-modulated audio carrier encoded with the analog waveform.

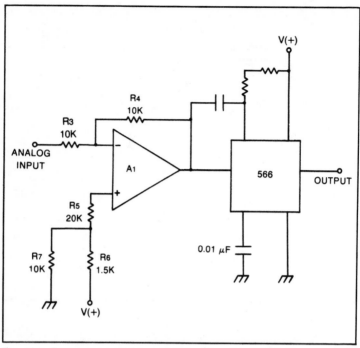

Fig. 18-4. Frequency modulator using the Signetics NE566 IC.

There are several approaches to demodulation of the FM carrier, but the most popular seem to be pulse-counting detection (PCD) and phase-locked loops (PLL). Envelope detection will not work, nor will product detection.

The block diagram for PCD demodulator is shown in Fig. 18-5. It consists of a Schmitt trigger, monostable multivibrator (i.e., one-shot multivibrator), integrator, and optional amplifier.

The modulated signal from the tape recorder is applied to the input of the Schmitt trigger circuit. This stage is used to square up the waveform. Recall that a Schmitt trigger output will snap high when the input signal voltage is greater than a certain threshold, and remains high as long as the signal remains above that threshold. If the signal voltage goes below the threshold, then the output snaps low again. This circuit action produces one square wave output for every input cycle.

The square wave signal is used to trigger the one-shot stage. The reason why this is done is that the square waves have variable width, and in the actual detection process I require pulses of constant amplitude and duration, which means that the area under all pulses is the same. Only the number of pulses varies with the modulation on the FM input signal.

These pulses are integrated, and that process recovers the original analog waveform that was recorded on tape.

The PCD method requires rigorous speed stability specifications for the machine used for playback. Speed variations tend to vary the frequency of the recorded signal, and this is seen as modulation by the detector. The detector cannot distinguish speed variation artifacts from real modulation. The PCD system does, however, allow quite a margin of error between the recording and playback speeds. This feature allows you to play back a recording made on another machine.

An example of PLL detector is shown in Fig. 18-6. This circuit is based on the Signetics 565 integrated circuit PLL device and a 741 operational amplifier. The operational amplifier acts as a buffer, level shifter, and low-pass filter.

This circuit suffers from the same dependence on tight wow and flutter specifications as does the PCD, but it also has the same ability to track playback signals that have a slightly different frequency than was originally recorded.

For applications with a high accuracy requirement, it is usually best to use a stereo recorder. Record the modulated carrier (i.e., that signal containing the analog waveform) on one channel and an unmodulated carrier on the other channel. The second channel signal serves as a reference for speed correction. An IC phase detector (Fig. 18-7) such as the Motorola MC4044 can be used to compare the signals from the two channels. Speed variations will affect

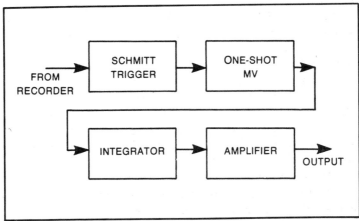

Fig. 18-5. Block diagram to a frequency demodulator for tape systems.

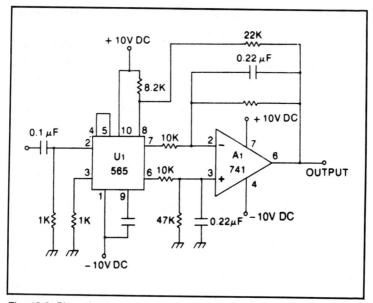

Fig. 18-6. Phase-locked loop frequency demodulator.

both signals equally, but frequency variations due to modulation affects only the one channel. The integrated output of the phase detector, therefore, will be the recovered analog waveform.

DIGITAL DATA ON AUDIO RECORDERS

Digital data can be recorded on an audiotape recorder in much the same manner, except that only two tones are used: one each for the two possible logic states. This type of modulation is sometimes called frequency-shift keying (FSK) after teletypewriter ter-

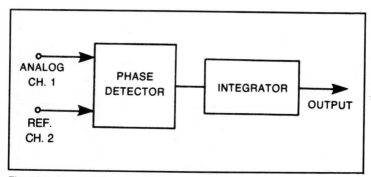

Fig. 18-7. Phase detector frequency demodulator.

minology, and is essentially the same as the circuits discussed in the chapter on serial data communications.

In this case, the output of the UART transmitter or other serial data source modulates the VCO, but the VCO output is reduced in amplitude to a level that is comparable with other audio sources used with the recorder, as low as − 40 to − 56 dBm.

The UART transmitter and receiver requires a 16 × baud rate clock. This clock can be provided for the transmitter only, and the signal recorded on the unused channel of a stereo tape recorder. It is then squared up in a Schmitt trigger and applied to the 16 × clock input on the receiver. This tactic allows speed variations to exist without fouling up the recovered data.

Consider Fig. 18-8, which shows a block diagram to a digital data recovery system used by many microcomputer manufacturers. In this system logic 1 is represented by a 2125-hertz tone, and logic 0 by a 2975-hertz tone.

Since the UART output is high (i.e., logic 1) when there is no transmission taking place, the 2125-hertz tone would be applied continuously. At the instant when a data transmission begins, the tone shifts to 2975 hertz indicating that the start bit is a logic 0, after which the tone wobbles back and forth according to the applied bit pattern.

At the receive end, the recorder output signal is passed through a bandwidth limited amplifier that passes only frequencies in the 2100- to 3000-hertz range, a tactic that tends to reduce noise problems.

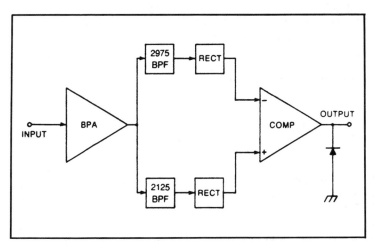

Fig. 18-8. Digital FSK decoder.

The signal is split into two paths at the output of the first amplifier, one path leads through a 2975-hertz filter and the other through a 2125-hertz filter. Each filter is sharply tuned so only signals of the proper frequency will pass through.

The output signals from the two filters are rectified in an operational amplifier *ideal* or *precision* rectifier circuit. The outputs of the rectifiers are applied to alternate inputs on a voltage comparator.

The inverting input of the comparator is connected to the output of the 2975-hertz (i.e., logic 0) tone, while the noninverting input of the comparator is connected to the output of the 2125-hertz (i.e., logic 1) filter.

If a logic 1 tone is present, then the comparator output will be positive, giving a logic 1 level.

Similarly, when the logic 0 tone is present, the input polarity seen by the comparator reverses, so the comparator output tries to go negative. The germanium diode across the comparator output clamps the negative excursion to approximately 0.2 volts, so the respective logic levels will be positive for logic 1 and 0 volts for logic 0. These levels are compatible with most CMOS systems, but clipping or level shifting will be required for operation with TTL devices.

DATA LOGGERS

Tape recorders are used to record data on a continuous basis, and it is usually deemed impractical to try using them for more than a few hours. Also, many scientific and engineering experiments do not require continuous, moment-for-moment recording of the data, but do require monitoring over a long period of time. Some are left running for 12 hours or even up to several days. Many applications only require a data point every minute, every 5 minutes, or even once per hour.

A *data logger* is a solid-state memory device that will store digitized data (i.e., binary). These instruments can be quite costly and complex if high speed or large word length are required, but are more reasonable in slow speed, eight-bit versions.

A simple data logger can be constructed (see Fig. 18-9A) using readily available microcomputer memory chips. The memory system can be constructed especially for this project, or (more reasonably) one of the 4K, 8K, 16K, or 32K eight-bit microcomputer memory boards can be adapted to this application. Some 4K boards sell for under $100, while several 8K boards sell for just a little over $125.

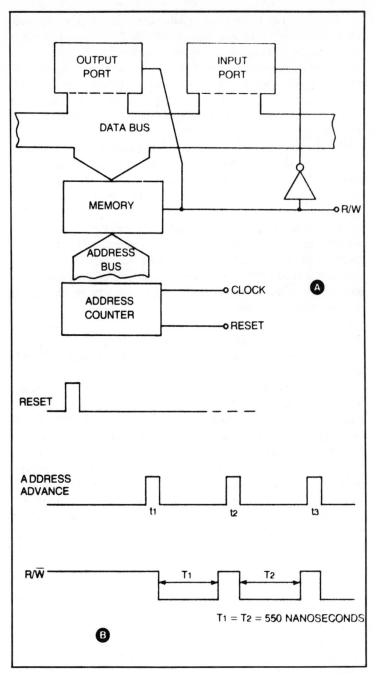

Fig. 18-9. (A) Digital data logger. (B) Timing diagram.

There are two approaches to data logger design. The approach shown in Fig. 18-9A uses ordinary CMOS or TTL logic chips, while the alternate scheme uses a microprocessor.

In the circuit of Fig. 18-9A a memory array is connected to an eight-bit data bus and an address bus. A binary counter or address generator is clocked at a rate equal to the desired sampling interval. This can be anything from a few milliseconds to once every hour.

The timing diagram is shown in Fig. 18-9B. A reset pulse sets the counter address to 00000000. Alternately, you might want to speed up the clock and advance the address to a starting location other than 00000000. A magnitude comparator such as the 7485 will tell the circuit when to shut off.

Clock pulses $t1$, $t2$, $t3$, . . . t_n advance the address counter by one location each. Following the memory address advance, on the trailing edge of the clock pulse, the R/W line goes low, enabling the input port and setting the memory in a WRITE condition. Data on the input port is then strobed into the memory location given by the address counter.

In order to read from memory, it is only necessary to invert the logic level applied to the R/W line following the clock pulse. It is good design practice to inhibit the input port during READ operations.

A simple data logger can be constructed from a microcomputer. Although table-top mainframe models can be used, only a so-called controller or minimum system is needed, i.e., some ROM, RAM (as much as required), one I/O port, and a data update clock.

The update clock could be wired to an interrupt line. A program is then written to keep the computer in a no-op loop until the clock interrupt line becomes active. The program would then increment the address counter and input (or output) the data as required.

A standard 8K (i.e., 8192-word) memory board costing about $125, when used with a once-per-5-minutes data sampling rate, will store 40,960 minutes (that's 683 hours, or 28 days) worth of data.

Chapter 19

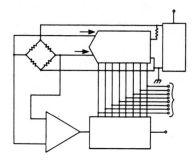

Telephone Dialer Circuits

ONE OF THE MORE AGGRAVATING TECHNOLOGICAL INNOVA-tions of the past few years is the automatic computer-controlled telephone dialer/solicitor. One application is to have the computer dial telephone numbers in sequence (an easy trick for a computer), deliver a prerecorded advertising message, then record the listener's response—which at my house would be unprintable.

Another application, which to me seems more socially accept-able, is for a computer to sense an alarm condition such as a fire or break-in at home or office, then dial a predetermined telephone number to warn a human to take an indicated action. Alarms could be connected to the interrupt lines of a microcomputer in this ap-plication. The computer could then take care of other tasks, or if a dedicated alarm system machine, just idle in an endless no-op loop until the alarm creates an interrupt situation. It would then branch to program the services that interrupt and dial the pro-grammed number.

In this chapter, I will talk about circuits that will dial either pulses or Touch-Tone telephones under program control. It is also useful at this point to review the telephone interconnection mate-rial given in Chapter 17.

This discussion is based on the Motorola MC14408/MC14409 binary-to-phone-pulse converter chips, and the MC14410 Touch-Tone generator chip. There are probably others on the market by the time this is published, so consult the CMOS special chip catalogues of Motorola and other semiconductor manufacturers for additional devices.

DIAL PULSE SYSTEM

The telephone can be represented by a model circuit such as Fig. 19-1A. A dc power supply potential (E) of around 60 volts will be measured across the line when the receiver is on the hook (i.e., when $S1$ is open). When the receiver is lifted off the hook, then the impedance of the telephone set loads the line, and the approximate voltage across the receiver is (see Fig. 19-1B):

$$E2 \ = \ 60 \ \times \ \frac{R2}{R1 \ + \ R2} \qquad (19.1)$$

This drop in level is often used by telephone-answering and signaling devices to assure that the receiver is off the hook before starting the message. It also signals bugging equipment of an active line.

Alternatively, an incoming call can be recognized before the receiver is lifted off the hook by the ringing signal, which is an ac tone in the 16- to 25-hertz range, that has an amplitude as high as 30 volts RMS.

Alternatively, an incoming call can be also recognized by the actual acoustical ringing of the bell. A microphone placed in close proximity to the telephone set will pick up the sound of the bell and turn on the equipment. If the microphone is placed close enough, and is not too sensitive, then only the ringing of the bell will set off the alarm circuitry, ambient room noise will cause only an occasional false alarm.

Many low-cost telephone-answering devices (TADs), especially those manufactured prior to the decisions that permitted foreign interconnects, were designed so that the telephone sits on top of the machine, immediately over the microphone. A solenoid-driven latch lifts the receiver off the hook after the bell has been recognized. It will then play a taped message to the caller.

Most of the more recent devices, however, take advantage of the looser regulations regarding interconnects, and connect directly to the telephone line. The telephone company uses a standard four-pin jack for remote or portable telephone connection. The manufacturer encourages the buyer of their equipment to have these installed, and will supply their equipment with a matching plug. One

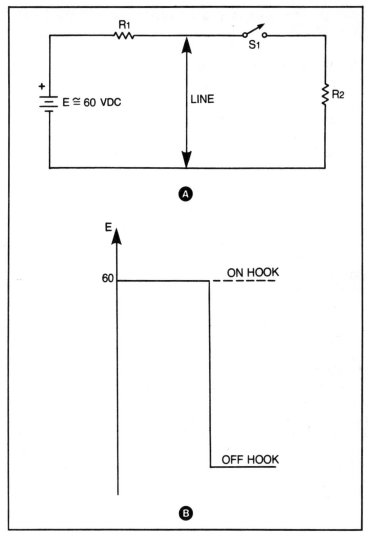

Fig. 19-1. Simplified representation of the phone line. (A) Circuit. (B) Voltage waveform.

could use almost any plug/socket combination that is handy, but if discovered, it will likely be viewed dimly by the phone company.

The MC14408/MC14409 Devices

The block diagram to the Motorola MC14408 and MC14409 IC binary-to-phone-pulse devices are shown in Fig. 19-2. These

365

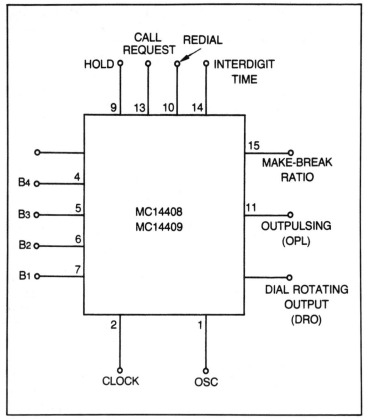

Fig. 19-2. Motorola MC14408/MC14409 telephone dial pulser IC.

MOS LSI (i.e., McMOS) chips will convert the four-bit binary word applied to their inputs to the number of serial output pulses indicated by the value of the applied four-bit code.

These devices are identical to each other in every respect, except for the dial-rotating output (DRO). On the MC14408 device the DRO remains high during continuous "outpulsing" (output pulsing), while on the MC14409 the DRO remains low during outpulsing.

The MC14408/09 devices can be used either alone, provided that suitable binary driving source is available for the inputs, or as a pushbutton dialing adapter for non-Touch-Tone systems. In the latter case, the companion MC14419 two-of-eight-to-binary converter chip is used as the BCD source.

These chips will dial numbers up to sixteen digits in length, and will redial the last number entered. They also have selectable dialing rates (i.e., 10 or 20 pps), interdigit time (i.e., 150 to 800 msec), and make-break ratio (i.e., 61% or 67%).

The output stage is a bipolar transistor that can accommodate a variety of loads including discrete transistor load drivers, TTL and CMOS digital logic devices. These features, incidentally, allow other applications not limited to telephone dialing systems.

MC14410 Two-of-Eight Tone Encoder

The Motorola MC14410 is an LSI CMOS chip that will generate two simultaneous tones on command from a standard two-of-eight contact closure key pad. The key pads are now widely available, especially from dealers supplying amateur radio operators, who use them (and the MC14410 or similar chips) to make VHF-FM (i.e., 2m) transceivers able to dial the telephone through a repeater.

The keyboard format has become standard, and requires that the pushbutton switches be used in a four-by-four matrix, that is to say a grid of four rows intersecting four columns. When a pushbutton is depressed, a switch is closed that shorts together one row and one column, creating a unique logic situation out of sixteen possible states.

The MC14410 contains its own clock circuit, controlled on-chip by a 1 MHz oscillator. The clock creates *two* simultaneous semi-sine waves by a special digital addition technique. The frequency tolerance is on the order of ±0.2%.

Figure 19-3 shows the basic connection scheme for the MC14410 chip. Note that very little extra circuitry is required external to the chip. The clock frequency is set by the 1000 kHz crystal ($Y1$) shunted by the 15M resistor.

The only connections required are the four row inputs, four column inputs, +5 volts dc, ground, and the high- and low-frequency outputs. The two outputs are separate from each other, allowing independent operation should it become necessary. In the example of Fig. 19-3, however, the two frequencies are summed as they should be in a Touch-Tone dialing system.

The example shown is from the Motorola applications literature, and is the simple case where a keyboard is used. The effect of the keyboard is to simultaneously ground a single row line and a single column line. This chip can be connected to a microcomputer output port to accomplish the same thing. In that case, the

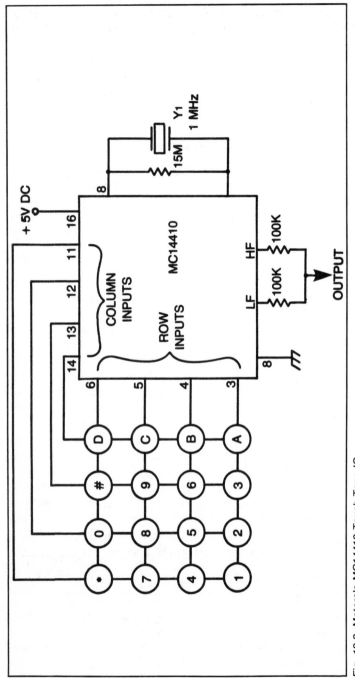

Fig. 19-3. Motorola MC14410 Touch-Tone IC.

dormant state would be for all eight bits to be high. To select a Touch-Tone digit one need only make two bits at a time low.

Example 19-1

The first of the following lists gives the frequencies generated by the MC14410 when each row or column is *zero,* while the second one gives the tone-pair required for each digit on the telephone dial. Note that each dial digit requires two tones, one each from high- and low-frequency groups.

Input Line	Low Group (Hz)	High Group (Hz)
*P*1	697	—
*R*2	770	—
*R*3	852	—
*R*4	941	—
*C*1	—	1209
*C*2	—	1336
*C*3	—	1477
*C*4	—	1633

Digit	Tone Pair	Digit	Tone Pair
0	*R*4*C*2	8	*R*3*C*2
1	*R*1*C*1	9	*R*3*C*3
2	*R*1*C*2	A	*R*1*C*4
3	*R*1*C*3	B	*R*2*C*4
4	*R*2*C*1	C	*R*3*C*4
5	*R*2*C*2	D	*R*4*C*4
6	*R*2*C*3	*	*R*4*C*1
7	*R*3*C*1	#	*R*4*C*3

How would you go about autodialing a telephone number, such as the almost universal emergency number 911, used in many cities for fire, police, and rescue?

Solution:

First, I must assign the rows and columns of the MC14410 to bits of a microcomputer output port. This is done in the following list:

Bit	Designation
1(LSB)	*C*1
2	*C*2
3	*C*3

Bit	Designation
4	$C4$
5	$R1$
6	$R2$
7	$R3$
8(MSB)	$R4$

Note that this is almost arbitrary, and you can make your own assignment protocol so long as each condition is unique to that system and is kept consistent.

Digit 9 is represented by the tone pair $R3C3$, while digit 1 is represented by tone pair $R1C1$. The output word from the computer when no numbers are being dialed should be 11111111. This word will set all lines to the MC14410 high.

To generate a *9*, I will want to set $R3C3$ low. $R3$ corresponds to bit 7, while $C3$ corresponds to bit 3. The output word that causes the MC14410 to create a digit 9 tone pair is

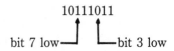

The digit 1 is represented by $R1C1$. The preceding list tells us that $R1$ is represented by bit 5, and $C1$ is bit 1. The binary word used to generate this condition at the MC14410 is

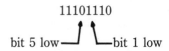

In an application such as a burglar alarm, where the computer would dial a telephone number, I would write a program that would dial a given number (such as 911 or your local security company), then activate a taped or computer-synthesized message. Note that most microcomputers are now available with a compatible voice synthesizer plug-in card that could handle a simple message. Also, be aware that in some jurisdictions the police and fire will not respond to an automatic message. In that case, have the computer call the security service, your home (if it is an office system), or a temporary number entered into the machine just before you left. Imagine being at a party and being the only one to receive a computerized message.

It is necessary to hold each number on the output port long enough for the telephone company central office equipment to recognize and act on that digit. This is a brief matter of a few milliseconds, but a computer can zip through all digits too fast!

Chapter 20

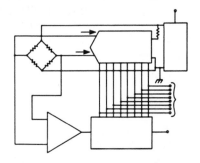

Analog Signal-
Processing Circuits

RARELY ARE SIGNALS OBTAINED FROM TRANSDUCERS AND electrodes suitable for direct processing in a computer or any other electronic instruments. Quite apart from the necessity of converting the signal to a digital form, some amount of predigitizing processing is often both advisable and desirable. It is usually necessary in many cases to "launder" the signal in some sort of intermediate processing to bring it to a state where it is more useful. In the simplest, almost trivial, case this intermediate processing will consist only of amplification or scaling. In other cases, though, the intermediate processing will be a lot more involved.

There are two approaches to almost every signal-processing job, and these can be denoted as the hardware and software concepts. Proponents of the hardware approach often contend that it is best to build electronic circuitry to do these jobs external to the computer or other digital instrument. The software buffs, on the other hand, contend that suitable computer programs can do the job in a superior manner.

The truth actually falls somewhere between these two views in most cases, there being good reasons to support *both* broad concepts. Proponents of the hardware approach often contend that it the hardware approach, be it digital or analog circuitry, where the computer's memory is limited to a small amount, where time constraints exist, or in those cases where adding a modest hardware package is a lot easier than doing the job in software. In short, the word is "trade-off" engineering jargon for compromise, albeit rational compromise.

Under the general rubric of signal-processing circuitry, I will cover a number of predominantly operational amplifier circuits such as integrators, differentiators, logarithmic and antilog amplifiers, analog multipliers, and active filters.

INTEGRATORS

A basic passive integrator circuit (see Fig. 20-1) consists of a resistor in series with the signal path, and a capacitor shunting the

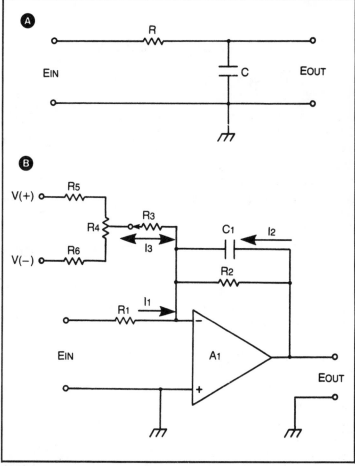

Fig. 20-1. Integrator circuits. (A) RC integrator. (B) Operational amplifier integrator.

373

signal path on the output side of the resistor. You will see this circuit later in the chapter when the low-pass filter is discussed, so keep it in mind.

The active integrator circuit is shown in Fig. 20-1B. For the moment, discount resistors $R2$ through $R6$; pretend for the moment that they do not exist. They are used to make the circuit actually work with real operational amplifiers. In the present discussion, I will make the presumptuous condition that the operational amplifier is ideal.

Considering the ideal case, the integrator consists of the operational amplifier $A1$, resistor $R1$, and feedback capacitor $C1$. From elementary capacitor theory I know that:

$$I2 = C1 \times \frac{d(E_{OUT})}{dt} \qquad (20.1)$$

Rearranging Eq. 20.1 gives us:

$$\frac{d(E_{OUT})}{dt} = \frac{I2}{C1} \qquad (20.2)$$

$$\int \frac{d(E_{OUT})}{dt}\, dt = \int \frac{I2}{C1}\, dt \qquad (20.3)$$

$$E_{OUT} = \int \frac{I2}{C1}\, dt \qquad (20.4)$$

$$E_{OUT} = \frac{1}{C1} \int I2\, dt \qquad (20.5)$$

But from the basic properties of the operational amplifier I know that

$$I1 = -I2 \qquad (20.6)$$

and,

$$I1 = E_{IN}/R1 \qquad (20.7)$$

So, substituting Eq. 20.7 into Eq. 20.6 gives us:

$$I2 = E_{IN}/R1 \qquad (20.8)$$

And substituting Eq. 20.8 into Eq. 20.5 gives me the transfer equation for operational amplifier integrators:

$$E_{OUT} = \frac{1}{C1} \int \frac{-E_{IN}}{R1} \, dt + C \qquad (20.9)$$

$$E_{OUT} = \frac{-1}{R1C1} \int E_{IN} \, dt + C \qquad (20.10)$$

Voltage E_{OUT} will change by the factor

$$\Delta E_{OUT} = \frac{-1 \text{ volt}}{R1C1 \text{ sec}} \qquad (20.11)$$

for each volt applied to the input.

Before going on to other topics, let me first examine Eq. 20.11 to see what type of performance is expected from ordinary operational amplifiers. Let me assume a case in which $R1 = 100K$, $C1 = 0.1 \, \mu F$, and $E_{IN} = 1$ volt. Further, assume that the maximum value of E_{OUT} allowed for the particular operational amplifier selected for $A1$ is ± 12 volts dc.

$$E_{OUT} = \frac{-E_{IN} \times \text{time (sec)}}{(10^5)(10^{-7})} \qquad (20.12)$$

$$12 \text{ volts} = \frac{-1 \text{ volt} + \text{time (sec)}}{10^{-2}} \qquad (20.13)$$

$$12 \text{ volts} = -100 \text{ volts} \times \text{time (sec)} \qquad (20.14)$$
$$(12 \text{ volts})/(-100 \text{ volts}) \times \text{time (sec)} \qquad (20.15)$$
$$0.12 \text{ seconds} = \text{time for } A1 \text{ to saturate} \qquad (20.16)$$

Next, see what happens if $C1$ is changed to $0.001 \, \mu F$:

$$E_{OUT} = \frac{-E_{IN} \times \text{time (sec)}}{(10^5)(10^{-9})} \qquad (20.17)$$

$$(12 \text{ volts})/(10^4) = \text{time (sec)} \qquad (20.18)$$
$$1.2 \text{ milliseconds} = \text{time to saturate } A1 \qquad (20.19)$$

The lesson in this comparison is to mind carefully the values of $R1$ and $C1$, relative to the expected integration time. Consider integrating a signal with a frequency of 1 kHz. The period of this signal is of the same order as the time to saturate the operational amplifier in our second example. The general rule to follow is that integrator time constant should be long compared with the period of the applied signal.

Figure 20-2 shows the effect of the integrator on different types of input waveform. In each case, the top waveform is the input to a circuit such as Fig. 20-1A, and the bottom trace is the output. Figure 20-2A shows the effect on a sine wave, namely phase shift. Recall from elementary calculus that this is the proper behavior when the waveform is a sine or, alternatively, a cosine. Each integrator stage produces a quadrature phase change in the applied signal.

Before leaving the subject of active integrators, let me explain the purpose of resistors $R2$ through $R6$, which had been deleted from our basic-theory-of-operation discussion. These components are required to overcome some defects found in real, as opposed to ideal, circuits. There is inevitably an offset associated with all operational amplifiers. Even premium-grade devices suffer from these problems. The output artifact created by these errors tends to charge the feedback capacitor. In experiments performed on several operational amplifiers, varying in quality from the 741 to a $10 premium type, it was found that the output would rise at a constant rate until the amplifier was saturated—hardly useful in real circuits.

An offset null circuit (resistors $R3$, $R4$, $R5$, and $R6$) is used to eliminate these problems. A counter current is created to discharge the capacitor at precisely the same rate as it is charged by the error current, thereby creating a null or equilibrium condition.

In most cases, the use of an operational amplifier with a MOS-FET input stage, and the high resolution null circuit shown will completely eliminate the problem.

Resistor $R2$ shunted across the integration capacitor is used to drain any accumulated charge due to unregenerate dc offsets appearing on a dynamic input waveform. Resistor $R2$ will typically have a value over 1 megohm, and a good rule of thumb is to use as high a value as will do the job in any particular case. The symp-

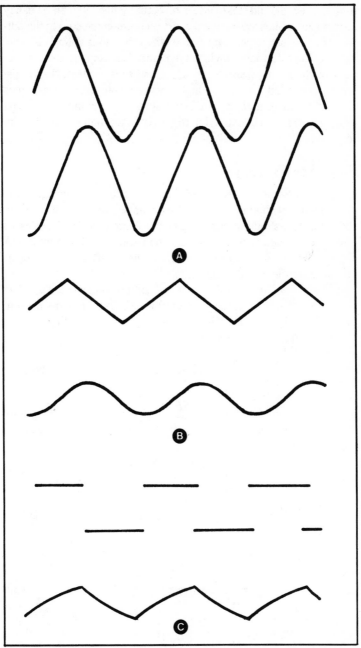

Fig. 20-2. Effect of integration. (A) On a sine wave. (B) On a triangular wave.
(C) On a square wave.

tom created, should this resistor be actually needed, can be easily seen on an oscilloscope or strip-chart recorder connected to the output. If the baseline remains stable when the input voltage is zero, yet climbs off the screen when a dynamic (i.e., nonconstant) waveform such as sine, triangle, or square wave is applied, then resistor $R2$ is needed, or has too high a value. Note that this assumes that the dc offset in the applied waveform is an artifact; if it is part of the information contained in the waveform, then it should be integrated.

DIFFERENTIATORS

An electronic differentiator can be made by rearranging the components of the integrator, a fact that one might suspect from the respective similar natures between the mathematical processes alluded to in their namesakes from the paper world. Figure 20-3A shows a simple RC differentiator, while Fig. 20-3B is an active operational amplifier differentiator.

Again, I will initially discount the components needed to bring the differentiator form the ideal world to the real. For the present let me consider only $R1$, $C1$, and the operational amplifier. In the circuit of Fig. 20-3B I know that:

$$I2 = -I1 \qquad (20.20A)$$

$$I1 = \frac{d(E_{IN})}{dt} \times C1 \qquad (20.20B)$$

$$I2 = \frac{d(E_{OUT})}{dt} \qquad (20.21)$$

I can obtain the transfer equation for the circuit by substituting Eqs. 20.20B and 20.21 into Eq. 20.20A:

$$\frac{E_{OUT}}{R1} = \frac{-C_{1d}(E_{IN})}{dt} \qquad (20.22)$$

$$E_{OUT} = \frac{-C1R_{1d}(E_{OUT})}{dt} \qquad (20.23)$$

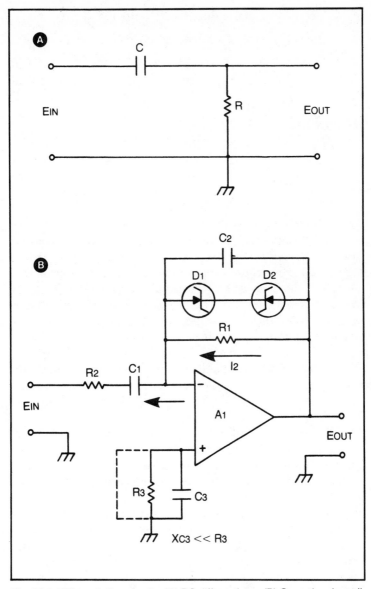

Fig. 20-3. Differentiation circuits. (A) RC differentiator. (B) Operational amplifier differentiator.

Equation 20.23 is the transfer equation for Fig. 20-3B.

Let me consider now the effects of the RC time constant ($R1C1$) on the gain of the circuit. Using the same parameter as for the integrator case presented earlier, set $R1 = 100K$ and $C1 = 0.1\ \mu F$:

$$A_V = R1C1 \qquad\qquad (20.24)$$
$$A_V = (10^5)(10^{-7}) \qquad\qquad (20.25)$$
$$A_V = 10^{-2} \qquad\qquad (20.26)$$
$$A_V = 0.01 \qquad\qquad (20.27)$$

Next set $R1 = 100K$ and $C1 = 0.001\ \mu F$.

$$A_V = R1C1 \qquad\qquad (20.28)$$
$$A_V = (10^5)(10^{-9}) \qquad\qquad (20.29)$$
$$A_V = (1)(10^{-4}) \qquad\qquad (20.30)$$
$$A_V = 0.0001 \qquad\qquad (20.31)$$

The rule for applying differentiators is to set the time constant to approximately 10% of the period of the applied signal, or less. Figure 20-4 shows the effects of electronic differentiation on various types of input waveform. It sometimes happens that the period used to set the differentiator time constants is not the actual waveform period, but that of the leading or trailing edges, or some other significant portion of the waveform. Consider Fig. 20-4B, for example. If the duration of a waveform is very long compared with the rise and fall times of its edges, and I want to differentiate the edges of the square wave, then it is necessary to set the time constant of the circuit relative to the rise time. This was done to make Fig. 20-4B.

A square wave applied to the input produces sharp spikes at the leading and trailing edges. Notice that these are the only times when the amplitude of the input signal was not a constant. A constant amplitude applied to the input of any differentiator will produce zero output. A positive spike is created on the leading edge, and a negative spike on the trailing edge, as befits the fact that one has a positive derivative and the other a negative derivative. Note that, if Fig. 20-3B had been used to make the oscilloscope photos, the spikes would have been inverted by the operational amplifier.

Figure 20-4C shows the effect of a properly selected time constant of differentiation on a triangle wave. The leading and trailing edges of a triangle waveform have constant but opposite slopes, that is to say that the rate of change (derivative) is constant. This shows up in Fig. 20-4C as a square wave at the output of the differentiator. Notice that the square wave is positive on the rising

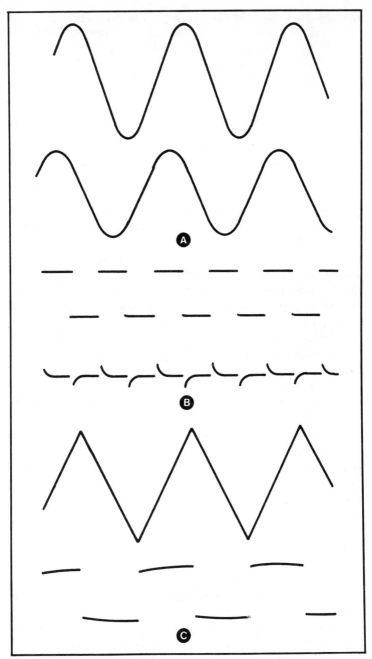

Fig. 20-4. Effect of differentiator. (A) On a sine wave. (B) On a square wave. (C) On a triangular wave.

(leading) edge, and negative on the falling (trailing) edge.

A differentiator can be calibrated using a ramp or triangle if you know the slope in volts/second; something that can be determined from an examination of the waveform on a calibrated oscilloscope. If, for example, the upper waveform in Fig. 20-4C was a 500-hertz triangle it has a period of 2 milliseconds. It will reach the 1-volt peak amplitude in 1 millisecond. The slope, then, is:

$$\frac{1 \text{ volt}}{1 \text{ msec}} \times \frac{1000 \text{ msec}}{1 \text{ sec}} = 1000 \text{ volts/sec} \qquad (20.32)$$

You now know that the amplitude of the square wave in the lower trace of Fig. 20-4C represents a derivative of 1000 volts/second. Of course, this calibration is only useful if you are able to infer significance from the numbers. Suppose, for example, a fluid pressure instrument equipped with a differentiator was calibrated so that the *dP/dt* was 100 torr/second. You could then find the rate of change of the applied pressure by examining the amplitude of the differentiator output signal.

The differentiator that contains only $R1$, $C1$, and the operational amplifier is a problem waiting to occur. Keep in mind that this is a high-frequency feedback circuit, so several anomalies are likely to exist, and they will cause problems.

Figure 20-5 shows the frequency response plot for the operational amplifier differentiator of Fig. 20-3B. A feedback amplifier is usually regarded as unstable if the feedback transfer plot intersects the open-loop response plot in the wrong region. Ordinarily, I want the plot of the feedback curve to intersect the open-loop gain curve in the 6 dB/octave segment. In the case of the differentiator, the plot of the RC network has a rising characteristic with frequency and intersects the open-loop response curve in the 12 dB/octave region, making it potentially unstable. By adding resistor $R2$, however, I produce a falling response characteristic to counter the curve of the feedback network. This resistor should have a value such that

$$R2 = 0.503/f3 \ C1 \qquad (20.33)$$

although in many practical cases a nominal value between 33 ohms and 150 ohms is used. This forces the plot of the RC network to effectively intersect with the open-loop response curve below the unity-gain crossover point.

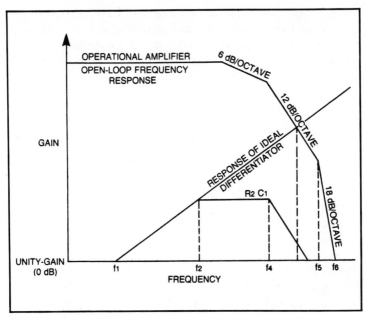

Fig. 20-5. Differentiator response plot.

Capacitor $C2$ forces the gain plot of the circuit to have a steeper slope. This capacitor usually has a value in the 10 pF to 560 pF range, and the rule of thumb for its selection is:

$$C2 = 1/2\pi f4R1 \qquad (20.34)$$

Resistor $R3$ is used just like the compensation resistor from the inverting follower circuit; it reduces the effect of the operational amplifier's input bias current on capacitor $C1$. If used, the resistor may require a parallel bypass capacitor. Normally, if $R3$ is less than about 10K, no bypassing is required. The reactance of $C3$ at frequency $f1$ should be about one-tenth of the value of $R3$. The purpose of $C3$ is to bypass the normal thermal noise potentials generated in $R3$.

A pair of back-to-back zener diodes ($D1$ and $D2$) are shunted across resistor $R1$ to limit the output excursions, and to prevent latchup of the operational amplifier.

COMPARATORS

A comparator (see Fig. 20-6A) is an operational amplifier that snaps to maximum output voltage when the differential input voltage ($E1 + E2$) exceeds a very small amount. The gain of the comparator is essentially the open-loop gain of the operational amplifier used, and that can be considerable. If this gain is 100,000 and the maximum allowable output voltage is 10 volts, the amplifier will saturate with an input voltage of

$$E_{IN} = E_{OUT}/A_V \qquad (20.35)$$
$$E_{IN} = 10^1/10^5 \qquad (20.36)$$
$$E_{IN} = 10^{-4} \text{ volts} = 0.1 \text{ mV} \qquad (20.37)$$

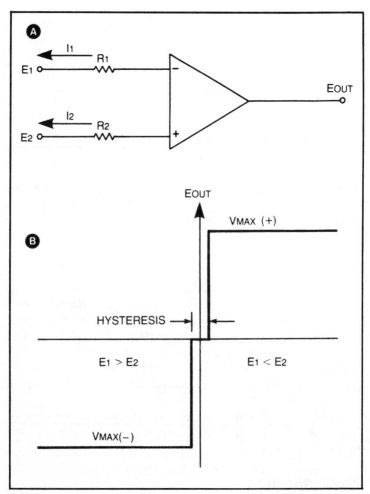

Fig. 20-6. Comparator. (A) Circuit diagram. (B) Transfer function.

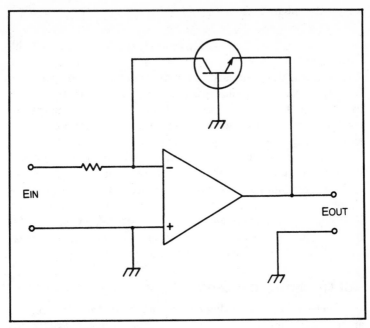

Fig. 20-7. Logarithmic amplifier.

The purpose of the comparator is to examine the input voltages (or if $R1$ and $R2$ are deleted, two currents) and issue an output logic level that indicates whether $E1 = E1$, $E1$ is greater than $E2$, or $E1$ is less than $E2$. The transfer function for a comparator is shown in graphical form at Fig. 20-6B.

LOGARITHMIC AMPLIFIERS

Often there is a need for converting an analog voltage function from a linear to algorithmic voltage. The base-emitter voltage (V_{BE}) characteristic of an ordinary transistor follows the relationship:

$$V_{BE} = (kT/q) \ln (I_C /I_S) \qquad (20.38)$$

where k is Boltzmann's constant ($1.38 \times 10^{-23}\ J/°K$), T is the absolute temperature in degrees kelvin (°K), q is the elementary electronic charge (1.6×10^{-19} coulombs), I_S is the theoretical reverse saturation current of 10^{-13} amperes, and I_C is the collector current.

At room temperature, normally taken to be about 300°K, Eq.

20.38 reduces to:

$$V_{BE} \approx (26 \text{ mV}) \ln (I_C /10^{13}) \qquad (20.39)$$

But note that this term is strongly dependent upon temperature. In fact, this same relationship is used as a temperature transducer in an earlier chapter. It is necessary to either hold the temperature constant, or temperature compensate the circuit. An example is shown in Fig. 20-7.

The inverse circuit is shown in Fig. 20-8, and this circuit will deliver an output that is the antilog of the input voltage. Very often the logarithmic/antilog circuits are used in a signal compression/expansion system to increase the dynamic range of an instrument. The linear input voltage is compressed by the logarithmic amplifier, is processed by additional circuitry, and is then expanded back to the linear realm by the antilog amplifier.

MULTIPLIERS & DIVIDERS

There are certain analog electronic circuits that will produce an output that is proportional to the product or quotient of two input voltages. Although some designers use discreet component circuits to perform these functions, and some use operational amplifier circuits, it is usually cheaper and more accurate to use a monolithic

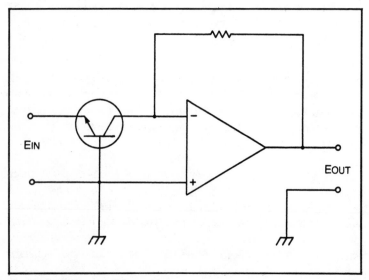

Fig. 20-8. Antilog amplifier.

integrated circuit multiplier, or a multiplier made into the form of an analog function module. Analog Devices, RCA, Motorola, Burr-Brown, and others manufacture products that obey the transfer function

$$E_{OUT} = kV1(V2/V3)^m \qquad (20.40)$$

where m can be any number between 0.2 and 5.

Although somewhat more expensive than integrated circuit multipliers, these modules are extremely flexible in that an external resistor network, or strapping certain pins, allows the various specific transfer functions.

ACTIVE FILTERS

A filter is a circuit that passes frequencies within its bandpass and rejects all others. Although the topic of filters is among the most complicated in the field, I will distill the subject down to a few basic fundamentals.

Figure 20-9 shows the frequency response characteristics of five types which we will consider: low-pass, high-pass, bandpass, sharp-bandpass or peaking, and band-reject or notch filters.

A low-pass filter, whose trace is shown in Fig. 20-9A, passes signals from dc to some cutoff frequency f_{CO}. At frequencies greater than the cutoff frequency, the response falls off at a given rate until it is essentially zero. The low-pass filter is eminently useful for removing high-frequency noise and other artifacts from an analog signal, or for the prevention of aliasing in A/D converters, and to smooth out the quantization ripple in the output of D/A converters.

The high-pass filter response is shown in Fig. 20-9B, and it has exactly the opposite shape as the low-pass filter of the previous example. This type of filter will pass frequencies *above* the cutoff frequency, and reject those from dc to the cutoff frequency. The response below the cutoff frequency falls off at a given rate until it is essentially zero.

Operational amplifier integrators will perform the function of low-pass filtering, and an operational amplifier differentiator can serve as a high-pass filter. The rolloff above the cutoff frequency, though, is only 6 dB/octave (20 dB/decade), so the filter action is not optimum.

A bandpass filter (see Fig. 20-9C) is designed to pass only a certain range of frequencies between upper and lower limits. A special case of the bandpass filter circuit is the peaking amplifier which

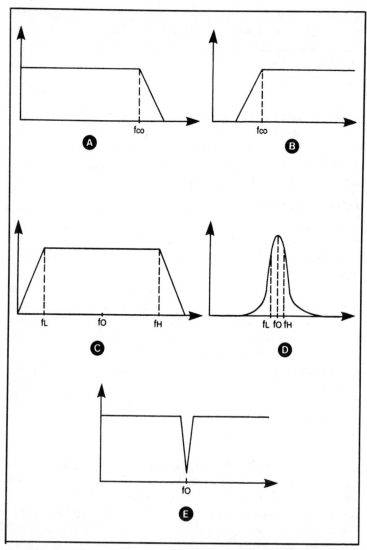

Fig. 20-9. Filter response curves. (A) Low-pass response. (B) High-pass response. (C) Bandpass response. (D) Narrow or sharp bandpass response. (E) Notch response.

has a response such as shown in Fig. 20-9D. The principal difference between two filters with these respective properties is expressed by their Q figures—figures of merit—which is given by:

$$Q = f_O / (f_H - f_L) \qquad (20.41)$$

388

where f_O is the center bandpass frequency, f_H is the upper cutoff frequency, and f_L is the lower cutoff frequency.

A more rigorous definition of Q depends upon the ratio of stored energy to the cyclically dissipated energy. Equation 20.41 and its attendant definitions assume the bandpass to be symmetrical, and that the rolloff slopes of the upper and lower cutoff frequencies be equal. For present purposes, though, Eq. 20.41 is entirely sufficient and adequate. The curve in Fig. 20-9C shows a low-Q characteristic, while that in Fig. 20-9D indicates a high-Q characteristic.

The notch filter, also known as the band-reject filter if the notch is broad enough, passes all frequencies *except* those in the immediate neighborhood of a specific center frequency. The response curve of this type of filter is shown in Fig. 20-9E. This class of filter is used to null out or reject unwanted single-frequency artifacts, such as 60-hertz power mains interference, from a circuit.

Before proceeding further, I should decide upon a convention so that my equations have at least some chance of working. Note that I am not being rigorous here, and it is suggested that those desiring rigor sign up for an engineering mathematical analysis course, then follow up with a few electrical engineering courses so that a few of the professional journal papers on active filters can be read.

Figure 20-10 shows a bandpass amplifier response that is per-

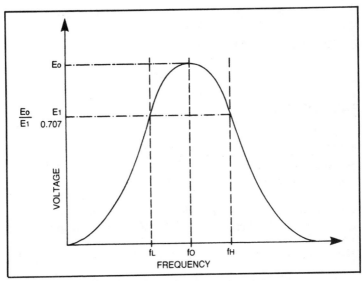

Fig. 20-10. Perfectly symmetrical bandpass response curve.

fectly symmetrical about center frequency f_O. The cutoff frequencies occur at the half-power points, signified by point $E1$ on the graph. Although referred to as a power point, this is usually specified in terms of easier to measure voltage units. The derivation of this notion depends upon the fact that power is expressed by E^2/R. At center frequency f_O power is given by:

$$P = E_O^2/R \qquad (20.42)$$

At f_H or f_L (again assuming symmetry) the power is given by:

$$P = 1/2(E_O)^2/R = E1^2/R \qquad (20.43)$$

so,

$$\frac{E_O^2}{2R} = \frac{E1^2}{R} \qquad (20.44)$$

From which we can deduce:

$$E_O^2/2 = E1^2 \qquad (20.45)$$
$$E_O^2/E1^2 = 2 \qquad (20.46)$$
$$E_O/E1 = 2^{1/2} = 1.414 \qquad (20.47)$$

By Eq. 20.47, then, I see that

$$E1 = 0.707E_O \qquad (20.48)$$

PASSIVE FILTERS

Simple RC integrators and differentiators can be used to make simple filters of modest performance. I have used these circuits on innumerable occasions to eliminate interference in electronic instrumentation circuits. Although the example chosen (Fig. 20-11A and B) uses the low-pass case for the sake of explanation, the presentation also applies for the high-pass case if you hold the book up to a mirror. The cutoff frequency of this filter is given by:

$$f_{CO} = 1/2\pi R1C1 \qquad (20.49)$$

The circuit can be viewed as a voltage divider with the output taken

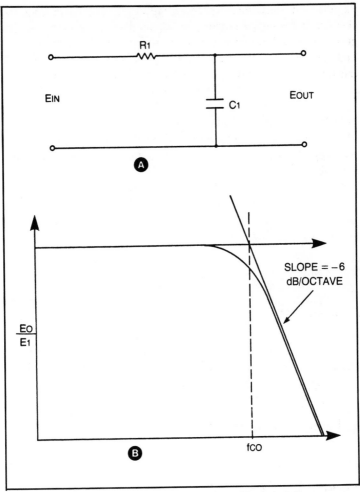

Fig. 20-11. Passive low-pass filter. (A) Circuit. (B) Response Curve.

across the reactance of capacitor $C1$. I can claim, then, that the output voltage will reflect the inverse dependence upon frequency normally exhibited by a capacitive reactance. The percentage of the input signal voltage that is delivered to the output decreases as the frequency increases. The rolloff curve for a single section RC filter such as Fig. 20-11A and B is 6 dB/octave or 20 dB/decade.

Greater rolloff factors can be obtained by cascading filter sections, but one is warned that this will greatly attenuate the bandpass signal also, unless certain steps are taken. Each section adds 6 dB/octave to the rolloff slope. A two-section filter, then, rolls off at 12 dB/octave, and a three-section filter rolls off at 18 dB/octave.

Passive *RC* filter sections are affected a great deal by the load resistance across their output terminals, and this is the reason why cascaded sections attenuate the bandpass signal (too much) as well as the stop band signal.

One solution is to buffer each *RC* section with a unity-gain, noninverting operational amplifier follower. The source impedance seen by each filter section input terminal will be low (i.e., an operational amplifier output impedance), and the load impedance across the capacitor is very high (i.e., the input impedance of a noninverting follower). Any attenuation in the circuit that falls within the bandpass can be made up using a gain follower at the output of the cascade chain or prior to the input.

A BETTER SOLUTION

The method outlined above is a sloppy way to obtain high-order filtering, and there exist certain techniques for obtaining the required rolloff properties without all of those extra stages: the so-called active filter.

In order to avoid becoming too rigorous, I will first evade the issue of defining the term "order," except that I will say order denotes the slope of the response rolloff beyond the f_{CO} point and that, in general, a first-order rolloff is 6 dB/octave, second-order rolloff is 12 dB/octave, and third-order rolloff is 18 dB/octave.

Also avoided are the various *types* of active filter response curve. For my discussion, I will limit the filter category initially to unity-gain, maximally flat bandpass, second-order types.

An assumption made in this treatment is that $Q = 2^{1/2}/2$, and that certain resistance and capacitor ratios are maintained. These will be specified for each case as it arises.

The general form for these circuits is shown in Fig. 20-12A. It consists of an operational amplifier connected in a unity-gain configuration and an input network consisting of several impedances.

The amplifier selected for $A1$ should be a premium-grade device if an attempt at high performance is made, but an ordinary 741-type operational amplifier will often suffice for most lower performance applications. Good choices are the LF156 series, and the RCA CA3130, CA3140, and CA3160 devices. The idea is to obtain a wide frequency response, and as high an input impedance as possible.

The impedances shown in Fig. 20-12A will be either resistances or capacitive reactances depending upon whether low-pass, high-

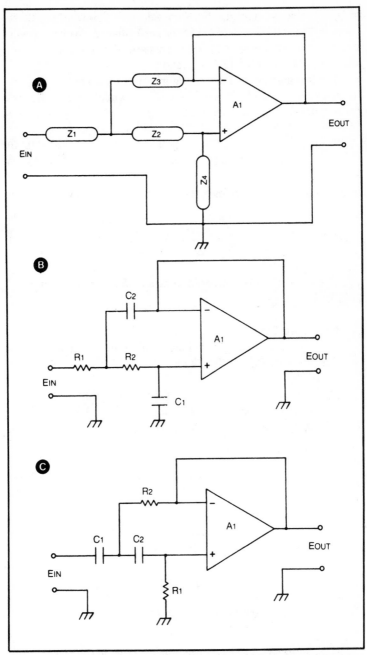

Fig. 20-12. Active filter. (A) General circuit. (B) Low-pass circuit. (C) High-pass circuit.

393

pass, or bandpass configurations are selected. The values of these components will be critical to proper performance, so either precision types or hand-selected components are required. These filters have a rolloff slope of 12 dB/octave.

The low-pass active filter configuration is shown in Fig. 20-12B. In this case $Z1$ and $Z2$ are resistances, while $Z3$ and $Z4$ are capacitances. It is specified that:

$$C2 \; = \; 2C1 \qquad\qquad (20.50)$$
$$R1 \; = \; R2 \qquad\qquad (20.51)$$

If these specifications are met, then the cutoff frequency is given by

$$f_{CO} \; = \; \frac{1}{2\pi R2 \,(C1C2)^{1/2}} \qquad\qquad (20.52)$$

where $C1$ and $C2$ are expressed in farads, $R2$ is in ohms, and f_{CO} is in hertz.

Example 20-1
Find the component values required of a 1 kHz low-pass filter in which $C1 = 0.001 \; \mu F$.
Solution:
1. Since $C2 = 2C1$, capacitor $C2$ will be $0.002 \; \mu F$.
2. Solve Eq. 20-52 for $R2$ assuming that $C1 = 0.001 \; \mu F$, $C2 = 0.002 \; \mu F$, and $f_{CO} = 1000$ hertz.

$$R2 \; = \; \frac{1}{2\pi f_{CO}(C1C2)^{1/2}} \qquad\qquad (20.53)$$

$$R2 \; = \; \frac{1}{(2)\,(3.14)\,[(10^{-9})\,(2 \times 10^{-9})]^{1/2}} \qquad\qquad (20.54)$$

$$R2 \; = \; 112.5K \qquad\qquad (20.55)$$

3. Since $R1 = R2$, let $R1 = 112.5K$ also.

In the high-pass configuration merely reverse the roles of impedances $Z1$ through $Z4$ making $Z1$ and $Z2$ capacitances, while $Z3$

and $Z4$ become resistances. The cutoff frequency is given by:

$$f_{co} = \frac{1}{2\pi C1(R1R2)^{1/2}} \qquad (20.56)$$

Assuming that

$$C1 = C2 \qquad (20.57)$$

$$R2 = R1/2 \qquad (20.58)$$

Example 20-2

Find the component values required for a 1 kHz high-pass filter if $R1 = 220K$.

Solution:

1. $R2 = R1/2 = 220K/2 = 110K$.
2. Solve Eq. 20.56 for $C1$.

$$C1 = \frac{1}{2\pi f_{co}(R1R2)^{1/2}} \qquad (20.59)$$

$$C1 = \frac{1}{(2)(3.14)(10^3)[(1.1 \times 10^5)(2.2 \times 10^5)]} \qquad (20.60)$$

$$C1 = 1.02 \times 10^{-9} \text{ farads} \qquad (20.61)$$
$$C1 = 0.00102 \ \mu F = 1020 \text{ pF} \qquad (20.62)$$

A bandpass filter can be made by cascading a low-pass stage with a high-pass stage. The cutoff frequency of the high-pass stage is set to the desired lower cutoff of the bandpass response, while the cutoff frequency of the low-pass section is set to the high-frequency cutoff point of the desired response. Once again, though, a superior solution exists in that I can use the bandpass version of the multiple feedback path filter circuit.

MULTIPLE FEEDBACK PATH FILTERS

Figure 20-13 shows the general form for the multiple feedback path active filter circuit. This particular circuit is a little more difficult to tame than the simpler circuit of the preceding examples, but generally yields better results. The general transfer equation

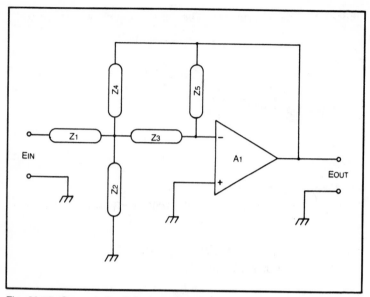

Fig. 20-13. General circuit for a multiple feedback path active filter.

for the circuit of Fig. 20-13 is:

$$\frac{E_{\text{OUT}}}{E_{\text{IN}}} = \frac{1/(-Z1Z3)}{(1/Z5)(1/Z1 + (1/Z2 + 1/Z3 + 1/Z4) + (1/Z3)(1/Z4)}$$

(20.63)

provided that the open-loop gain of the operational amplifier is extremely high, so that several $1/A_{VB}$ (gain at bandpass frequencies) terms approach zero.

The low-pass version of this filter is shown in Fig. 20-14. Note that the impedances at $Z1$, $Z3$, and $Z4$ have become resistances, while the impedances at $Z2$ and $Z5$ are capacitances. In this case:

$$\omega_O = 2\pi f_{CO} = (1/R2R3C1C2)^{1/2}$$

(20.64)

Solving Eq. 20.64 for f_{CO} gives the cutoff frequency:

$$f_{CO} = \frac{1}{(2\pi)} \left(\frac{1}{R2R3C1C2^{1/2}} \right)^{1/2}$$

(20.65)

or, in the more commonly encountered form

396

$$f_{CO} = \frac{1}{2\pi\sqrt{R2R3C1C2}} \qquad (20.66)$$

I can simplify the design of this type of circuit by assuming that $Q = 2^{1/2}/2$, which is approximately 0.707. The gain inside of band-pass is approximated by

$$A_{VB} = R3/R1 \qquad (20.67)$$

If I permit the ratio $C1/C2$ to be a constant, denoted by k, and let

$$k = 4Q^2(A_{VB} + 1) \qquad (20.68)$$

I can claim that

$$C1 = kC2 \qquad (20.69A)$$
$$C1 = C2[4Q^2(A_{VB} + 1)] \qquad (20.69B)$$

Substituting my given assumption concerning Q,

$$C1 = C2(4)(\sqrt{2}/2)^2(A_{VB} + 1) \qquad (20.70)$$

$$C1 = 2C2(A_{VB} + 1) \qquad (20.71)$$

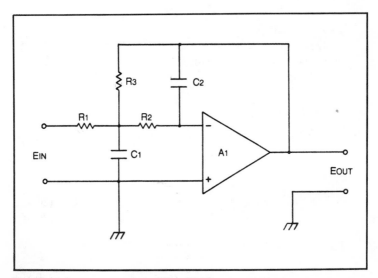

Fig. 20-14. Low-pass version of Fig. 20-13.

The value of $R2$ is given by:

$$R2 = \frac{1}{\omega_o^2 C1^2 R3[4Q^2(A_{VB} + 1)]} \qquad (20.72)$$

$$R2 = \frac{1}{8\pi^2 f_{CO} C1^2 R3(A_{VB} + 1)} \qquad (20.73)$$

And, finally, the values of the other resistors are given by:

$$R1 = R3/A_{VB} \qquad (20.74)$$

$$R3 = 1/4\pi f_{CO} QC2 \qquad (20.75)$$

$$R3 = 1/2\sqrt{2f_{CO} C2} \qquad (20.76)$$

The protocol for component selection sequence is:

1. Set $C1$ to a convenient value.
2. Compute $C2$.
3. Compute $R3$.
4. Compute $R1$.
5. Compute $R2$.

Example 20-3

Design a 1 kHz second-order filter of the multiple feedback path type. Assume the following: $Q = 0.707$, $A_{VB} = 1$.

Solution:

1. Set $C1$ to 0.004 μF.
2. From Eq. 20.71:

$$C1 = 2C2(A_{VB} + 1) \qquad (20.77)$$

so,

$$C1C2 = 2(1 + 1) \qquad (20.78)$$
$$C1/C2 = 4 \qquad (20.79)$$

Therefore,

$$C2 = C1/4 \qquad (20.80)$$

$$C2 = 0.004/4 \tag{20.81}$$
$$C2 = 0.001 \ \mu\text{F} \tag{20.82}$$

This last computation can be performed several times until a pair of standard-value capacitors are obtained.

3. Compute $R3$ using Eq. 20.76.

$$R3 = 1/2\sqrt{2}f_{\text{CO}}C2 \tag{20.83}$$
$$R3 = 1/(2)\,(\sqrt{2})\,(1000)\,(10^{-9}) \tag{20.84}$$
$$R3 = 354\text{K} \tag{20.85}$$

4. Compute $R1$ from Eq. 20.67.

$$A_{VB} = R3/R1 \tag{20.86A}$$
$$1 = R3/R1 \tag{20.86B}$$
$$R1 = R3 \tag{20.87}$$
$$R1 = 354\text{K} \tag{20.88}$$

5. Compute $R2$ from Eq. 20.72.

$$R2 = \cfrac{1}{8\pi^2 f_{\text{CO}}{}^2 C1^2 R3(A_{VB} + 1)} \tag{20.89}$$

$$R2 = \cfrac{1}{(8)\,(3.14)^2(1000)^2(4 \times 10^{-9})^2(3.54 \times 10^5)\,(1 + 1)} \tag{20.90A}$$

$$R2 = 1.119\text{K} \tag{20.90B}$$

The high-pass multiple feedback path filter is shown in Fig. 20-15. Notice that, once again, the principal difference between the high-pass and low-pass designs is that the resistors and capacitors exchange places.

In the high-pass circuit the gain is set by the capacitor ratio

$$A_{VB} = C1/C3 \tag{20.91}$$

and if a flat bandpass response is desired,

$$C1 = C2 \tag{20.92}$$

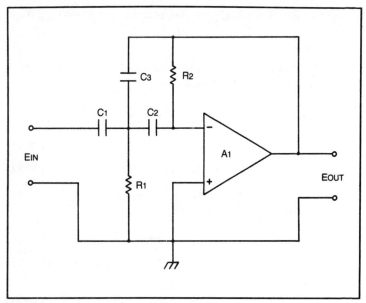

Fig. 20-15. High-pass version of Fig. 20-13.

By an argument similar to the low-pass case, the cutoff frequency is given by

$$f_{CO} = 1/2\pi\sqrt{R1R2C2C3} \qquad (20.93)$$

the resistors are selected from

$$R1 = 1/2\pi f_{CO}QC1(2A_{VB} + 1) \qquad (20.94)$$

and,

$$R2 = Q(2A_{VB} + 1)/(2\pi f_{CO}C1) \qquad (20.95)$$

The design sequence is:

1. Select $C1$ arbitrarily.
2. Find $C2$ and $C3$.
3. Compute $R1$.
4. Compute $R2$.

Example 20-4

Design a maximally flat, second-order high-pass filter with

400

a cutoff frequency of 1 kHz. Assume that bandpass gain A_{VB} is unity.

Solution:

1. Set $C1 = 0.001\ \mu F$.
2. $C1 = C2 = 0.001\ \mu F$.
3. Compute the value of $R1$ from Eq. 20.94.

$$R1 = 1/2\pi f_{CO}QC1(2A_{VB} + 1) \qquad (20.96)$$

$$R1 = 1/2(3.14)\,(10^3)\,(0.707)\,(10^{-9})\,(2 + 1) \quad (20.97)$$

$$R1 = 75K \qquad (20.98)$$

4. Compute the value of $R2$ from Eq. 20.95.

$$R2 = Q(2A_{VB} + 1)/2\pi f_{CO}C1 \qquad (20.99)$$

$$R2 = (0.707)\,(3)/(2)\,(3.14)\,(10^{-9}) \quad (20.100)$$

$$R2 = 338M \qquad (20.101)$$

An example of the bandpass multiple feedback path filter circuit is shown in Fig. 20-16. The bandpass gain is given by

$$A_{VB} = 1/(R1/R3)\,(1 + C2/C1) \qquad (20.102)$$

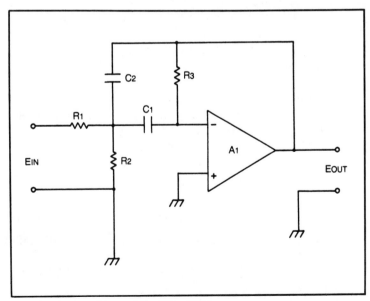

Fig. 20-16. Bandpass version of Fig. 20-13.

401

But in the maximally flat bandpass version of the circuit, I let
$C1 = C2$, so Eq. 20.102 becomes

$$A_{VB} = 1/(2R1/R3) \qquad (20.103)$$

$$A_{VB} = R3/2R1 \qquad (20.104)$$

The circuit Q is found in the manner given earlier [$f_O/(f_H - f_L)$], and the center frequency from:

$$f_O = \frac{1}{2\pi C1} \sqrt{\frac{(R1 + R2)}{R1R2R3}} \qquad (20.105)$$

The resistor values are found from

$$R1 = Q/A_{VB}2\pi f_O C2 \qquad (20.106)$$

$$R2 = Q/(2Q^2 - A_{VB})2\pi f_O C2 \qquad (20.107)$$

$$R3 = 2Q/\omega_O C2 \qquad (20.108)$$

$$R3 = Q/\pi f_O C2 \qquad (20.109)$$

The component selection sequence is:

1. Determine Q, f_O, f_L, and f_H.
2. Set $C1$ and $C2$ ($C1 = C2$) to a convenient value.
3. Compute $R1$.
4. Compute $R2$.
5. Compute $R3$.

Example 20-5

Design a band-pass filter with a Q of 20, centered about
1 kHz. Assume a bandpass gain of unity.
Solution:
1. If $Q = 20$ and $f_O = 1000$ hertz, then

$$f_H - f_L = f_O/Q \qquad (20.110)$$

$$f_H - f_L = 1000/20 \qquad (20.111)$$

$$f_H - f_L = 50 \text{ hertz} \qquad (20.112)$$

If the bandpass is symmetrical f_H is $f_O + 25$ Hz, or 1025 hertz,
and $f_L = f_O - 25$ hertz or 975 hertz.
2. Set $C1 = C2 = 0.001 \ \mu F$

3. Compute the value of $R1$ from Eq. 20.106.

$$R1 = Q/A_{VB}2\pi f_O C2 \qquad (20.113)$$
$$R1 = (20)/(1)\,(2)\,(3.14)\,(10^3)\,(10^{-9}) \qquad (20.114)$$
$$R1 = 3.18M \qquad (20.115)$$

4. Compute the value of $R2$ from Eq. 20.107

$$R2 = Q/(2Q^2 - A_{VB})2\pi f_O C2 \qquad (20.116)$$
$$R2 = (20)/[(2)\,(20)^2 - 1]\,(2)\,(3.14)\,(10^3)\,(10^{-9}) \qquad (20.117)$$
$$R2 = (20)/(799)\,(2)\,(3.14)\,(10^{-6}) \qquad (20.118)$$
$$R2 = 3.98K \qquad (20.119)$$

5. Compute the value of $R3$ from Eq. 20.109.

$$R3 = Q/\pi f_O C2 \qquad (20.120)$$
$$R3 = (20)/(3.14)\,(10^3)\,(10^{-9}) \qquad (20.121)$$
$$R3 = 6.37M \qquad (20.122)$$

In all of these examples, you might have to try various capacitor values in order to see which will allow the use of standard components. Keep in mind when making your choices that filter performance is best when precision resistors and low-drift (i.e., silver mica, poly-carbonate, polyethylene, etc.) capacitors are used. It is also part of the standard wisdom to opt for standard capacitors and nonstandard value resistors if such a question arises because it is easier to use a potentiometer to trim a value than it is to use a variable capacitance.

Chapter 21

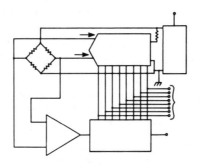

Building Fixed and
Adjustable Power Supplies

THE DC POWER SUPPLY IS ONE OF THE MOST IMPORTANT
parts of any electronic construction project, yet it is often also
the least considered portion of the design. The dc supplies used
by hobbyists include both integral power supplies mounted inside
of a project cabinet, and "universal" bench power supplies used
for testing, adjusting, troubleshooting or otherwise messing with
an electronic circuit. In this article, I will discuss the methods for
"designing" simple dc regulated power supplies; both fixed voltage and variable voltage models will be considered.

FIXED-VOLTAGE TYPES

The problem of designing a fixed output voltage regulator is
made much simpler these days by the three-terminal integrated circuit voltage regulators that are now available. There are quite a
few of these devices on the market, but they all share certain characteristics. For one thing, the output voltage is fixed at some standard value; the actual value is usually identified from the type number. For example, a 7805 is a 5-volt regulator, while the 7812 is
a 12-volt model. There are exceptions to the numbering rule, but
the general characteristics are summarized as follows:

1. LM-309 series are 5-volt regulators at 100-mA or 1-ampere,
   depending upon case style.

2. LM-323 is a 5-volt 3-ampere regulator in a "K" package (i.e., "same-as-TO3" transistor package.

3. LM-340n-xx are positive output voltage regulators; the "n" in the type number denotes the package type, while "xx" is the voltage rating. For example, LM-340K-12 is a 12-volt regulator in a "K" package (so it will pass 1 ampere).

4. 78xx is a family of regulators that are similar to the LM-340n-xx series; the letters "xx" denote output voltage (7812 is a 12-volt regulator). Package style determines output current.

5. LM-320n-xx is a negative output version of the LM-340n-xx, while 79xx is a negative output version of the 78xx. (Note: the input and ground terminals on the LM-320 and 79xx are reversed from the pinouts of the LM-340 and 78xx; failure to observe this convention will result in destruction of the regulator!)

6. The package designations are as follows: "H" indicates a TO-5 case, or its plastic equivalent, and a current of 100-mA; "K" indicates a TO-3 case and a current of 1 ampere (1.5 amperes if properly heatsinked); "T" indicates a TO-220 plastic power transistor-type case, and a current rating of 750-mA in free air or 1 ampere if properly heatsinked.

Regarding the current ratings, I don't think that it makes good sense to use the maximum rating if accompanied with the admonition ". . . if properly heatsinked." Heat and high currents are the twins that destroy electronic circuits. If these devices are routinely operated at or near maximums, then I believe that you will experience a higher than usual failure rate.

Figure 21-1 shows the basic circuit for a three-terminal, fixed voltage IC regulator. The transformer (T1), rectifier (BR1), and filter capacitor (C1) are selected according to the usual rules for any dc power supply. The transformer steps the 115-volt ac line voltage down to the level required for the input of the regulator. Normally, there will be a 2.5 volt difference required between the rated regulator output, and the minimum allowable input voltage. For a + 5-volt regulator, therefore, a minimum of + 7.5-volts is needed.

The rule for selecting a transformer is to provide at least the minimum voltage required for proper operation of the regulator. Keep in mind that the voltage across the regulator input, which is also the voltage across filter capacitor, will be approximately 0.9 times the peak ac voltage across the transformer secondary. Since the secondary voltage is specified in RMS values, I must multiply the rated value by 1.414 in order to find the peak voltage. If all terms are accounted for, then, the output voltage will be approximately 0.9 × 1.414 × V_{rms}, or 1.27 × V_{rms}. The minimum RMS value of the secondary voltage should be the minimum value of dc input required to the regulator, divided by 1.26. For the + 7.5-volts required for a + 5-volt regulator, then, I need an RMS rating of 7.5/1.26, or 5.95-volts. Since 6.3-volts RMS is the next highest standard value, I would select a 6.3-Vac transformer for this application.

The current rating of the transformer should be the highest expected dc value, plus a margin for safety. In addition, be aware that most transformers that have a center-tapped secondary are rated for regular fullwave rectification, not fullwave bridge rectification. The current available when the bridge circuit is used is *one-half* that rated value. This is true because the voltage is twice, and we still do not want to exceed the "volts times amperes" rating of the transformer. Some transformers will take a higher current when it is drawn, but it is not good practice to make the transformer work so hard. The current rating must be at least the current rating of the regulator, and preferably more. There is a general design rule that requires the use of only about 75 percent of capacity on the average.

There are two ratings on the rectifier that need attention: the forward current and the peak inverse voltage (PIV). The forward current is simply the amount of current that the bridge rectifier will pass normally in the forward direction without suffering a "heart attack." For most cases, the rating should be equal to or greater than the regulator forward current rating. Again, having some excess capacity so the rectifier never operates for long at its maximum rating, is good design practice and will make the circuit more reliable.

The peak inverse voltage rating (PIV) is the maximum reverse bias voltage that the rectifier will withstand without breaking down. If this voltage is exceeded, the rectifier can be destroyed. The normal rule of thumb is to use a minimum PIV that is 2.83 times (heck, make it 3 to be sure!) the applied RMS. The reason for this rule is that the normal PIV seen by the rectifier is 1.414 × RMS plus the voltage on the capacitor (C1 in Figure 21-1), which is also 1.414 × RMS; thus 2 × 1.414 × RMS is 2.83 × RMS. This rule does not mean much when dealing with 6.3 Vac transformers, because the minimum available PIV rating is 25 volts (with those being a lot harder to find than 50-volt PIV units!), which is above the max-

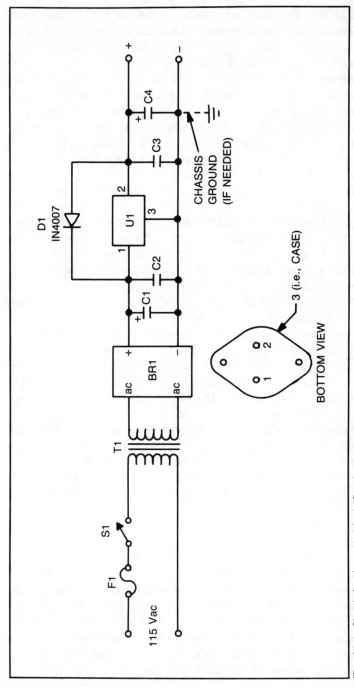

Fig. 21-1. Circuit for three-terminal IC voltage regulators.

407

imum reverse voltage generated. But the rule becomes more and more important when the voltage increases. I can recall a famous amateur radio transceiver that had a habit of popping the bridge rectifier diodes frequently. The problem was that the 2.83 rule was violated. When three 1000-volt PIV diodes were connected in series (along with 470-kohm balancing resistors in parallel with each diode), the problem went away. The 2.83 rule is *not trivial*.

The filter capacitor ($C1$) in Fig. 21-1 is selected to provide enough ripple reduction to make the regulator happy, but does not have to provide all of the ripple reduction needed by the external circuitry powered by the regulator (the regulator adds considerable ripple reduction). Most authorities recommend a capacitance for $C1$ equal to 1000-μF/ampere of current drawn, with some recommending 2000-μF/ampere. If the latter recommendation is followed, then a 1-ampere regulator (the most common type) requires a 2000-μF capacitor. Another "rule" requires me to keep a capacitance of at least 500-μF in the circuit, even when the forward current is less than 500 milliamperes.

The working voltage rating ("WVdc") of the filter capacitor must be somewhat higher than the maximum expected voltage. Keep in mind that most electrolytic capacitors have a 20-percent tolerance, and normally voltages vary 15 percent. I will thus require a 35 percent margin of error on the WVdc rating. For example, if I have a 12-volt regulator that inputs 18 volts, the filter capacitor will normally see 18 volts. Using the 35 percent rule, we would specify 18-volts $\times$ 1.35, or 24.3 volts (or more). Since 25 WVdc is a standard rating, I can use that value as the minimum value for the WVdc rating of $C1$. I personally prefer to use 35 WVdc or 50 WVdc if practical; again, safety is a consideration.

The capacitors marked $C2$ and $C3$ in Fig. 21-1 are used for noise immunity/protection. These capacitors have a value of 0.1-μF to 1.0-μF. They are normally mounted as close as possible to the regulator device; in fact, many designers mount them on the regulator itself.

The output capacitor $C4$ is optional and is used to improve the transient response of the regulator. When external current demand increases very rapidly, it will take a certain amount of time for the regulator to catch up (microseconds). During this time, the external circuit will draw current from the capacitor, thereby preventing a "glitch" in the power supply voltage. The value of capacitor $C1$ is 100-μF/ampere, with a WVdc rating of not less than 1.35 times the rated output voltage of the regulator.

If capacitor C4 is used, then it is advisable to also use diode D1. The purpose of this diode is to dump the charge in C4 when the circuit is turned off. Otherwise, the charge can be dumped back into the circuit through the regulator, which can cause damage. Any diode in the 1N4002 through 1N4007 series will suffice.

ADJUSTABLE-VOLTAGE TYPES

Adjustable voltage regulators used to be somewhat harder to design than fixed types, but today there are several three- and four-terminal devices on the market that will serve nicely. In this chapter, I will limit my discussion to the LM-317 and LM-338 devices, since those two are readily available to hobbyists through mail order and walk-in retail outlets that sell parts (Jim-Paks sell both). The LM-317 and LM-338 devices are similar to each other in function, except that the LM-317 device handles 1.5-amperes, while the LM-338 device is rated at 5 amperes. Information given below for the LM-338 device is generally applicable for the LM-317 also.

Figure 21-2A shows a circuit based on the LM-338K device, while Fig. 21-2B shows the bottom view of the LM-338K package. Note well in Fig. 21-2B, that the output terminal is the case, which is exactly the opposite from fixed-voltage, three-terminal devices. It also means that the builder must insulate the case and/or the heatsink it rests on from chassis, especially if chassis is used as the common ground!

The transformer, rectifier, and filter capacitor are selected from the same criteria as given earlier, so will not be discussed here.

The LM-338K can accept an input voltage up to 35 volts and will produce a maximum output voltage of several volts less than that figure.

The exact output voltage is set by the ratio of two resistors, R1 and R2. In Fig. 21-2A, resistor R2 is a variable resistor made from a potentiometer with the wiper and one end terminal shorted together. The output voltage will be approximately:

$$V_\mathrm{o} = 1.25 \mathrm{\ V} \left[\frac{R2}{R1} + 1 \right] + (R2 \times I_{adj})$$

Normally, the term $(R2 \times I_{adj})$ is so small as to be ignored. With the values given in Fig. 21-2A, the output voltage can be varied from 1.25 volts to over 35 volts (if the input would allow it). In some cases, where the voltage is set and then forgotten, the

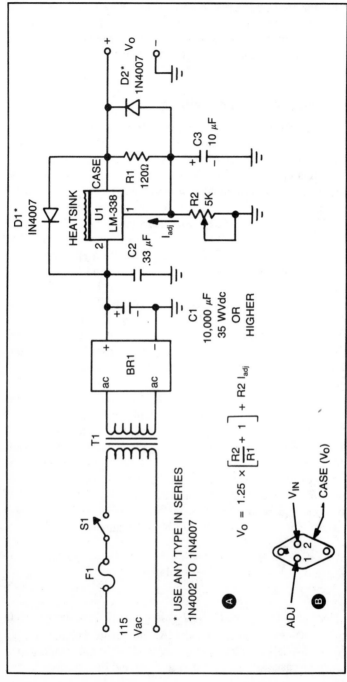

$$V_O = 1.25 \times \left[\frac{R2}{R1} + 1 \right] + R2\ I_{adj}$$

* USE ANY TYPE IN SERIES 1N4002 TO 1N4007

Fig. 21-2. (A) LM-338 circuit is used to 5-amperes. (B) LM-338 pinouts.

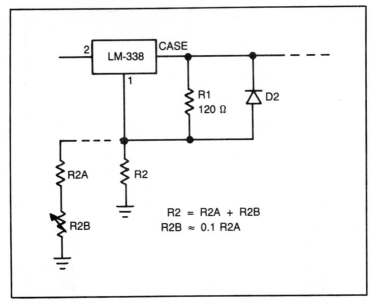

Fig. 21-3. Fixed and adjustable versions of the circuit.

potentiometer (*R2*) will be a trimmer pot and mounted on the power supply printed wiring board. In other cases, it will be a front panel control so that the operator can adjust it.

Diodes *D1* and *D2* serve the same protection function in this circuit as in the previous circuits. Once again, any diode in the series 1N4002 through 1N4007 will suffice.

If you want to make the LM-338K (or the LM-317) into a fixed voltage regulator, then adopt one of the two circuit modifications shown in Fig. 21-3. In the one case, two fixed resistors are used for *R1* and *R2*. Normally, *R1* will be 120 ohms and *R2* will be selected to set the output voltage at the required level. In the other case, *R2* is broken into two components, *R2A* and *R2B*. The value of *R2B* is roughly 10 to 15 percent of the value of *R2A*, and is set to trim the output to a precise value. This arrangement has the advantages of fixed operation, while allowing trimming of the output voltage to the exact value required.

My final circuit is shown in Fig. 21-4. Here I have two LM-338K devices connected together to form a 1.2 to 16-volt dc, 10-amperes (approx.) regulated power supply. Since the cases of the LM-338K devices are connected together, I can use the same heatsink for both. Potentiometer *R2* sets the output voltage, and is adjustable to just over 16 volts dc.

411

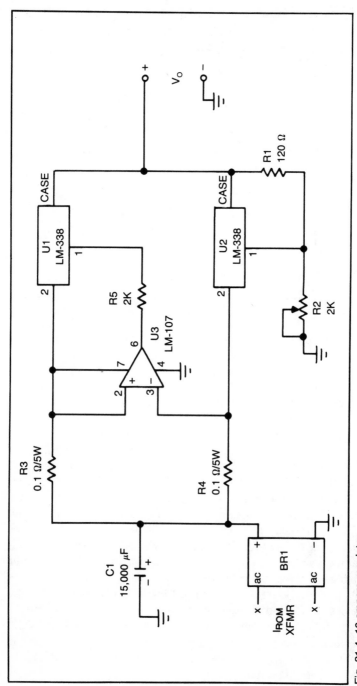

Fig. 21-4. 10-ampere regulator.

412

The input voltage for this circuit can be acquired from rectifying a 12.6 Vac (RMS) transformer, and filtering it with 15,000 μF, or more, of capacitance.

CONCLUSION

The dc power supply is very important to the success of your electronic project. Indeed, dc supplies are also very important to have on the workbench. With the information presented in this article you will be able to design and build most of the elementary supplies needed by hobbyists and small labs using solid-state electronics.

Chapter 22

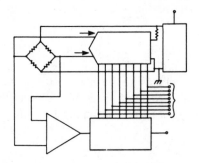

Elements of
Computer Interfacing

THE ART OF INTERFACING IS ESSENTIALLY THE ART OF CON-
necting the computer to some devices. Under the term "interfacing" falls a variety of activities that include, among others, connection of memory and I/O devices, keyboards, displays,
peripherals, A/D converters, D/A converters and so forth. I have
to consider certain details of interfacing. First, for example, I must
learn how to increase the drive capacity of the microprocessor chip
(most will only drive a couple of LS-TTL loads!). Second, I must
learn how to decode addresses. Whether an I/O-based machine like
the Z80, or a memory-mapped machine like the 6502, the address
bus will carry the addresses of both memory locations and I/O ports.
It is, therefore, critical to be able to decode the address when a
location is called. Finally, I will learn to generate system signals
such as INput and OUTput. These signals can be used to chip select I/O ports, memory chips, and other devices.

BUS BUFFERING

The output lines of all microprocessors are limited in driving
capacity. If you recall from Chapter 4 my discussion of the TTL
family of devices, then you may recall the terms "fan-in" and "fan-
out." These refer to the drive requirements and drive capacity of
TTL inputs and outputs respectively. The basic unit of measurement is the "standard TTL load." In TTL jargon, this means a

current source of approximately 1.8 mA in the standard +2.4 to +5.0 volt range that is normal to TTL devices. A fan-in of one means that the load requirement is one standard TTL input. A typical TTL device has a fan-in of one for the inputs and a fan-out of ten for the outputs. Thus, a standard TTL device can drive up to ten standard TTL inputs. The low-power Schottky devices (74LSxx devices) require less driving capacity, so have a fan-in of less than one. To keep things in perspective, therefore, I will rate microprocessor chip driving capacity in terms of TTL fan-in/fan-out figures. Some microprocessors have a fan-out of less than two, being capable of driving only two LS-TTL loads. Others, on the other hand, *have* a fan-out of two, which means that they can drive two standard TTL loads (or 3.2 mA at TTL logic levels). Unfortunately, this limitation places a severe constraint on the microcomputer designer.

If I want to drive more than a few TTL loads, then it is necessary to create a bus buffer, as shown in Fig. 22-1. A bus buffer

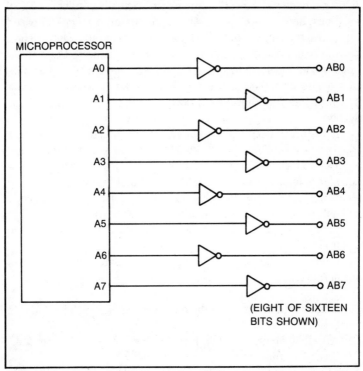

Fig. 22-1. Increasing the drive capacity of microprocessor using bus driver stages.

can be used on either address bus or data bus, although the one shown here is for the data bus. It consists of a set of high power, noninverting buffer stages. These devices will have a fan-out of 10 to 100, depending upon type. In some cases, they are available as six buffers in a single chip package, in which event you will need two IC packages and will have four unused stages in one of the packages. In other cases, you might select a bidirectional octal bus transceiver. Such a chip will have the drive capacity, while at the same time it is bidirectional, allowing input reads as well as output writes.

Some computers contain the bus buffer/driver, while others do not. In the cases where the computer does not have a buffered data/address bus, then you might want to consider constructing one.

ADDRESS DECODING

The address of any memory location within the 64K range of permissible addresses is defined by the binary word applied to the sixteen-bit address bus (2^{16} combinations actually results in 65,536 locations, but this is called "64K" in computerese). On Z80-based machines, the address of the I/O port will appear on the lower eight bits of the address bus, which means that the discrete I/O instructions will address 255 locations numbered from port 0 to port 255. In order to let a port or memory location know when it is being hailed by the CPU therefore, I need to have some form of address decoder.

An address decoder is a circuit or device that will produce a true output, if and only if, the correct eight- or sixteen-bit address is present on the address bus. At all other times, the output of the decoder is false. Please note that "true" and "false" are relative terms here, definable as I choose. In some cases, for example, I will define true as high, and in other cases, true will be low. In Fig. 22-2, for example, the SELECT signal is active low, so the true condition is low.

Figure 22-2 shows an eight-input NAND gate. Recall the rules of operation for NAND gates:

1. If any input is low, then the output is high (i.e., "false" in this case).
2. All eight inputs must be high for the output to be low (i.e., "true" in this case).

If I connect the inputs of the NAND gate to lines from the ad-

416

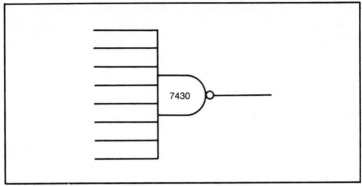

Fig. 22-2. 7430 NAND gate.

dress bus, then I will generate a low, if and only if, the correct address is present on the bus. In the configuration shown, the correct address must be 11111111 (which is FF in hexadecimal, or "FFh"). Thus, only one location or I/O port (no. 255) can be addressed by this circuit. Figure 22-3 solves the problem.

The idea in using the 7430 device as an address decoder is to conspire to force all bits of the address bus high when the correct address is on the line. This requirement means that I must invert all lows to make them highs. The circuit in Fig. 22-3 has inverters on bits B3, B4, and B7. This circuit will therefore issue an active-

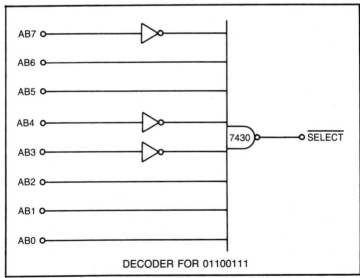

Fig. 22-3. Sample 7430 eight-bit address decoder.

417

low SELECT signal when the address present is 01100111 (or 67h). When this address is present, all inputs of the 7430 are high, so the 7430 output is low. If any other address is present instead, then at least one bit of the group A0 - A7 will be incorrect, so its corresponding input of the 7430 will be low—and the output remains high (the inactive state).

There is still something of a problem with the circuit in Fig. 22-3: it is not easily reconfigured for different locations. In some projects I might want to reassign the memory location or port number of the decoder without having to re-wirewrap or butcher a printed wiring board. In those cases, it might be nice to have a switch to select whether an inverter is present or not. There are several approaches to this problem.

First, I could place an inverter at each location, and then connect the input of the 7430 to the pole terminal of a single-pole, double-throw PC DIP switch. At one setting, the switch would place the address bus bit directly on the 7430 input line; in the other setting it would place the output of the inverter on the 7430 input line (Fig. 22-4A).

Another arrangement is the circuit of Fig. 22-4B. Here, just one input of the 7430 is treated with an exclusive-OR (XOR) gate (only one input is shown for sake of simplicity and to keep the artist from quitting). Recall the rules for operation of an XOR gate:

1. If either, but not both, inputs are high, then the output is high.
2. If both inputs are high, then the output is low.
3. If both inputs are low, then the output is low.

In other words, as long as the two inputs see different signal levels, then the output of the XOR is high, but if the same signal is applied to both then the output is low. In tabular form it appears as follows:

INPUTS		OUTPUT
A	**B**	
LOW	LOW	LOW
HIGH	LOW	HIGH
LOW	HIGH	HIGH
HIGH	HIGH	LOW

In Fig. 22-4B, one input of the XOR gate is connected to a pull-up resistor and a grounding switch. When the switch (S1) is open,

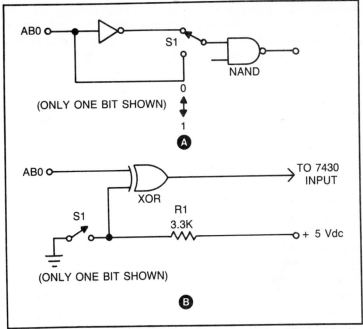

Fig. 22-4. (A) High/low selector for address decoder. (B) XOR gate circuit.

then the input of the XOR gate is high, and when S1 is closed the input of the XOR gate is low. The other input of the XOR gate is connected to a bit of the address bus. I will consider both situations.

When S1 is open, the control input of the XOR gate is high. In this case, a low applied on the A3 bit of the address bus will cause the output of the gate to be high. This is because the two inputs are different. In this case, the bit is inverted. Similarly, when A3 is high, the XOR gate sees both inputs high so produces a low output. Again, the address bus bit was inverted.

When S1 is closed, then the control input of the XOR gate is low. In this case, the circuit is noninverting. When A3 is high, the XOR gate sees different inputs, so produces a high output (no inversion). Similarly, when A3 is low, the XOR gate sees both inputs are the same and produces a low output (again, there is no inversion).

My final method is actually a modification of the first two methods. Both of the previous methods involved having an inverter on each input of the 7430 device. But I rarely require an inverter on each input. The more prudent method is to use from one to six inverters (*three* is a common number) on the printed circuit board, and then connect them such that they can serve any bit of the address bus at will. In many cases, the designers use a wire jumper system on the printed circuit board to make this work. A pair of jumpers will be used to connect the inputs and outputs of the inverter.

I could also use a 7485 (or its CMOS equivalent) to make the address decode function work. This IC is a four-bit magnitude comparator. It examines a pair of four-bit binary words ("A" and "B"), and then issues outputs that indicate whether $A > B$, $A < B$ or $A = B$. I will normally use the active-high, $A = B$ output to indicate that the correct address is present on the bus. The address bus bits are connected to one word input (e.g., the A inputs), while the other inputs (B in our hypothetical example) are programmed with the correct address. I could use a variety of methods for programming the B inputs: switches like in Fig. 22-4B, a four-bit latch (for programmable reassignment), or hard wire jumpers. Note that the 7485 devices can be cascaded. There are cascading inputs on each 7485, which can be connected to the outputs of the lower-order stages. Thus, I can create address decoder circuits in four-bit chunks.

Figure 22-5 shows the use of a 7442 chip for address decoding. This chip is a TTL device, and was never intended by the original designer as an address decoder. It was, instead, a BCD-to-one-of-ten-decoder. It would examine a four-bit, binary-coded decimal (BCD) word at its inputs, and then cause one of ten unique outputs to go low. Thus, if the binary word 0100 was present on the inputs, the "4" decimal output would be low and all others would be high. The 7442 was used to decode BCD to drive lamp columns of high-voltage *Nixie* tube switching transistors.

The decoder circuit of Fig. 22-5 uses the 7442 to select which of ten (of a possible sixteen) unique addresses are being called for by the lower four bits of the address bus (*A*0 - *A*3). Thus, on a Z80 machine, I could call for I/O ports 0 through 9 with this circuit (plus an IN or OUT signal, of which more later). The outputs 0, 1, 2, 3, 4, 5, 6, 7, 8, and 9 are shown in Fig. 22-5 and below:

A0	A1	A2	A3	Active (LOW) Output	7442 Pin No.
0	0	0	0	0	1
0	0	0	1	1	2
0	0	1	0	2	3
0	0	1	1	3	4

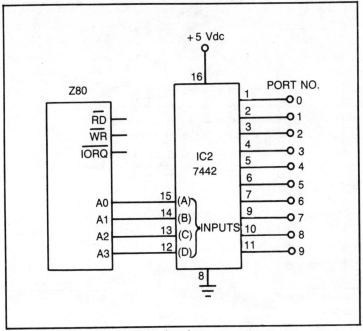

Fig. 22-5. 7442 I/O port address decoder.

A0	A1	A2	A3	Active (LOW) Output	7442 Pin No.
0	1	0	0	4	5
0	1	0	1	5	6
0	1	1	0	6	7
0	1	1	1	7	9
1	0	0	0	8	10
1	0	0	1	9	11

In a simplified computer that does not have a large amount of memory, I might want to consider a simplified decoder circuit such as Fig. 22-6. The elements of decoding include not just the address, but also an INput or OUTput signal. These functions are combined into one three-input NOR gate in Fig. 22-6. If I want to make an input port, then the Z80 active-low read (RD) signal is connected to the NOR gate input. If an output port is desired, use the write (WR) input instead. I have either RD or WR, and the active-low input/output request (IORQ) line connected to the NOR gate. Normally, the third input would be connected to the active-low output of a decoder such as discussed above. But in this simplified case,

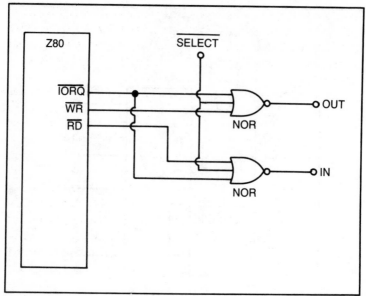

Fig. 22-6. IN/OUT decoder.

I am using just a single bit of the address bus to address the I/O port. If there is less than 32K worth of memory, then I can use bit $A15$ of the address bus singly as the I/O address. The decimal address of this port is 32,768. Since $A15$ is on for this address and all addresses higher, I cannot use this circuit if there are address locations or I/O ports higher than 32,768.

Most of the devices that I would normally select to form address decoder circuits are eight bit or less in length. Thus, I need some means of combining these outputs to form 16-bit or greater addresses. Figure 22-7 shows a means for combining eight- or four-bit address decoders to make higher order decoders. In this specific case, a pair of eight-bit decoders are combined to form a sixteen-bit decoder. This circuit works for any address decoder that has an active-low output for the SELECT signal. In this case, the SELECT1 and SELECT2 signals are combined in a two-input NOR gate. Recall the rules for a NOR gate:

1. If any input is high, then the output is low.
2. All inputs must be low for the output to be high.

Examine Fig. 22-7 in the light of these rules. Since both SELECT1 and SELECT2 will be low if and only if the correct sixteen-bit address is on the bus, the output of the NOR gate is high only when the correct address is present. Although the sense of the SELECT signal is changed (it is now active-high), the signal is valid for a sixteen-bit address. I could, if an active-low signal is needed, invert the SELECT signal.

Before moving on to another topic, I will consider one other little bit of miscellanea. Inverters come in sixes: the hex inverter is a standard IC in TTL and CMOS lines. Therefore, if I need less than six inverters in some design, the balance is wasted. Thus, if I need three inverters, then there will be three more in the package to be used otherwise. By combining these inverters as in Fig. 22-8, however, I can make a three-input NOR gate. This circuit depends upon using open-collector inverters, which require pull-up resistors to the V + (in this case + 5 Vdc) power supply. If I "wired-OR" all outputs together and used a common pull-up resistor, then the circuit becomes a three-input NOR gate (or as many inputs as there are inverters). In this manner, I am able to conserve chip count in my design.

SYSTEM IN-AND-OUT SIGNALS

The system IN and OUT signals indicate when INput and OUTput operations (respectively) are taking place. In Fig. 22-9, I use Z80 terminology to show how these signals are generated. There

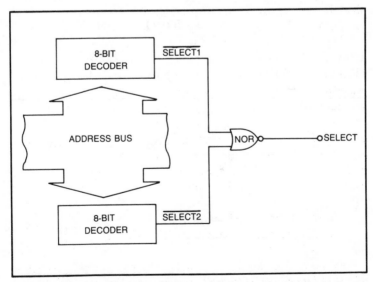

Fig. 22-7. 16-bit device select circuit.

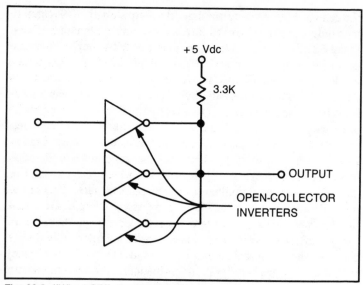

Fig. 22-8. "Wired-OR" circuit using inverters.

are three signals involved: IORQ (input/output request), RD (read) and WR (write). During read operations, RD and IORQ will be simultaneously low and WR is high. During write operations, WR and IORQ are low and RD is high. In tabular form it appears as follows:

	IORQ	RD	WR
INPUT OPERATION	LOW	LOW	HIGH
OUTPUT OPERATION	LOW	HIGH	LOW

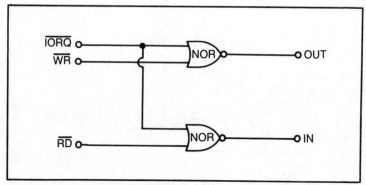

Fig. 22-9. NOR version of IN/OUT decoder.

In Fig. 22-9, I use these signals and a pair of NOR gates to generate IN and OUT signals. During a read operation, the RD and IORQ signals applied to the two inputs of NOR gate $G1$ are simultaneously low, so the output of $G1$ is high. This signal becomes our active-high INput signal. Similarly, when a write operation takes place, WR and IORQ applied to the inputs of NOR gate $G2$ are low simultaneously, so the output of $G2$ goes high forming my OUTput signal.

The signals generated in Fig. 22-9 are system signals. That is, they are active whenever an input (IN) or output (OUT) operation is taking place. There is no discrimination between ports in this case. Unless there is to be only one port forever, then I will need some additional logic to make the thing work. I could, for example, combine an active-high address decoder signal with either of these signals in a NAND gate, and produce an active-low device select signal that is unique to that lone address. Similarly, I could invert either or both of these signals and then combine them with an active-low SELECT signal in a NOR gate, to form an active-high unique select signal. Alternatively, I could use the discrete IN and OUT methods shown below.

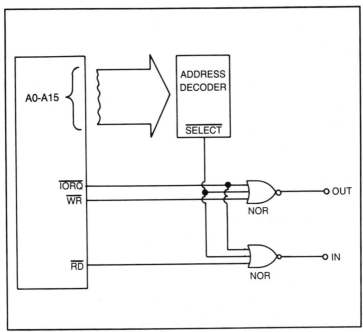

Fig. 22-10. Typical Z80 IN/OUT decoder with address selection.

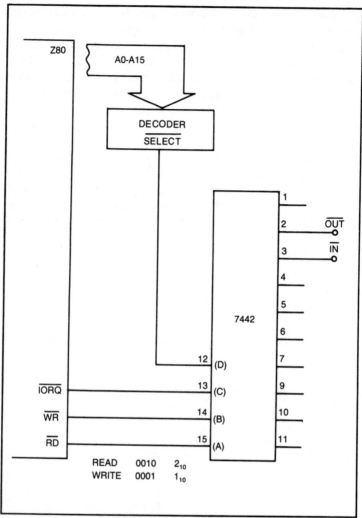

Fig. 22-11. 7442 IN/OUT decoder.

DISCRETE IN-AND-OUT SIGNALS

A discrete IN or OUT signal is one that is unique to a specific port or memory location. In Fig. 22-10, I have a circuit that could be used on an I/O board that is plugged into the microcomputer. It would generate an IN signal only when the address of the port is called, and the Z80 is calling for an input read. It would generate an OUT signal only when the address of the port is called, and the Z80 is calling for an output write.

This circuit is based on the same concepts as previously shown, except that a three-input NOR gate is required due to the addition of the active-low output of an address decoder circuit. Thus, all three inputs must be low for the output to be high. For an INput operation then, the address decoder output must be low (indicating that the Z80 is calling for the correct address), IORQ must be low and RD must be low simultaneously. On an OUTput operation, the same situation pertains except that WR is low instead of RD.

I may also use the 7442 as a discrete decoder, as shown in Fig. 22-11. In this case, I use the highest order input of the 7442 to examine the active-low output of the address decoder. If the correct binary word is applied, indicating an INput, then the corresponding 7442 output will go low—similarly for the OUTput operation.

Chapter 23

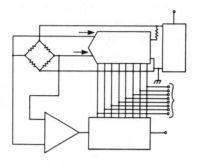

Single-Board
And Small Computers

THE INSTRUMENTATION AND CONTROL FIELD HAS EXPERI-
enced nothing less than a revolution, now that cheap, small
but powerful computers are available on the market. These computers are able to perform a large number of different instrumentation or measurement functions that were once performed by either
analog electronic circuits, or discrete (non-computer) digital circuits.
The use of computers to control instrumentation and control systems is certainly not new by any means, but it is now widespread
because of the microcomputer. Once, it was limited to those applications where a minicomputer could be justified, at very high cost.
Now, a simple KIM-1, AIM-65, Apple *II*e or Timex 1000 can perform chores that once required a much larger arsenal—and budget.
In this section, I am going to look at some of these computers with
an eye for interfacing. Each model selected for inclusion here has
either an interface connector on the rear, or a plug-in socket inside
of the machine (e.g., Apple II) that permits some flexibility in interfacing chores. The machines covered are AIM-65 and other
KIM-1 family machines, Timex 1000, 1500 and 2068, Apple *II*e
and the Commodore 64.

AIM-65 AND RELATED MACHINES

One of the first microcomputers used by many now-experienced
designers, technical people of assorted descriptions, and just plain

old "hobby hackers" was the MOS Technology KIM-1. Although intended as a trainer to introduce engineers to the 6502 microprocessor produced by MOS Technology, it soon became what a publisher might call a "run-away best seller." At a price of $149.95, it became the best buy in town. Many an engineer cut his microcomputer teeth on the little KIM-1.

The KIM-1 begat a cult of devotees that published numerous articles, designed an endless array of 6502-based software, and produced a lot of add-on extras for the little KIM-1 single-board computer. It seems that the KIMmy-cult people were exceptionally given to humor in their presentations, and attached a somewhat feminine gender to the neuter computer. For example, when one chap designed a cabinet for the KIM-1 microcomputer, he titled the article about it, "A Dress for KIM."

The KIM-1 is hard to find anymore, but its legacy lives on in other computers. The SYM-1 by Synertek Systems Corporation is a directly compatible "KIMmy," while the Rockwell AIM-65 is an advanced KIMmy. Even the Ohio Scientific Superboard II is arguably a derivative of the KIM-1, even though considerably more advanced than the early virgin KIM-1.

There are also two I/O ports available on the KIM-series of computers. The I/O channel was a 6522 integrated circuit called a versatile interface adapter (VIA). This device offers two ports, which can be configured as either input or output on a bit-for-bit basis. In other words, it is not necessary to make either port all input or all output. I can, for example, make bit 0 of port A an input line, while bit 1 (or any other bit) of port A is configured as an output. The difference is programmable and can be reconfigured at will. The 6522 contains two *data direction registers*, one each for Port A and Port B (these are labeled DDRA and DDRB). Writing a "1" to a bit in the DDR makes the corresponding I/O port bit an output, while a "0" renders it an input.

The 6522 registers are all memory mapped in a 6502-based machine. In the Rockwell AIM-65, for example, the following 6522 functions are mapped at the indicated locations in page A0 of memory:

Location	Function
A000	Port B Output Data Register (i.e., Port B I/O port)
A001	Port A Output Data Register (i.e., Port A I/O port)
A002	Port B Data Direction Register (DDRB)

Location	Function
A003	Port A Data Direction Register (DDRA)
A004	Timer T1 write T1L-L
A005	Timer T1 write T1L-H & T1C-H, clear interrupt flag
A006	Timer T1 Write T1L-L
A007	Timer T1 Write T1L-H
A008	Timer T2 Write T2L-L
A009	Timer T2 Write T2C-H, clear T2 interrupt flag
A00A	Shift Register (SR)
A00B	Auxiliary Control Register (ACR)
A00C	Peripheral Control Register (PCR)
A00D	Interrupt Flag Register (IFR)
A00E	Interrupt Enable Register (IER)
A00F	Port A Output Data Register (ORA)—no handshake

Let me consider an example using some 6502 programming. Suppose I wanted to configure Port A as an input port. Recall from above, that an input port is obtained by writing a "0" (i.e., low) to the corresponding bit of the DDR. In this case, in order to make all of Port A into an input, I need to write all zeroes to DDRA. I need a program something like the following:

    LDA #$00
    STA $A001

In this two line program, the 6502 is told to load the accumulator with the hexadecimal number 00 (which is 00000000 in binary). It is then told to store the number loaded into the accumulator at location A001, which is the memory-mapped address of the DDRA. After this program is executed, I can use Port A as an input. In most 6502 programs, the data to be output would first be loaded into the accumulator, and then stored at location $A003, which is the address of the output register for Port A (i.e., ORA). A program similar to the above is typically used:

    LDA ($nnnn)
    STA $A003

What the above code is telling the 6502 is to load the accumulator with the data stored at memory location $nnnn (a dummy number used for illustration), and then store those contents at the memory location of the ORA. When this program is executed, that data will appear on the output terminal. Of course, these simple examples are simplistic because they only take into account two of the 6502 addressing modes.

There are two interface connectors on the KIM-1 bus machines. These are called the Applications Connector and the Expansion Connector. The Applications Connector is basically the I/O ports, while the Expansion Connector is the data and address buses. The Expansion Connector would be used to add memory, video drives, and so forth. These connectors are defined below. Both connectors are printed circuit card-edge connectors printed onto the board.

KIM-Bus Applications Connector (Numbered pins are on top of board, letter pins are on the bottom)

Pin No.	Designation	Function
1	GRND	Ground/Power Supply Common
2	PA3	Bit 3 of Port A
3	PA2	Bit 2 of Port A
4	PA1	Bit 1 of Port A
5	PA4	Bit 4 of Port A
6	PA5	Bit 5 of Port A
7	PA6	Bit 6 of Port A
8	PA7	Bit 7 of Port A
9	PB0	Bit 0 of Port B
10	PB1	Bit 1 of Port B
11	PB2	Bit 2 of Port B
12	PB3	Bit 3 of Port B
13	PB4	Bit 4 of Port B
14	PA0	Bit 0 of Port A
15	PB7	Bit 7 of Port B
16	PB5	Bit 5 of Port B
17	KBR0	Keyboard, Row 0
18	KBCF	Keyboard, Column F
19	KBCB	Keyboard, Column B
20	KBCE	Keyboard, Column E
21	KBCA	Keyboard, Column A
22	KBCD	Keyboard, Column D
A	PWR5	+5 Vdc from main board
B	K0	Memory Bank Select

Pin No.	Designation	Function
C	K1	Memory Bank Select
D	K2	Memory Bank Select
E	K3	Memory Bank Select
F	K4	Memory Bank Select
F	K4	Memory Bank Select
H	K5	Memory Bank Select
J	K7	Memory Bank Select
K	Decode	Memory decode signal. Used to increase memory size with external
L	AUDIN	Audio input from cassette player
M	AUDOUTL	Audio output to cassette. Low-level signal for microphone
N	PWR12	+12 Vdc power from main board
P	AUDOUTH	High level audio output for use with cassette recorder that has a . . . /?
R	TTYKBD+	Positive terminal of 20-mA TTY current loop (serial input), keyboard
S	TTYPNT+	Positive terminal of 20-mA TTY current loop (serial output), printer
T	TTYKBD-	Negative terminal of "R" above
U	TTYPNT-	Negative terminal of "S" above
V	KBR3	Keyboard, Row 3
W	KBCG	Keyboard, Column G
X	KBR2	Keyboard, Row 2
Y	KBCC	Keyboard, Column C
Z	KBR1	Keyboard, Row 1

KIM-Bus Expansion Connector (Numbered pins are on top of board, letter pins are on bottom)

Pin No.	Designation	Function
1	SYNC	Output that is high during clock phase 1
2	RDY	Active-low input that inserts WAIT state into program execution
3	P1	Phase 1 clock signal
4	IRQ	Active-low interrupt request

432

Pin No.	Designation	Function
5	RO	Reset overflow (resets the overflow flip-flop inside CPU)
6	NMI	Active-low nonmaskable interrupt input
7	RST	RESET. Goes to reset line of 6502
8	DB7	Data bus bit 7
9	DB6	Data bus bit 6
10	DB5	Data bus bit 5
11	DB4	Data bus bit 4
12	DB3	Data bus bit 3
13	DB2	Data bus bit 2
14	DB1	Data bus bit 1
15	DB0	Data bus bit 0
16	K6	Active-high output that indicates that the CPU addresses a location in the range
17	SSTOUT	Single-Step Output
18	(no connection)	
19	(no connection)	
20	(no connection)	
21	PWR5	+5 Vdc power from main board
22	GRND	Ground
A	AB0	Address Bus Bit 0
B	AB1	Address Bus Bit 1
C	AB2	Address Bus Bit 2
D	AB3	Address Bus Bit 3
E	AB4	Address Bus Bit 4
F	AB5	Address Bus Bit 5
H	AB6	Address Bus Bit 6
J	AB7	Address Bus Bit 7
K	AB8	Address Bus Bit 8
L	AB9	Address Bus Bit 9
M	AB10	Address Bus Bit 10
N	AB11	Address Bus Bit 11
P	AB12	Address Bus Bit 12
R	AB13	Address Bus Bit 13
S	AB14	Address Bus Bit 14
T	AB15	Address Bus Bit 15
U	P2	Clock phase 2
V	R/W	High for read operations, low for write operations

Pin No.	Designation	Function
W	R/W	Low for read operations, high for write operations (complement of "V" above)
X	PLLTST	Phase locked loop test for FM-audio signal used to record data on audio
Y	P2	Complement of clock phase-2 ("U" above)
Z	RAMRW	RAM read/write control signal to activate RAM during phase-2 clock periods

TIMEX T/S-1000 & T/S-1500

The Timex 1000 microcomputer was based on the Sinclair ZX-80, and is very similar (maybe identical) to the ZX-81 device. This little computer seems to have taken the world by storm as more than 550,000 were sold in only a few months. Although since replaced with the T/S-1500, and the much more powerful T/S-2068 machines, the T/S-1000 retains a loyal following of enthusiasts.

All of the Timex computers have a built-on keyboard. The 1000 uses membrane keys, which some people find difficult to adjust to—although others love them. These are the so-called "flat keys," that are merely painted onto the case of the computer. The keys seem to be long-lasting, and I have used mine for almost a year with no ill effects or lack of reliability. The keys on the 1500 keyboard are rubber "chicklet" keys. While they provide a certain amount of tactile feedback, some people (not myself) dislike them for their rubbery feel. The Timex 2068 microcomputer, on the other hand, uses real, honest-to-gosh hard keys.

T/S-1000 & T/S-1500 Interfacing

Probably, your main intent in reading this book is to use the microcomputer in some practical project. For this purpose, interfacing is needed—which explains the heavy emphasis on interfacing in the text material. The T/S-1000 and T/S-1500 microcomputers are based on the Z80 microprocessor chip, so one is expected to become familiar with the control signals of that device. The interface connector on the rear of the 1000 and 1500 follows the definitions given below.

434

SIDE-A (TOP)

Pin No.	Function	Description
1	D7	Data bus bit 7
2	RAM-CS	Active-low input to deselect internal RAM and select external RAM bank
3	(slot - no connection)	
4	D0	Data bus bit 0
5	D1	Data bus bit 1
6	D2	Data bus bit 2
7	D6	Data bus bit 6
8	D5	Data bus bit 5
9	D3	Data bus bit 3
10	D4	Data bus bit 4
11	INT	Interrupt Request (maskable), active-low
12	NMI	Nonmaskable Interrupt Request, active-low
13	HALT	Active-low input to halt CPU execution
14	MREQ	Active-low memory request line
15	IORQ	Active-low input/output request line
16	RD	Active-low read line
17	WR	Active-low write line
18	BUSAK	Active-low bus acknowledge output
19	WAIT	Active-low indicates wait state
20	BUSRQ	Active-low input requests control of data bus
21	RESET	Active-low reset line forces Program Counter register to 0000H
22	M1	Active-low output to indicate M1 machine cycle
23	REFSH	Active-low refresh signal (used to refresh dynamic memory)

SIDE-B (Bottom)

Pin No.	Function	Description
1	5V	5 Volt dc regulated

Pin No.	Function	Description
2	9V	9 Volt dc unregulated
3	(slot—no con-nection)	
4	GND	GROUND
5	GND	GROUND
6	clock	
7	A0	Address bus A0
8	A1	Address bus A1
9	A2	Address bus A2
10	A3	Address bus A3
11	A15	Address bus A15
12	A14	Address bus A14
13	A13	Address bus A13
14	A12	Address bus A12
15	A11	Address bus A11
16	A10	Address bus A10
17	A9	Address bus A9
18	A8	Address bus A8
19	A7	Address bus A7
20	A6	Address bus A6
21	A5	Address bus A5
22	A4	Address bus A4
23	ROM-CS	Active-low input that disconnects internal ROM to allow use of external ROM

The TTL-compatible outputs of the Timex 1000 and 1500 interface connector will not support large numbers of TTL inputs hanging on the bus. The interfacer is expected to supply buffers/drivers for this purpose if more than a few TTL inputs must be driven. Similarly, the dc power supply is not the world's largest, so if there are more than a few TTL chips on the external circuit, you should provide a separate +5 volt dc TTL-compatible dc power supply (regulated, of course).

APPLE II

The *Apple II* microcomputer is one of the "standards," and has been around in one model or another for almost as many years as any microcomputer. This machine was one of the first really small, desktop microcomputers and has been extremely successful. This machine comes with 48K of memory, with the memory space between 48K and 64K taken up by internal programming. The *Apple II* is well known for simple peripheral driver devices because it replaces hardware in these circuits with software. The Apple is based on the 6502 microprocessor chip.

There have been many Apple "look-alikes" on the market, all using the 6502 chip and claiming Apple compatibility. Be a little careful when selecting an Apple look-alike that claim compatibility, because some are "compatible—*except . . .*" and that "except" can be profound. There are also Apple counterfeits on the market. Most of these are manufactured in Southeast Asia, and are offered for sale as if they were Apples—not merely Apple compatible. They look like Apples, carry Apple logos, and are offered as Apples. But most of them are second-quality machines. Even where they are made well, with good quality components, there is always the problem of obtaining service for the machine: I doubt that the Apple service shops would be helpful when presented with a counterfeit of their product!

The Apple II machine uses outboard 5.25-inch disk drives, while the Apple III contains a single built-in 5.25-inch drive. Most Apple dealers offer a package that includes one disk drive, and offer the second drive as an option.

The main reason for including the Apple computers in a book on single-board computers is that it is one of the most popular machines in use. This machine is a low-cost, highly viable alternative for laboratories, industrial process controls, and other similar applications that are often the province of the single-board computers. The Apple II is also often recommended as the development system for 6502-based, single-board computers. The idea is that you will develop the assembly language program for the single-board computer on the Apple II, and then transfer it to a read only memory (ROM) for insertion into the ROM socket of the SBC. This system allows a flexible system that can easily be changed to reflect new situations.

One reason that the Apple is so popular is that it contains eight printed circuit card-edge connectors on the motherboard that accepts plug-in printed circuit cards. These sockets are used by the interfacer, the customizer, and so forth. The sockets have 50 pins, 25 on each side of the board. Pins 1 through 25 are on the component side, while 26 through 50 are on the "wiring side." There are a number of non-Apple accessory cards that can be plugged into Apple II's. These include extra ROM cards, I/O cards of one sort or another, A/D or D/A converters, and so forth. The Apple II can be custom configured with plug-in cards to meet your particular need.

The pinouts of the Apple II plug-in slots are given below.

Pin No.	Description	Function
1	I/O SELECT	Active-low signal that is low when the unique address (of 16 possible addresses) for that particular connector is called for in a program. All 16 of these addresses are in the range $C800 to $C8FF.
2	A0	Address bus bit 0
3	A1	Address bus bit 1
4	A2	Address bus bit 2
5	A3	Address bus bit 3
6	A4	Address bus bit 4
7	A5	Address bus bit 5
8	A6	Address bus bit 6
9	A7	Address bus bit 7
10	A8	Address bus bit 8
11	A9	Address bus bit 9
12	A10	Address bus bit 10
13	A11	Address bus bit 11
14	A12	Address bus bit 12
15	A13	Address bus bit 13
16	A14	Address bus bit 14
17	A15	Address bus bit 15
18	R/W	6502 read/write signal. This signal is low for WRITE operations, and high for READ operations.
19	(no connection)	
20	I/O STR	Active-low signal that indicates an I/O operation is taking place. This signal is activated when any address between $C800 and $C8FF is called in the program.
21	RDY	Active-low input signal that will add a WAIT state to the CPU (e.g., stop program execution) if low during clock Phase 1.

Pin No.	Description	Function
22	DMA	Active-low input that allows an external device to gain control over the data bus for direct access to the memory.
23	INTOUT	Interrupt Output signal that permits prioritizing of the interrupts. The INTOUT signal is connected to the INTIN signal of the next higher order plug-in printed circuit card.
24	DMAOUT	Direct memory access prioritization.
25	+5	Five-volt power supply from main board.
26	GND	GROUND
27	DMAIN	Direct memory access prioritization input.
28	INTIN	Interrupt prioritization input
29	NMI	Nonmaskable interrupt input signal to 6502 (active-low)
30	IRQ	Interrupt Request (active-low)
31	RES	Active-low reset line
32	INH	Active-low input that will disconnect on-board monitor ROMs to allow custom programmed ROMs to be used.
33	−12	−12 volt dc power supply from pc board
34	−5	−5 volt dc power supply from pc board
35	(no connection)	
36	7MHz	7-MHz clock from main pc board
37	Q3	2-MHz clock from main pc board
38	01	Phase-1 clock signal
39	USER1	Disables all ROMs and locations $C800 thru $C8FF.
40	02	Phase-2 clock signal
41	DEVICESEL	Active-low signal indicates that one of the 16 address of that connector is present on address bus.

Pin No.	Description	Function
42	D7	Data bus bit 7
43	D6	Data bus bit 6
44	D5	Data bus bit 5
45	D4	Data bus bit 4
46	D3	Data bus bit 3
47	D2	Data bus bit 2
48	D1	Data bus bit 1
49	D0	Data bus bit 0
50	+12	+12 volt power supply from main pc board.

COMMODORE 64

The Commodore 64 machine is a small desk-top computer that is often mistaken for a toy—but it is far from a toy, it is a professional machine in a small package. The Commodore 64 is based on the 6510 chip, which is a variant of the 6502. As a result, most machine language programs written for 6502 machines (like AIM-65 or Apple II) will run on the Commodore 64, even though it is necessary to manually enter the instructions (Commodore will not read an Apple disk or AIM-65 audio tape). The connector pinouts for the Commodore 64 are shown below:

USER PORT

The USER PORT on the Commodore 64 is connected to the 6526 CIA chip, and is used to connect to printers, modems and similar devices. It is a 24-pin connector of the card-edge type.

Numbered pins are on top side,
lettered pins on the bottom side of the board

Pin No.	Designation	Function
1	GROUND	Ground (0 volts)
2	+5 Volts	dc power
3	RESET	Resets all functions (cold start)
4	CNT1	Serial Port Counter #1
5	SP1	Serial Port #1
6	CNT2	Serial Port Counter #2
7	SP2	Serial Port #2

Pin No.	Designation	Function
8	PC2	Handshaking (CIA #2)
9	SERIAL ATN	Serial bus ATN line
10	9 Vac+	Positive phase of 9 Vac line
11	9 Vac–	Negative phase of 9 Vac line
12	GND	Ground
A	GND	Ground
B	FLAG2	CIA handshaking
C	PB0	Data bit
D	PB1	Data bit
E	PB2	Data bit
F	PB3	Data bit
H	PB4	Data bit
J	PB5	Data bit
K	PB6	Data bit
L	PB7	Data bit
M	PA2	Data bit
N	GND	Ground

Serial Bus connector

The serial bus connector is a DIN-type connector that provides the serial data stream to printers and the like, specifically the Commodore graphics printer and the disk drive.

Pin No.	Designation
1	Serial SRQ input
2	Ground
3	Serial ATN line IN/OUT
4	Serial CLK line IN/OUT
5	Serial DATA line IN/OUT
6	No Connection (RESET on some models)

Expansion Port Connector

The Expansion Port Connector is the 44-pin card-edge connector on the back of the Commodore 64 that allows total access to the data bus, address bus, control signals and other signals needed to interface the computer with a wide range of products, memory, I/O ports, and so forth.

Pin No.	Designation	Function
1	GND	Ground
2	+ 5 Vdc	(dc power, combined $<$ = 450 mA both pins)
3	+ 5 Vdc	
4	IRQ	Active-low interrupt Request line
5	R/W	Read/Write signal (high for read, low for write).
6	CLOCK	8.18-MHz system clock
7	I/O1	I/O Block #1 @ $DE00-$DEFF (active-low)
8	GAME	Active-low game signal input
9	EXROM	Active-low external ROM signal
10	I/O2	I/O Block #2 @ $DF00-$DFFF (active-low)
11	ROML	8K decoded RAM/ROM @ $8000
12	BA	Bus Available
13	DMA	Active-low, Direct Memory Access input
14	D7	Data bus bit
15	D6	Data bus bit
16	D5	Data bus bit
17	D4	Data bus bit
18	D3	Data bus bit
19	D2	Data bus bit
20	D1	Data bus bit
21	D0	Data bus bit
22	GND	Ground
A	GND	Ground
B	ROMH	8K decoded ROM/RAM @ $E000
C	RESET	Active-low input connected to 6510 RESET line
D	NMI	Active-low nonmaskable interrupt input
E	P2	Phase-2 clock output
F	A15	Address Bus Bit
H	A14	Address Bus Bit
J	A13	Address Bus Bit
K	A12	Address Bus Bit
L	A11	Address Bus Bit

Pin No.	Designation	Function
M	A10	Address Bus Bit
N	A9	Address Bus Bit
P	A8	Address Bus Bit
R	A7	Address Bus Bit
S	A6	Address Bus Bit
T	A5	Address Bus Bit
U	A4	Address Bus Bit
V	A3	Address Bus Bit
W	A2	Address Bus Bit
X	A1	Address Bus Bit
Y	A0	Address Bus Bit
Z	GND	Ground

Chapter 24

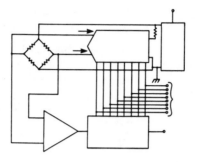

Biopotentials Amplifiers

BIOPOTENTIALS ARE ELECTRICAL SIGNALS GENERATED IN living things. Most familiar are signals such as electrocardiogram (heart signals), electroencephalogram (brain signals), electromyograph (skeletal muscle signals), and so forth. These signals are generated at the cell level, and are due to differing electrolytic concentrations between inside and outside of the cell. For example, most cells are rich in potassium inside of the cell, and rich in sodium outside the cell. (Note: both potassium and sodium are found on both sides of the cell wall, but the relative concentration is different.) The different ionic potentials of these elements produces a voltage of about – 80 millivolts (i.e., inside more negative than outside the cell). When the cell discharges—for example, when the heart contracts—the cell wall breaks down a little bit and becomes permeable to the external sodium. The ionic charge on both sides of the wall equilibrates, and the "action potential" snaps to as much as + 20 millivolts for a short while. The cell wall then reasserts itself, and restores the resting potential (– 80 mV) concentrations as before.

It is possible to pick up the potentials from internal muscles like the heart from surface electrodes. Unfortunately, not all is easy in this respect, for there are some severe problems. Fig. 24-1 shows some of these problems.

One of the first problems is that the signal is at a very low level. While EMG signals may look like 100-millivolt powerhouses, other

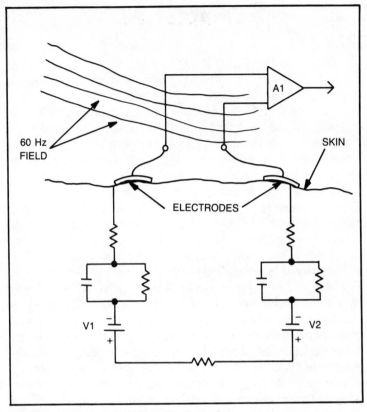

Fig. 24-1. Equivalent circuit for skin electrodes.

signals are not so strong. The typical ECG waveform, for example, is on the order of 1 millivolt, with significant components a lot less. The EEG signal is a small signal indeed: on the order of 5 to 100 microvolts. The amplifier must have a high gain in order to correctly display these signals. If I have an oscilloscope or strip-chart (paper) recorder with a 1-volt full-scale input, then I will need the following gains:

Electrocardiogram (ECG)
1000 millivolts = 1 Volt, so
$A_V = 1000/1$
$A_V = 1000$
Electroencephalogram (EEG)
Assuming 50 microvolts signals, and 1 uV = 1/1,000,000 volts

$$A_V = \text{1-volt/0.000050-volt}$$
$$A_V = 20{,}000$$

With gains required of 1000 to 20,000, we clearly need some high gain! Of course, not all display devices require a 1-volt input, but the principle is the same: A_V = output/input.

The second problem is that biopotentials signals tend to be subject to intense 60-Hz ac fields from the surrounding ac power lines. These fields can introduce a tremendous ac component onto the signal, that either shows up as a 60-Hz baseline component or totally obliterates the signal. The solution is to make amplifier A1 a differential input amplifier. Given that the two electrode wires are in the same field, they will be affected similarly by the field, and produce equal inputs to the amplifier. The nature of the differential amplifier is that it will cancel equal potential inputs, so the 60-Hz interference goes away—and remains away so long as the balance is maintained.

Of course, it is sometimes difficult to maintain that balance, but perhaps never more so than when the operators of the equipment, typically nurses, physicians or scientists, subvert the intent of the design. I recall one urgent summons to an intensive care unit in the hospital where I worked. The Monitoring Technician could not get the 60-Hz interference to go away, so had me summoned—the Repairman! The problem was that the patient tended to sweat a lot, and would not keep the electrodes on. The karaya adhesive used on the back of the electrodes kept coming loose, and the electrode would fall off. The nurse, being quick-witted and resourceful, painted the area where the electrode would go with a solution called Betadyne—great disinfectant, but also an insulator when it dried. Hence, she made the observation that the ECG monitor worked fine after she made the change, but that it deteriorated within ten minutes and was now all 60-Hz and no ECG! The problem was that the Betadyne was a conductor when it was wet, but an insulator when it dried. The electrodes were essentially free-floating on the patient's body—which acted like an antenna for the 60-Hz field.

At this point, I know of two parameters for the biopotentials amplifier: high gain and differential inputs. What else?

Refer again to Fig. 24-1, and note the complex RC circuit beneath the skin. Researchers are generally agreed that the electrode-skin contact is such a complex network, rather than a simple ohmic contact as between two similar metals. There are actually two problems here: dc offset and impedance. The impedance of the electrode is anything from 1000 ohms to 20,000 ohms, so the input of the amplifier must be a high impedance input. Thus, my amplifier must be a high input impedance, high gain, differential amplifier (it's getting more complicated!).

The second problem is the dc offset potentials, V1 and V2. Since the electrodes are metallic, and the skin is electrolytic, there will be a half-cell potential between the electrode and the skin when the two are placed in contact with each other. This potential can be many times higher than the signal potential, so will easily saturate the high gain amplifier needed for the signal. Of course, if the potentials V1 and V2 are exactly equal, then they will cancel out because of the differential inputs of the amplifier. Unfortunately, this situation does not exist very often, especially since a 1 millivolt difference out of a 1-volt potential can cause the amplifier to saturate!

The solution to the offset voltage problem is to ac-couple the amplifier (of which more later). Now what do I know about our amplifier? It must be:

1. Ac-coupled
2. High input impedance
3. Differential
4. High gain

The general diagram for a biopotentials amplifier is shown in Fig. 24-2. Here is seen what might be required. The input amplifier A1 is the biopotentials amplifier discussed above. The next stage is for frequency shaping and signal processing. In general, the frequency shaping function must do any or all of the following: set low-pass frequency limit, set high-pass frequency limit, and set any notch-filter frequencies (e.g., 60 Hz to take out residual 60-Hz

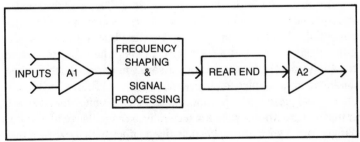

Fig. 24-2. Instrument block diagram.

signals not handled by the differential input). Signals processing might be nothing but additional gain (it isn't always wise to put all gain in the same stage), or might include things like differentiation, integration, analog multiplication, and so forth (although many signal processing functions are now done in software rather than hardware). The "Read End" is merely a stage that provides display control over the signal: position, span, dc balance, etc. (see Chapter 16).

Finally, we have the output buffer amplifier. This stage might be part of the Rear End section, or provided separately. It has the function of isolating the amplifier from the external world.

AC COUPLING

The dc offset potential of the input electrodes makes it necessary to ac-couple the signal amplifier. But there is also a requirement for low-end frequency response. The ECG signal, for example, normally has a frequency component spectrum of 0.05 Hz to 100 Hz. A signal of 0.05 Hz is almost dc, so one might be tempted to use a dc amplifier. Unfortunately, I must not only provide both 0.05 Hz as a low-end, -3 dB point in the amplifier response, but also ac-couple it to prevent saturation of the amplifier. There are several ways to do this job, as shown in Fig. 24-3 and Fig. 24-4.

The circuit in Fig. 24-4 is not often used because it requires the input amplifier to have low gain. It does offer the advantage of bipolar operation with high value electrolytic capacitors. Such capacitors are not well regarded for bipolar operation, so a pair of them in series (back-to-back) are needed. The diodes are used to switch capacitors in opposite halves of the ac input signal. When the signal is positive going, then capacitor C1 is used, D1 is reverse biased, and D2 is forward biased (shorting out C2). Similarly, when the input signal is negative going, C2 is used, D1 is forward biased (shorting out C1) and D2 is reverse biased.

Another solution is shown in Fig. 24-4. This circuit is more widely used. The amplifiers A1 and A2 are merely the input amplifiers of a differential instrumentation amplifier (or its ICIA version). Resistors $R1$ and $R2$ are used to keep the capacitors from charging from the bias currents, and are sometimes deleted when those currents are picoampere in range (e.g., on BiMOS or BiFET operational amplifiers). Do not use less than 10 megohms for this application, as these resistors set the input impedance of the amplifier. Also, make sure to balance them either with hand-selection with an ohmmeter, or, by using 1-percent tolerance resistors.

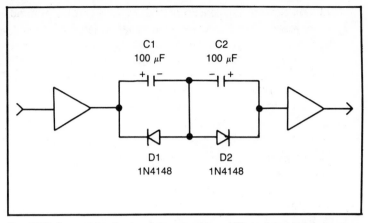

Fig. 24-3. Ac-coupling for low-frequency bioamplifiers (interstage).

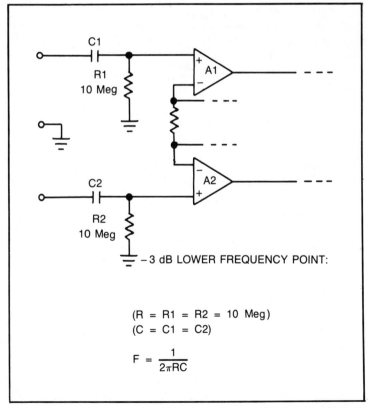

−3 dB LOWER FREQUENCY POINT:

(R = R1 = R2 = 10 Meg)
(C = C1 = C2)

$$F = \frac{1}{2\pi RC}$$

Fig. 24-4. Ac-coupling for input of instrumentation amplifier used in biomedi-cal applications.

The capacitors are selected according to the desired lower end, – 3 dB frequency response limit. The equation is:

$$C = \frac{1,000,000}{6.28\ R\ F}$$

Where:

C is the capacitance in microfarads
F is the frequency in Hertz
R is the resistance in Ohms

EXAMPLE:

$$C = \frac{1,000,000}{6.28\ R\ F}$$

$$C = \frac{1,000,000}{(6.28)\ (10,000,000)\ (0.05\ \text{Hz})}$$

$$C = 0.32\ \mu\text{F}$$

Because 0.33 is the next higher "standard value," one would normally select 0.33-μF for this application: $C1 = C2 = 0.33$-μF.

The upper – 3 dB frequency limit is set by other means. If the operational amplifiers have compensation terminals, then capacitors or RC networks across these terminals are used. Otherwise, select a capacitor to shunt the feedback resistor in any stage. The value of the capacitor is found from the same formula as above.

Appendix

Appendix A

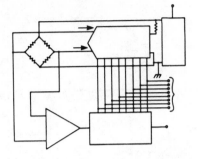

Transient Suppressor for
Small Computers

HIGH VOLTAGE TRANSIENTS, CALLED "GLITCHES" IN THE JAR-gon, are capable of interrupting the operation of digital circuits. While most analog circuits tend to simply absorb the transients, digital circuits often see them as valid signals, and will respond accordingly. I can recall working in a hospital where the high voltage power supply for the X-ray machines tended to throw transients out on the ac power lines—only to raise Hob with the digital word processors in the Medical Records department.

There are two basic ways to solve the transient problem. The most expensive (and in hard cases, the only) method is to use a special isolation transformer between the ac power lines and the computer or other digital equipment. The transformer has a 1:1 turns ratio, so the output voltage is the same as the input voltage. The transformer is only efficient at low frequencies however, so I find that it will attentuate the high-frequency components of the transient pulse. The only hooker in the plan is the fact that these transformers are expensive. Both Sola and Topaz make adequate models.

The cheap solution, which works often enough to be worth trying, is to use a General Electric *metal-oxide varistor* (MOV) device as shown in Fig. A-1. The MOV device is somewhat like a pair of high-voltage zener diodes connected back-to-back. It will clip all potentials over the threshold limit, but pass without attention all lower potentials. Thus, the MOV device will be ineffective when

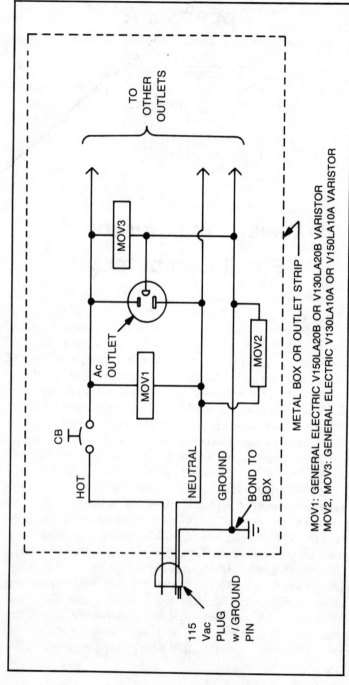

METAL BOX OR OUTLET STRIP

MOV1: GENERAL ELECTRIC V150LA20B OR V130LA20B VARISTOR
MOV2, MOV3: GENERAL ELECTRIC V130LA10A OR V150LA10A VARISTOR

Fig. A-1. Transient suppressor circuit using MOVs.

the line potential is normal, but clips high voltage transients that might come along.

In Fig. A-1 there are three MOV devices. Strictly speaking, only the one across the hot and neutral lines (i.e., MOV1) is necessary. The others are highly recommended, however.

You will be surprised at how less costly this constructed device is than the equivalent ready-built. I found a six outlet "power-strip," complete with the circuit breaker, at a Harry and Harriet Homeowner store for $7 on sale, and the MOV devices were $1.60 each. My $12 investment would cost $50 in a computer supplies store!

Index

457

464
Edited by Roland S. Phelps

Other Best Sellers From TAB

Other Best Sellers From TAB